CATACLYSMIC VARIABLES AND RELATED OBJECTS

ASTROPHYSICS AND SPACE SCIENCE LIBRARY

A SERIES OF BOOKS ON THE RECENT DEVELOPMENTS
OF SPACE SCIENCE AND OF GENERAL GEOPHYSICS AND ASTROPHYSICS
PUBLISHED IN CONNECTION WITH THE JOURNAL
SPACE SCIENCE REVIEWS

VOLUME 101
PROCEEDINGS

CATACLYSMIC VARIABLES AND RELATED OBJECTS

PROCEEDINGS OF THE 72nd COLLOQUIUM
OF THE INTERNATIONAL ASTRONOMICAL UNION
HELD IN HAIFA, ISRAEL, AUGUST 9–13, 1982

Edited by

MARIO LIVIO

and

GIORA SHAVIV

Faculty of Physics, Technion Israel Institute of Technology, Haifa, Israel

D. REIDEL PUBLISHING COMPANY
DORDRECHT : HOLLAND / BOSTON : U.S.A.
LONDON : ENGLAND

Library of Congress Cataloging in Publication Data

CIP

International Astronomical Union. Colloquium (72nd : 1982 : Haifa, Israel)
Cataclysmic variables and related objects.

(Astrophysics and space science library ; v. 101. Proceedings.)
Includes index.
1. Cataclysmic variable stars–Congresses. 2. Stars–Congresses.
I. Livio, Mario, 1945- . II. Shaviv, Giora, 1937- .
III. Title. IV. Series.
QB835.I59 1982 523.8'446 83-3276
ISBN 90-277-1570-X

Published by D. Reidel Publishing Company,
P.O. Box 17, 3300 AA Dordrecht, Holland.

Sold and distributed in the U.S.A. and Canada
by Kluwer Boston Inc.,
190 Old Derby Street, Hingham, MA 02043, U.S.A.

In all other countries, sold and distributed
by Kluwer Academic Publishers Group,
P.O. Box 322, 3300 AH Dordrecht, Holland.

D. Reidel Publishing Company is a member of the Kluwer Group.

Printed in The Netherlands

TABLE OF CONTENTS

EDITORS' FOREWORD

Colloquium No. 72 of the International Astronomical Union covered many observations and theoretical developments in the field of cataclysmic variables and related objects. Much time was devoted to discussions and we made an effort to include as much of the discussions material as possible in the proceedings.

The Local Organizing Committee would like to thank:

The International Astronomical Union for travel grants

The Israel Academy of Sciences for financial support

The Technion-Israel Institute of Technology for financial support and assistance

Bank Leumi Le-Israel for a generous support

We also thank the Dean of the Faculty of Physics, our colleagues and students for their assistance.

MARIO LIVIO

GIORA SHAVIV

SCIENTIFIC ORGANIZING COMMITTEE

B. Warner (Chairman), G.T. Bath, D. Crampton, J.E. Pringle, E.L. Robinson, G. Shaviv, R.E. Williams, J. Smak

LOCAL ORGANIZING COMMITTEE

G. Shaviv (Chairman), A. Finzi, M. Livio, H. Netzer, D. Sadeh

LIST OF PARTICIPANTS

BATH, Geoffrey, T.	Dept. of Astrophysics, Oxford, England
BIANCHINI, Antonio	Osservatorio Astronomico, Padova, Italy
BROWNLEE, Robert, R.	Los Alamos, New Mexico, U.S.A.
CHANMUGAM, Ganesh	Dept. of Physics and Astronomy, Louisiana State University, U.S.A.
COLVIN, Jeff	EG&G, Los Alamos, U.S.A.
COWLEY, Anne, P.	Dominion Astrophysical Observatory, Canada
CRAMPTON, David	Dominion Astrophysical Observatory, Canada
EGGLETON, Peter P.	Institute of Astronomy, Cambridge, England
EVANS, A.	Dept. of Physics, University of Keele, United Kingdom
FEINGOLD, Susan J.	Dept. of Physics, Technion, Israel
FINZI, A.	Dept. of Mathematics, Technion, Israel
FRIEDJUNG, Michael	Institut d'Astrophysique, Paris, France
GILLILAND, Ronald, L.	High Altitude Observatory, Boulder, U.S.A.
HARPAZ, Amos	Dept. of Physics, Technion, Israel
HAYLI, Avram	Observatoire de Lyon, France
HEISE, John	Space Research Lab., Utrecht, The Netherlands
KIEBOOM, Karin	Dept. of Physics, Technion, Israel
KING, Andrew	Astronomy Dept., University of Leicester, England
KITAMURA, M.	Tokyo Astronomical Observatory, Japan
KOZLOVSKY, Ben Zion	Dept. of Physics and Astronomy, Tel Aviv University, Israel
LAMB, Don, Q.	Harvard-Smithsonian Center for Astrophysics, U.S.A.
LANGER, Steven	Dept. of Astronomy, University of Illinois, U.S.A.
LA DOUS, Constanze	Universitats Sternwarte, Munchen, West Germany
LIVIO, Mario	Dept. of Physics, Technion, Israel
MACDONALD, Jim	Dept. of Astronomy, University of Illinois, U.S.A.
MATTEI, Janet, A.	AAVSO, U.S.A.
MAZEH, Tsevi	Dept. of Physics and Astronomy, Tel Aviv University, Israel
MEYER, Friedrich	Max Planck Institut, Munchen, West Germany
MITCHELL, R.M.	Dept. of Applied Mathematics, Queen Mary College, London, England
MIYAJI, Shigeki	College of Arts and Sciences, Chiba, Japan
MOUCHET, Martine	Observatoire de Paris-Meudon, France
NETZER, Hagai	Dept. of Physics and Astronomy, Tel Aviv University, Israel
PIIROLA, Vilppu	University of Helsinki Observatory, Helsinki, Finland
RAMATY, Reuven	Goddard Space Flight Center, U.S.A.

REGEV, Oded	Dept. of Physics, Technion, Israel
RITTER, Hans	Max Planck Institut, Munchen, West Germany
ROBINSON, Edward L.	Dept. of Astronomy, University of Texas, Austin, U.S.A.
SCHATZMAN, Evry	Observatoire de Nice, France
SHAVIV, Giora	Dept. of Physics, Technion, Israel
SHLOSMAN, Isaac	Dept. of Physics and Astronomy, Tel Aviv University, Israel
SION, Edward, M.	Dept. of Astronomy, Villanova University, U.S.A.
STERKEN, C.	Astrophysical Institute, University of Brussels, Belgium
SUGIMOTO, Daiichiro	Dept. of Earth Science and Astronomy, University of Tokyo, Japan
TAPIA, Santiago	Steward Observatory, University of Arizona, U.S.A.
WARNER, Brian	Dept. of Astronomy, University of Cape Town, South Africa
WILLIAMS, Bob	Steward Observatory, University of Arizona, U.S.A.

THE PERIODS OF CATACLYSMIC VARIABLE STARS

Edward L. Robinson
Department of Astronomy and McDonald Observatory
University of Texas, Austin, Texas, USA

I. INTRODUCTION

To understand the structure and evolution of the cataclysmic variables, we will need accurate values for their masses, dimensions, mass transfer rates, and other physical properties. Unfortunately, despite an abundance of observational data on these systems, there is a severe dearth of reliable, quantitative information about their fundamental physical properties. Only two cataclysmic variables, U Gem and EM Cyg, are simultaneously eclipsing binaries and double-lined spectroscopic binaries, and only for these two systems can masses and dimensions be determined with a minimum of assumptions (Stover 1981a; Stover, Robinson, and Nather 1981). Even if there were more systems like U Gem and EM Cyg, it is not obvious that our information would be any more reliable, because observers are often unable to agree on the values of the directly measured quantities used to determine physical properties. Thus, the radial velocity curve of the brightest dwarf nova, SS Cyg, has been measured independently 5 times in the last 30 years. The agreement among the measurements is unsatisfactory, and the reasons for the disagreement are not completely understood (Joy 1956; Kiplinger 1979; Stover et al. 1980; Cowley, Crampton, and Hutchings 1980; Walker 1981). The physical properties may still be unreliable when the disagreements are understood and eliminated, because there is considerable uncertainty about the proper way to extract physical properties from observational data. For example, the observed radial velocity curves of cataclysmic variables are believed to be different from the true radial velocity curves of their component stars, but the amount of difference and ways to correct for the difference are unknown.

There is only one physical property known accurately for a large number of cataclysmic variables: the orbital period. The orbital periods have, therefore, an importance out of proportion to their immediate information content. The purpose of the present paper is to gather together into one place the data on the orbital periods of cataclysmic variables; to discuss the selection effects distorting

M. Livio and G. Shaviv (eds.), Cataclysmic Variables and Related Objects, 1–15.

Table 1

The Orbital Periods of Cataclysmic Variable Stars

Star	Class	Normal Mag (V)	Orbital Period (hr)	References
RX And	UG	13.7	5.08	Hutchings and Thomas (1982)
AE Aqr	MV	11.5	9.88	Chincarini and Walker (1981)
V603 Aql	N	11.9	3.48	Haefner (1981)
TT Ari	NL	10.6	3.2	Smak and Stepien (1975)
T Aur	N	15.4	4.91	Mumford (1976)
SS Aur	UG	14.8	4.33	Kraft and Luyten (1965)
KR Aur	NL	13.5	3.91	Shafter (1982)
Z Cam	UG	14.5	6.96	Kraft, Krzeminski, Mumford (1969)
YZ Cnc	SU	14.6	2.21(S)	Patterson (1979a)
AC Cnc	NL	13.8	7.21	Kurochkin and Shugarov (1981)
OY Car	SU	16	1.51	Vogt et al. (1981)
QU Car	NL	11.6	10.9	Gilliland and Phillips (1982)
HT Cas	UG	16.5	1.77	Patterson (1981)
BV Cen	UG	13.2	14.66	Gilliland (1983)
V436 Cen	SU	15.5	1.50	Gilliland (1982a)
WW Cet	NL	14.5	3.8	Kraft and Luyten (1965)
Z Cha	SU	15.2	1.79	Cook and Warner (1981)
T CrB	RN	10.8	227.5 d	Paczynski (1965)
SS Cyg	UG	12.2	6.60	Stover et al. (1980)
EM Cyg	UG	14.0	6.98	Mumford (1980)
V1500 Cyg	N	>20	3.35	Patterson (1979b)
V1668 Cyg	N		10.54?	Campolonghi et al. (1980)
HR Del	N	12.5	4.10	Hutchings (1979)
EF Eri	AM	15.3	1.35	Schneider and Young (1980a)
U Gem	UG	14.5	4.25	Arnold, Berg, Duthie (1976)
AM Her	AM	12.9	3.09	Young and Schneider (1979)
DQ Her	N	14.6	4.65	Africano and Olson (1981)
EX Hya	NL	13.5	1.64	Gilliland (1982b)
VW Hyi	SU	13.9	1.78	Vogt (1974)
WX Hyi	SU	14.9	1.80	Schoembs and Vogt (1981)
AY Lyr	SU	18.3	1.81(S)	Patterson (1979a)
MV Lyr	NL	12.5	3.21	Schneider, Young, Shectman (1981)
TU Men	SU	17	2.82	Stolz and Schoembs (1981)
BT Mon	N	15.4	8.01	Robinson, Nather, Kepler (1982)
V2051 Oph	NL	14	1.50	Bond (1977)
CN Ori	UG	14.8	3.91?	Schoembs (1982)
BD Pav	N	>16.4	4.30	Barwig and Schoembs (1981)
RU Peg	UG	13.1	8.99	Stover (1981b)
GK Per	N	14.0	1.99d	Crampton (1982)
RR Pic	N	12.3	3.48	Vogt (1975)
VV Pup	AM	15	1.67	Schneider and Young (1980b)
RZ Sge	SU	17.4	1.68(S)	Bond, Kemper, Mattei (1982)
WY Sge	N	19	3.69	Shara et al. (1982)
WZ Sge	SU?	15.3	1.36	Robinson, Nather, Patterson (1978)
V3885 Sgr	NL	10.4	4.94	Cowley, Crampton, Hesser (1977a)
VZ Scl	NL	15.6	3.47	Warner and Thackeray (1975)
LX Ser	NL	14.0	3.80	Africano and Klimke (1981)
RW Sex	NL	10.6	5.93	Cowley, Crampton, Hesser (1977b)
RW Tri	NL	12.8	5.57	Longmore et al. (1981)
EK TrA	SU	>17	1.56(S)	Vogt and Semeniuk (1980)
UX UMa	NL	13.5	4.72	Kukarkin (1977)
AN UMa	AM	16.5	1.91	Liebert et al. (1982)
0139-68	AM	15.6	1.83	Visvanathan and Pickles (1982)
0526-328	MV?	13.5	5.49	Hutchings et al. (1981b)
0643-1648	UG	13.2	4.3 or 5.3	Hutchings et al. (1981a)
1012-03	NL	14.5	3.23	Williams and Ferguson (1982)
1013-477	AM	17	1.72	Mason et al. (1982a)
1103+254	AM	16.2	1.90	Stockman et al. (1982)
1115+18	AM	18.2	1.50	Biermann et al. (1982)
1148+719	NL?	16	3.9	Patterson et al. (1982)
1405-451	AM	15.5	1.69	Mason et al. (1982b); Tapia (1982)
1550+191	AM	15.4	1.89	Liebert et al. (1981)
2129+47	NL	16.9	5.24	Thorstensen et al. (1979)
2215-086	MV	13.5	4.85	Shafter and Targon (1981)
2252-035	MV	13.3	3.59	Warner et al. (1981)
Lanning 10	NL	14.2	7.71	Horne et al. (1982)

the sample of orbital periods; and to determine the properties of the period distribution.

II. OBSERVATIONAL DATA

a) The Known Orbital Periods

The cataclysmic variables whose orbital periods are known with reasonable certainty are listed in Table 1 along with their class, their normal V magnitude, their orbital period, and a reference to the source of the orbital period. Binary systems that probably contain neutron stars instead of white dwarfs, such as 2A1822-371, have been excluded from the table. The two binary white dwarfs AM CVn and G61-29 (= GP Com), the peculiar variable V Sge, and binary systems without mass transfer such as V471 Tau have also been excluded. Orbital periods are now known for 66 cataclysmic variables.

The dwarf novae have been divided into two classes in Table 1: the normal dwarf novae (UG), and the SU UMa stars (SU) (for a description of the SU UMa stars, see Vogt 1980). The other sub-classes of the dwarf novae have been put into one or the other of these two classes; the Z Cam stars, for example, have been included in the UG class. The novae (N) have not been further subdivided, nor have the recurrent novae (RN). The remaining variables have been classified as nova-like variables (NL) except for the nova-like variables with magnetized white dwarfs, which have been placed in the AM Her class (AM) if the white dwarf rotates synchronously, and in the magnetic variable class (MV) if it does not.

We have used the quantity "Normal" V magnitude in Table 1 instead of the minimum magnitude or magnitude range. Neither the minimum nor the maximum magnitudes are appropriate quantities for discussing the statistics of orbital periods of cataclysmic variables because cataclysmic variables usually spend most of their time at some intermediate magnitude. Thus, the dwarf nova Z Cha fades to near magnitude 17 at the bottom of its deep eclipse and can reach magnitude 11.9 during its eruptions. These are the minimum and maximum magnitudes given by the General Catalogue of Variable Stars. However, Z Cha spends most of its time between eruptions near magnitude 15.2. I have defined the normal magnitude of Z Cha to be 15.2, and have entered that value in Table 1. Defining appropriate normal magnitudes for the cataclysmic variables requires some subjective interpretation of their light curves, so the normal magnitudes I give in Table 1 could differ by perhaps 1/2 magnitude from estimates for the same quantity made by others.

Some of the periods given for the SU UMa stars are superhump periods, not true orbital periods. There is an (S) appended to these periods in Table 1. If there is any doubt about the reliability of the period, a question mark has been appended. Finally, the references given in Table 1 are the most recent careful determinations of the orbital periods, not necessarily the most complete studies of the general properties of the systems.

b) Selection Effects

The distribution of cataclysmic variables as a function of orbital period is shown in Figure 1, where the orbital periods have been grouped together into one hour bins. Although there are some striking features to the distribution, the selection effects at work in Figure 1 are significant and must be dealt with before the distribution can be interpreted properly.

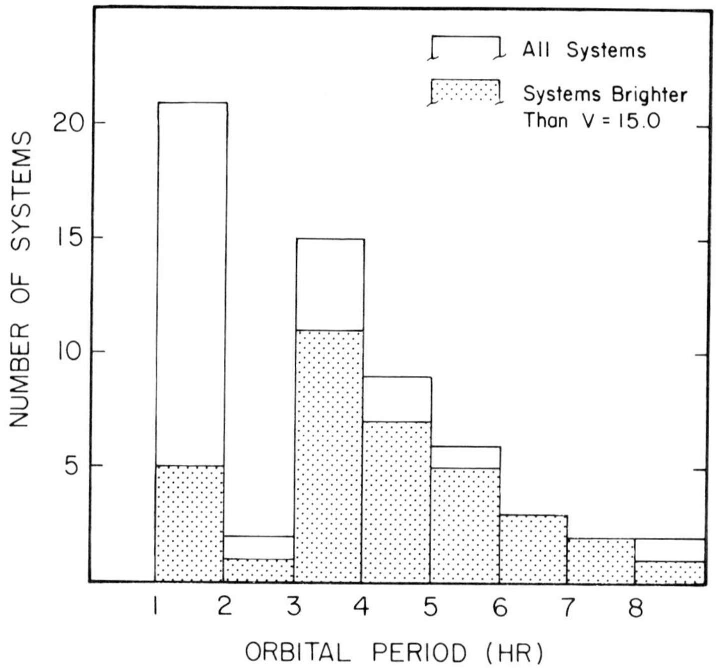

Figure 1 - The distribution of cataclysmic variables as a function of orbital period. Two samples of the known orbital periods have been graphed: the sample of all periods, and the sample of periods for cataclysmic variables brighter than magnitude 15. The latter sample is 70 percent complete.

The most serious selection effects fall into three groups. (1) Because different classes of cataclysmic variables have different light curves and luminosities, lists of cataclysmic variables are more complete for some classes than for others. For example, dwarf novae are easy to discover because of their frequent eruptions, but the nova-like variables are difficult to discover unless they have magnetized white dwarfs and are strong X-ray sources. As a result, dwarf novae and systems with magnetized white dwarfs are probably over-represented in Table 1, whereas nova-like variables without magnetized white dwarfs are under-represented. (2) Information is more complete for those classes of cataclysmic variables that have attracted the attention of the observers than for those that have not. The AM Her class is an example of this selection effect: the orbital period of <u>every</u> AM Her star has been measured. (3) The specific technique used to observe a cataclysmic variable will be more effective at detecting orbital modulation at some orbital periods than at others. Thus, photometry will detect orbital light variations more readily if the orbital period is near 2 hours than if it is near 10 hours.

The last two selection effects can be reduced or eliminated by taking a magnitude-limited, complete sample of the orbital periods. It is possible to achieve a good approximation to a magnitude-limited, complete sample by taking a subset of the periods given in Table 1. The upper half of Table 2 gives both the total number of cataclysmics known and the number with measured orbital periods in 1 magnitude intervals from magnitude 10.0 to 15.0. The sample is highly incomplete for cataclysmic variables fainter than magnitude 15.0, but even for the relatively faint variables with magnitudes between 14.0 and 15.0 the sample is now more than 50 percent complete. The sample is 70 percent complete for all variables brighter than magnitude 15.0, so by restricting ourselves to the variables brighter than magnitude 15.0 we obtain nearly the magnitude-limited, complete sample we need. This limited sample still has 40 systems, enough to give reasonable reliability to our results. The lower half of Table 2 lists the bright cataclysmic variables whose orbital periods have not yet been determined.

Figure 1 compares the distribution of orbital periods in the magnitude-limited sample to the distribution of periods in the entire sample. At orbital periods longer than about 3 hours the shapes of the two distributions are similar, but in the period range from 1 to 2 hours the distributions are markedly different. The difference is caused by the absence from the magnitude-limited sample of many SU UMa stars and AM Her stars that appear in the entire sample. These two groups have attracted considerable attention from observers in the past few years, so they are over-represented in the entire sample.

Table 2

The Completeness of the Orbital Period Sample

Magnitude Range	Systems With Known Orbital Periods	Total Number of Systems Known	Cumulative Percent With Periods
<10.0	0	0	
10.1 - 11.0	6	6	100
11.1 - 12.0	2	2	100
12.1 - 13.0	5	7	87
13.1 - 14.0	14	20	77
14.1 - 15.0	13	22	70

Bright Cataclysmic Variables Without Orbital Periods:

AE Ara	T Pyz
SY Cnc	V426 Oph
V751 Cyg	V841 Oph
DN Gem	V630 Oph
AH Her	V1016 Sgr
DI Lac	V2562 Sgr
TU Leo	VY Scl
RS Oph	SU UMa
KT Per	

Although use of the magnitude-limited sample reduces the importance of the last two selection effects, the selection effects in the first group are still acting, and are distorting the distribution, perhaps severely. For example, novae, which have mean absolute visual magnitudes near +4.5, are much brighter than dwarf novae, which have mean absolute visual magnitudes near +7.5 (McLaughlin 1960; Kraft and Luyten 1965). Because of the 3 magnitude difference in their absolute magnitudes, the volume of space included in a magnitude-limited sample is 60 times larger for novae than for dwarf novae. If all other selection effects were the same for the two groups (they are not), the classical novae would be over-represented in the magnitude-limited sample by a factor of 60 compared to the dwarf novae. All novae have orbital periods longer than 3 hours and most novae have orbital periods between 3 and 5 hours, so there is an excess or novae in the magnitude-limited sample in this period range. If they were removed from the sample, the large peak in the period distribution at these periods would be reduced.

There is another, even larger selection effect acting on the novae. No nova has ever been recognized to be a cataclysmic variable before its eruption; and even if a pre-nova were to appear in a list of cataclysmic variables, it is unlikely it would be recognized as a nova. Therefore, a nova only appears in Table 1 if it has erupted within the past few hundred years. As the eruptions of novae are thought to recur every 10^5 years or so, only 1 out of every 10^3 novae in the solar neighborhood has been found, and the novae are under-represented in Table 1 by this large factor.

There may be enough information available to estimate the true space density of dwarf novae and correct all the selection effects acting on their period distribution, but the necessary information is not available for the other classes. Thus, calculation of the true space density of novae requires knowledge of their mean recurrence time, but their recurrence time can only be estimated from theoretical arguments and could easily be in error by more than a factor of 10. Except for the dwarf novae, the selection effects involve such large and uncertain factors that any attempt to correct for them would risk introducing other even larger biases into the period distribution. Therefore, we will not correct for the remaining selection effects; and therefore, we must refrain from interpreting the fine details in the distribution of orbital periods.

III. RESULTS

a) The 80 Minute Period Limit

The period distribution has several features too strong to be caused by selection effects. Perhaps the most important of them is that the distribution terminates at an orbital period near 80 minutes. There are excellent reasons for believing that the 80 minute limit is not due to an observational selection effect, and that cataclysmic variables with orbital periods less than 80 minutes are non-existent or extremely rare. The first reason is that the period distribution does not fade gradually to zero at the limit, it terminates abruptly. There are 9 cataclysmic variables with orbital periods between 80 minutes and 100 minutes. It is difficult to imagine any selection effect that would permit so many systems to be found in such a narrow range of orbital periods, but render invisible all systems with orbital periods just shorter than 80 minutes. The second reason is that four binary stars with orbital periods shorter than 80 minutes are known, but none of them are cataclysmic variables. AM CVn and GP Com have orbital periods of 17.5 min and 46.5 min respectively, but are binary white dwarfs (Faulkner, Flannery, and Warner 1972; Nather, Robinson, and Stover 1981); 4U 1626-67 and 4U 1915-05 have orbital periods of 41.5 min and 50 min respectively, but the accreting stars in these systems are neutron stars not white dwarfs (White and Swank 1982; Walter *et al.* 1982; Middleditch *et al.* 1981). Therefore, there is nothing preventing observers from finding binary systems with short orbital periods.

Paczynski and Sienkiewicz (1981), and Rappaport, Joss, and Webbink (1982) have provided a theoretical explanation of the minimum period that appears to be at least qualitatively correct. The mass of the late type star in a cataclysmic variable decreases as it transfers mass to its white dwarf companion. While the mass of the late type star is large enough to support hydrogen burning in its core, it is near the main sequence mass-radius relation, and its radius decreases as its mass decreases. However, once the late type star has lost so much mass that it cannot burn hydrogen in its core, it must switch to the mass-radius relation for hydrogen white dwarfs, and thereafter its radius increases as its mass decreases. Thus, there is a minimum radius for the late type stars in cataclysmic variables. The minimum radius translates directly into a minimum orbital period, because the radius of the late type star must equal the radius of its Roche lobe to have mass transfer, and the radius of the Roche lobe sets the orbital period of a cataclysmic variable to within narrow limits.

While the late type star is switching from the main sequence mass-radius relation to the white dwarf mass-radius relation, it does not really fit either relation. During the switch the time scale for mass transfer becomes comparable to the thermal time scale of the late type star, and the late type star is thrown out of thermal equilibrium. The radius of the late type star becomes larger than its equilibrium radius; the greater the mass transfer rate, the larger its radius becomes. In the theoretical calculations by Rappaport, Joss, and Webbink (1982) and Paczynski and Sienkiewicz (1981), the rate of mass transfer, and thus the minimum orbital period, was set by the rate at which gravitational radiation carries angular momentum away from the binary system. Their calculations gave minimum periods between 60 and 70 minutes depending on the parameters of their models. Although the theoretical minimum period is close to the observed minimum period, Rappaport, Joss, and Webbink were unable to raise the theoretical minimum to 80 minutes without altering their models beyond reason. As the observed 80 minute limit is firmly established, the source of the discrepancy appears to be in the theory, not the observations.

b) The Period Gap

The gap in the distribution of orbital periods between 2 and 3 hours is striking. Figure 2 divides the period distribution into narrower period bins than Figure 1 to show more detail near the edges of the gap. The data used to make Figure 2 is not exactly the same as that for Figure 1. The periods of the four cataclysmic variables for which only superhump periods are available have been reduced by a few percent according to the formula of Stolz and Schoembs (1981) to give a better approximation to their true orbital periods. In particular, we have taken periods of 2.10 hr for YZ Cnc 1.75 hr for

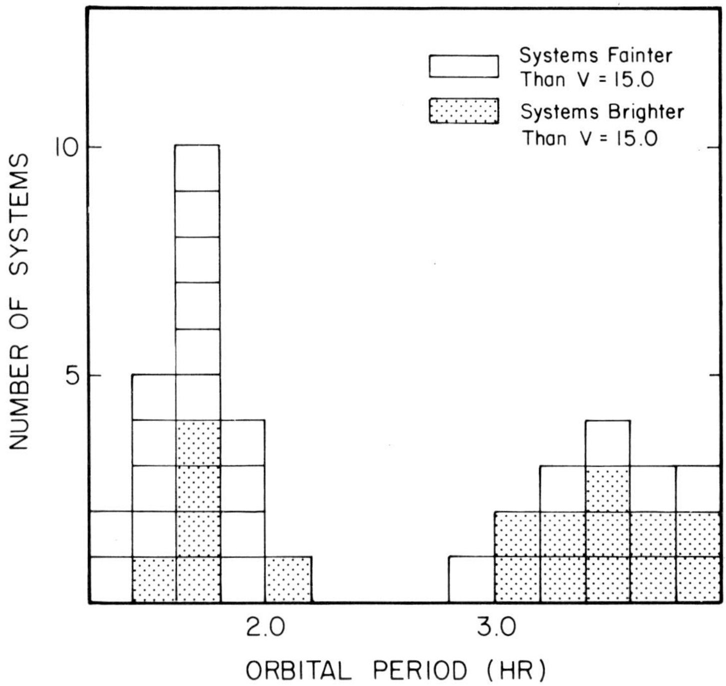

Figure 2 - The distribution of cataclysmic variables as a function of orbital period at orbital periods between 1.2 and 4.0 hours. The period distribution has been divided into narrower period bins than in Figure 1 to show more detail near the edges of the period gap. As described in the text, the periods used to make Figure 2 are not exactly the same as those tabulated in Table 1.

AY Lyr, 1.64 hr for RZ Sge, and 1.53 hr for EK TrA. The period gap runs from YZ Cnc at 2.10 hr to AM Her at 3.09 hr in the magnitude-limited sample and is nearly a full hour wide. The gap is more than 40 minutes wide even in the sample of all orbital periods. There are no other gaps from 80 minutes to 4.33 hours in the sample of all periods, and no gap of comparable width between 80 minutes and 6 hours. Figure 2 shows that even though the magnitude-limited sample has fewer stars, its distribution is also nearly unbroken except for the 2 to 3 hour gap. The statistical significance of the period gap is possibly not as high as the human eye would have it, but its existence is not in doubt.

The existence of the period gap implies two facts about the evolution of cataclysmic variable stars. First, cataclysmic variables do not form initially in the gap, either because it is impossible for close binary stars of any kind to form with periods between 2 and 3 hours, or because they are initially detached and cannot begin to transfer mass until they have evolved out of the gap. Webbink (1979) has suggested that the gap is caused by a difference in the evolution of cataclysmic variables with helium white dwarfs, which have masses less than $\sim$0.45 $M_\odot$, and cataclysmic variables with carbon-oxygen white dwarfs, which have masses greater than $\sim$0.55 $M_\odot$. The low mass systems form below the gap, and the high mass systems form above the gap. However, the recent discovery that the central star of the planetary nebula Abell 41 is a detached binary with an orbital period of 2.72 hr suggests that binary systems can form in the gap, but do not begin mass transfer until they have evolved out of the gap (Grauer 1982).

The existence of the gap also requires that cataclysmic variables do not evolve after their formation in a way that fills the gap, either because they do not evolve into the gap (e.g. Whyte and Eggleton 1980), because they change to a different kind of binary within the gap and are no longer cataclysmic variables (e.g. Robinson, et al. 1981), or because they evolve through the gap so rapidly that we are unlikely to find them in the gap (Rappaport, Joss, and Webbink 1982). The mass transfer in a few cataclysmic variables, AM Her and MV Lyr among them, can cease for extended periods of time, up to several years in MV Lyr. Robinson et al. (1981) noted that these systems are found preferentially near the edges of the period gap, and suggested on these grounds that cataclysmic variables do evolve through the period gap, but they cease transferring mass while in the gap, thus becoming extremely difficult to detect. The recent calculations by D'Antona and Mazzitelli (1982) support this suggestion.

c) Other Correlations

The various types of cataclysmic variables are segregated nearly perfectly according to orbital period, with an orbital period near 3.2 hr being a universal dividing line between groups. All the SU UMa dwarf novae have extremely short orbital periods; the shortest period in the group is 1.50 hr for V436 Cen and the longest 2.82 hr for TU Men. In contrast, with the sole exception of HT Cas, all the normal (UG) dwarf novae have orbital periods longer than 3.91 hr. Thus, except for HT Cas, the periods of the SU UMa stars and the normal dwarf novae fall into non-overlapping ranges. No explanation for the segregation of the dwarf novae into two period groups has been suggested. As the orbital period of HT Cas is 1.77 hr, placing it firmly in the middle of the SU UMa period range, perhaps further observations may show it is an SU UMa star not a UG star.

The magnetic variables show a segregation by period similar to that of the dwarf novae. All the synchronously rotating magnetic variables (the AM Her stars) have periods less than 3.09 hr, whereas all the non-synchronously rotating variables (the MV stars) have orbital periods longer than 3.59 hours. This correlation survives if systems like DQ Her and V533 Her, that have short period, rapid coherent oscillations, are included in the MV class; but the correlation may have an exception in EX Hya, which has a period of 1.64 hr and could possibly be a non-synchronous rotator (Gilliland 1982). Once again, the explanation for the segregation by period is unknown.

Finally, the classical novae all have periods above the 3.2 hr dividing period, with the shortest period being 3.35 hr for V1500 Cyg.

One other correlation betweent the type of a cataclysmic variable and its orbital period is known. As noted originally by Warner (1976), the recurrent novae tend to have giants instead of main sequence stars for their mass losing stars, and must have orbital periods of hundreds of days. This last correlation is not perfect. The recurrent nova T Pyx has colors too blue, B-V = 0.12 and U-B = -1.04 (Mumford 1971), to contain a giant; and recent observations of the recurrent nova U Sco show that it is too faint at minimum light to contain a giant.

IV. SUMMARY

Moderately or highly reliable orbital periods have been measured for 66 cataclysmic variables. The sample of periods is 70 percent complete for the known cataclysmic variables brighter than magnitude 15. Despite the large selection effects in the sample, even in the magnitude-limited sample, several reliable conclusions can be drawn about the distribution of the periods.

1. The distribution terminates abruptly at a period of 80 minutes. Cataclysmic variables with orbital periods less than 80 minutes are rare or non-existent.

2. There is a gap in the distribution between 2.1 hr (YZ Cnc) and 2.82 hr (TU Men). Normal cataclysmic variables with orbital periods in this range are rare or non-existent.

3. Most of the classes of cataclysmic variables are found in restricted period ranges, and a period near 3.2 hr is a universal boundry line between ranges. All but one of the normal dwarf novae (the UG stars in Table 1) have orbital periods longer than 3.9 hr, and all the SU UMa type dwarf novae have orbital periods less than 3.0 hr. The AM Her stars all have orbital periods less than 3.1 hr, and the non-synchronously rotating magnetic variables all have orbital periods longer than 3.5 hr. The novae all have periods longer than 3.3 hr. The recurrent novae tend to have long periods.

Theoretical explanations have been given for the first two of these properties, but the theoretical calculations do not yet agree quantitatively with the observed distribution. The correlations between the orbital periods and the types of the cataclysmic variables remain unexplained.

This work was supported in part by NSF grant AST-8108691.

REFERENCES

Africano, J. L., and Klimke, A. 1981, IAU Inf. Bull. Var. Stars, No. 1969.
Africano, J. L., and Olson, E. C. 1981, P.A.S.P., 93, 130.
Arnold, S., Berg, R. A., and Duthie, J. G. 1976, Ap. J., 206, 790.
Barlow, M. J., Brodie, J. P., Brunt, C. C., Hanes, D. A., Hill, P. W., Mayo, S. K., Pringle, J. E., Ward, M. J., Watson, M. G., Whelan, J. A. J., and Willis, A. J. 1981, M.N.R.A.S., 195. 61.
Barwig, H., and Schoembs, R. 1981, IAU Inf. Bull. Var. Stars, No. 2031.
Biermann, P., Kuhr, H., Liebert, J., Stockman, Strittmatter, P., and Tapia, S. 1982, IAU Circulars, No. 3680.
Bond, H. E. 1977, private communication.
Bond, H. E., Kemper, E., and Mattei, J. A. 1982, Ap. J., in press.
Campolonghi, F., Gilmozzi, R., Guidoni, U., Messi, R., Natali, G., and Well, J. 1980, Astron. Ap., 85. 14.
Chincarini, G., and Walker, M. F. 1981, Astron. Ap., 104, 24.
Cook, M. C., and Warner, B. 1981, M.N.R.A.S., 196, 55p.
Cowley, A. P., Crampton, D., Hesser, J. E. 1977a, Ap. J., 214, 471.
Cowley, A. P., Crampton, D., Hesser, J. E. 1977b, P.A.S.P., 89, 716.
Cowley, A. P., Crampton, D., and Hutchings, J. B. 1980, Ap. J., 241, 269.
Crampton, D. 1982, private communication.
D'Antona, F., and Mazzitelli, I. 1982, Ap. J., 260, in press.
Faulkner, J., Flannery, B., and Warner, B. 1972, Ap. J. (Letters), 175, L79.
Gilliland, R. L. 1982a, Ap. J., 254, 653.
Gilliland, R. L. 1982b, Ap. J., 258, 576.
Gilliland, R. L. 1983, Ap. J., in press.
Gilliland, R. L., Phillips, M. M. 1982, Ap. J., 261, in press.
Grauer, A. D. 1982, private communication.
Haefner, R. 1981, IAU Inf. Bull, Var. Stars, No. 2045.
Horne, K., Lanning, H. H., and Gomer, R. H. 1982, Ap. J., 252, 681.
Hutchings, J. B. 1979, Ap. J., 232, 176.
Hutchings, J. B., Cowley, A. P., Crampton, D., Williams, G. 1981a, P.A.S.P., 93. 741.
Hutchings, J. B., Crampton, D., Cowley, A. P., Thorstensen, J. R., and Charles, P. A. 1981b, Ap. J., 249, 680.
Hutchings, J. B., and Thomas, B. 1982, P.A.S.P., 94, 102.
Joy, A. H. 1956, Ap. J., 124, 317.
Kiplinger, A. L. 1979, A.J., 84, 655.
Kraft, R. P., Krzeminski, W., and Mumford, G. S. 1969, Ap. J., 158, 589.
Kraft, R. P., and Luyten, W. J. 1965, Ap. J., 142, 1041.
Kukarkin, B. V. 1977, M.N.R.A.S., 180, 5p.
Kurochkin, N. E., and Shugarov, S. Yu. 1981, Astron. Tsirk., No. 1154.

Liebert, J., Stockman, H. S., Williams, R. E., Tapia, S., Green, R. F., Rautenkranz, D., Ferguson, D. H., and Szkody, P. 1982, Ap. J., 256, 594.

Liebert, J., Tapia, S., Bond, H. E., and Grauer, A. D. 1982, Ap. J., 254, 232.

Longmore, A. J., Lee, T. J., Allen, D. A., and Adams, D. J. 1981, M.N.R.A.S., 195, 825.

Mason, K., Middleditch, J., Cordova, F., Jensen, K., Reichert, G., Bowyer, S., Murdin, P., and Clark, D. 1982a, IAU Circulars, No. 3685.

Mason, K. O., Middleditch, J., Cordova, F. A., Jensen, K. A., Reichert, G., Murdin, P. G., Clark, D., Bowyer, S. 1982b, Ap. J., in press.

McLaughlin, D. B. 1960, in Stars and Stellar Systems, Vol. 6, Stellar Atmospheres, ed. J. Greenstein (Chicago: University of Chicago Press), p. 585.

Middleditch, J., Mason, K. O., Nelson, J. E., and White, N. E. 1981, Ap. J., 244, 1001.

Mumford, G. S. 1971, Ap. J., 165, 369.

Mumford, G. S. 1976, Ap. J., 210, 416.

Mumford, G. S. 1980, A.J., 85, 748.

Nather, R. E., Robinson, E. L., and Stover, R. J. 1981, Ap. J., 244, 269.

Paczynski, B. 1965, Acta Astr., 15, 197.

Paczynski, B., and Sienkiewicz, R. 1981, Ap. J. (Letters), 248, L27.

Patterson, J. 1979a, A.J., 84, 804.

Patterson, J. 1979b, Ap. J., 231. 789.

Patterson, J. 1981, Ap. J. (Suppl.), 45, 517.

Patterson, J., Schwartz, D. A., Bradt, H., Remillard, R., McHardy, I. I., Pye, J. P., Williams, G., Fesen, R. A., and Szkody, P. 1982, Bull. Amer. Ast. Soc., 14, 618.

Rappaport, S., Joss, P., and Webbink, R. F. 1982, Ap. J., 254, 616.

Robinson, E. L., Barker, E. S., Cochran, A. L., Cochran, W. D., and Nather, R. E. 1981, Ap. J., 251, 611.

Robinson, E. L., Nather, R. E., and Kepler, S. O. 1982, Ap. J., 254, 646.

Robinson, E. L., Nather, R. E., and Patterson, J. 1978, Ap. J., 219, 168.

Schneider, D. P., and Young, P. 1980a, Ap. J., 238, 946.

Schneider, D. P., and Young, P. 1980b, Ap. J., 240, 871.

Schneider, D. P., Young, P., and Shectman, S. A. 1981, Ap. J., 245, 644.

Schoembs, R. 1982, IAU Inf. Bull. Var. Stars, No. 2116.

Schoembs, R., and Vogt, N. 1981, Astron. Ap., 97, 185.

Shafter, A. 1982, IAU Circulars, No. 3689.

Shafter, A. W., and Targon, D. M. 1981, Bull. Amer. Ast. Soc., 13, 802.

Shara, M., Moffat, A. F. J., McGraw, J., Dearborn, D. S., Bond, H. E., and Kemper, E. 1982, IAU Circulars, No. 3707.

Smak, J., and Stepien, K. 1975, Acta Astr., 25, 379.

Stockman, H., Foltz, C., Tapia, S., Schmidt, G., and Grandi, S. 1982, IAU Circulars, No. 3696.

Stolz, B., and Schoembs, R. 1981, IAU Inf. Bull. Var. Stars, No. 2029.

Stover, R. J. 1981a, Ap. J., 248, 684.
Stover, R. J. 1981b, Ap. J., 249, 673.
Stover, R. J., Robinson, E. L., Nather, R. E. 1981, Ap. J., 248, 696.
Stover, R. J., Robinson, E. L., Nather, R. E. and Montemayor, T. J. 1980, Ap. J., 240, 597.
Tapia, S. 1982, IAU Circulars, No. 3685.
Thorstensen, J., Charles, P., Bowyer, S., Briel, U. G., Doxsey, R. E., Griffiths, R. E., and Schwartz, D. A. 1979, Ap. J. (Letters), 233, L59.
Visvanathan, N., and Pickles, A. 1982, Nature, 298, 41.
Vogt, N. 1974, Astron. Ap., 36, 369.
Vogt, N. 1975, Astron. Ap., 41, 15.
Vogt, N. 1980, Astron. Ap., 88, 66.
Vogt, N., Schoembs, R., Krzeminski, W., and Pedersen, H. 1981, Astron. Ap., 94, L29.
Vogt, N., and Semeniuk, I. 1980, Astron. Ap., 89, 223.
Walker, M. F. 1981, Ap. J., 248, 256.
Walter, F. M., Bowyer, S., Mason, K. O., Clarke, J. T., Henry, J. P., Halpern, J., and Grindlay, J. E. 1982, Ap. J. (Letters), 253, L67.
Warner, B. 1976, in IAU Symposium 73, Structure and Evolution of Close Binary Systems (Dordrecht, Holland: Reidel), p. 85.
Warner, B., O'Donoghue, D., and Fairall, A. P. 1981, M.N.R.A.S., 196, 705.
Warner, B., and Thackeray, A. D. 1975, M.N.R.A.S., 172, 433.
Webbink, R. F. 1979, in IAU Colloquium 46, Changing Trends in Variable Star Research (Hamilton, N. Z.: University of Waikato Press), p. 102.
White, N. E., and Swank, J. H. 1982, Ap. J. (Letters), 253, L61.
Whyte, C., and Eggleton, P. 1980, M.N.R.A.S., 190, 809.
Williams, R. E., and Ferguson, D. H. 1982, Ap. J., 257, 672.
Young, P., and Schneider, D. P. 1979, Ap. J., 230, 502.

DISCUSSION FOLLOWING E. ROBINSON'S TALK

BATH: How well defined are the classes of dwarf novae in the sense of distinctions between SU Ursa Majoris systems and what we call normal, long period, dwarf novae? I ask this question because data that the BAA have obtained indicates that U Gem itself shows two classes of outbursts. What criteria are you using to distinguish SU UMa systems from normal dwarf novae?

ROBINSON: The classical SU UMa star has in addition to rather short eruptions, whose interval is only statistically predictable, eruptions, that are perhaps ten times longer in separation, in interval, than the short eruptions. They are much more predictable in the interval and the amplitudes are perhaps a magnitude or so higher and the life of the eruptions is maybe, two to three times longer. The classical SU UMa stars I think are pretty distinctive and I don't think would include a U Gem star.

BATH: I will show the light curve of U Gem itself and it is very difficult to distinguish between the two classes.

ROBINSON: I will have to see that before I would agree with that.

LAMB: Fred Lamb and I have recently completed some work on the magnetic torque in AM Her systems that I shall describe later. We find that AM Her is very much on the edge of being able to be synchronously rotating. The torque we discuss is due to magnetic coupling, but it is quite different from the one that has been proposed by Joss, Katz and Rappaport. We find a critical binary period beyond which synchronization cannot be achieved within the evolutionary lifetime of the system. It is sensitive to the white dwarf's magnetic field but is about 3 hours for 2×10^7 gauss. It may explain the distinction you mention.

MATTEI: I want to make a comment concerning SU UMa stars. I think when we compare the light curve of SU UMa stars with other dwarf novae, there is no question in distinguishing SU UMa type light curves from other types of dwarf novae, in fact just from the light curve we were able to determine another addition to the members of SU UMa stars, RZ Sagittae and Howard Bond and co workers have been able to definitely confirm that and recently I think I have another member of SU UMa stars just from the light curve Ty Pisces.

ROBINSON: Perhaps, you can tell me, has SU Ursa Majoris yet proved itself to be an SU UMa star?

MATTEI: Well, in fact, I would like to speak about SU Ursa Majoris because it has actually stopped being even a dwarf nova. It is just not doing anything.

ROBINSON: What about HT Cas?

MATTEI: HT Cas is not doing anything either.

WARNER: The intermediate polars I am going to talk about, there are five of them with orbital periods known and three certainly have periods very close to your 3.1 hours which is possibly suggestive and possibly not. In such a small sample it is unusual to find so many just at the point where you say there is a break in the type of object.

PHOTOMETRIC AND SPECTROSCOPIC OBSERVATIONS OF THE CATACLYSMIC VARIABLE AC CANCRI

M. Kitamura
Tokyo Astronomical Observatory
University of Tokyo, Mitaka, Tokyo 181, Japan

A. Okazaki
Tsuda College, Kodaira, Tokyo 187, Japan

A. Yamasaki
Dept. Earth Science and Astronomy
University of Tokyo, Meguro-ku, Tokyo 153, Japan

ABSTRACT

The eclipsing variable AC Cancri, recently identified as a cataclysmic variable by Shugarov (1981) and Okazaki, Kitamura and Yamasaki (1982), was further studied in detail on the bases of UBV-colours, multi-channel photoelectric light curves and image-tube spectrograms which were newly obtained at Tokyo Astronomical Observatory. The result of analysis of these materials for the nature of the cataclysmic variable is presented.

AC Cnc was discovered to be a variable star by Kurochkin (1960). Subsequently, Kurochkin and Shugarov (1980) found that this variable is an eclipsing binary with the orbital period of P = $0\overset{d}{.}30047768$. From UBV photoelectric light curves Shugarov (1981) suggested that AC Cnc is an ex-nova or a nova-like object. Okazaki, Kitamura and Yamasaki (1982) have identified AC Cnc as a cataclysmic variable from spectroscopic and photometric observations made in April 1981.

In order to study the nature of this cataclysmic variable in detail, we further made photometric and spectroscopic observations at Okayama and Dodaira Stations of Tokyo Astronomical Observatory as follows:

- UBV photoelectric photometry with the 0.9-m reflectors at Okayama and Dodaira during October 1981 to January 1982,
- multi-channel photometry with the 1.9-m reflector at Okayama in January 1982,
- spectroscopic observations using a Cassegrain image-tube spectrograph attached to the 1.9-m reflector at Okayama in April 1982.

M. Livio and G. Shaviv (eds.), Cataclysmic Variables and Related Objects, 17–20.

From UBV photoelectric observations the average magnitude and colour indices of AC Cnc outside eclipses during October 1981 to January 1982 were obtained as

$$V = 14.2$$
$$B - V = +0.4$$
$$U - B = -0.5.$$

Compared with this period, the star was brighter in April 1981 by $0\overset{m}{.}6$, $0\overset{m}{.}5$ and $0\overset{m}{.}4$ in U, B and V, respectively. Kurochkin and Shugarov (1980) already reported occasional increases in brightness of AC Cnc, and we have confirmed the erratic fluctuations in brightness outside eclipses. The colour indices for AC Cnc yield a position in the region where the majority of the cataclysmic variables fall in the (U - B, B - V) diagram.

A multi-channel photoelectric photometry of AC Cnc was made in the following six channels:

Channel 1 λλ3330 - 3700, Channel 2 λλ3820 - 4190,
3 λλ4300 - 4670, 4 λλ4780 - 5150,
5 λλ5265 - 6005, 6 a narrow pass band at Hβ.

As an example, the observed light curve in Channel 3 is shown in Fig. 1. A hot spot may be present but not so strong since a hump seen just before the primary minimum is rather weak.

In the multi-channel light curves outside eclipses flickering can be seen with a time scale of 10 - 30 min, rather long compared with flickering occurred in many other cataclysmic variables, but such flickering with a long time scale as seen in AC Cnc is also reported in the case of Lanning 10 (Horne, Lanning and Gomer, 1982).

We have obtained altogether eight Cassegrain image-tube spectrograms of AC Cnc, seven in blue and one in red, with a dispersion of 110 A mm^{-1}. Broad emission lines of Balmer series Hα - Hδ were found superimposed on a blue continuum spectrum. He II λ4686, and possible C III - N III λ4650, appear in emission in the spectrograms which were taken on April 30, 1981 (Fig. 2), when AC Cnc was bright. The profile of Hβ was found to vary with the orbital phase. This may indicate a non-uniform distribution of the temperature on the accretion disk (a hot spot).

In order to analyse the light curve, we combined photon counts of the multi-channel photoelectric observations in Channels 1, 2 and 3 to construct a wide-band light curve. Then, this light curve was analysed by a synthetic method (e.g., Young and Schneider, 1980). In the analysis we employed a binary model which consists of a circular accretion disk lying on the orbital plane and a spherical red dwarf star whose radius is equivalent to its critical Roche radius. The brightness distribution of the disk is assumed to be given by a power law of $r^{-\beta}$. The orbit is also assumed to be circular.

The solution of the light curve analysis could not be uniquely determined, but it is given as a function of the mass ratio q = M(red dwarf) /M(white dwarf). For example, in the case of q = 1 other system parameters are given by

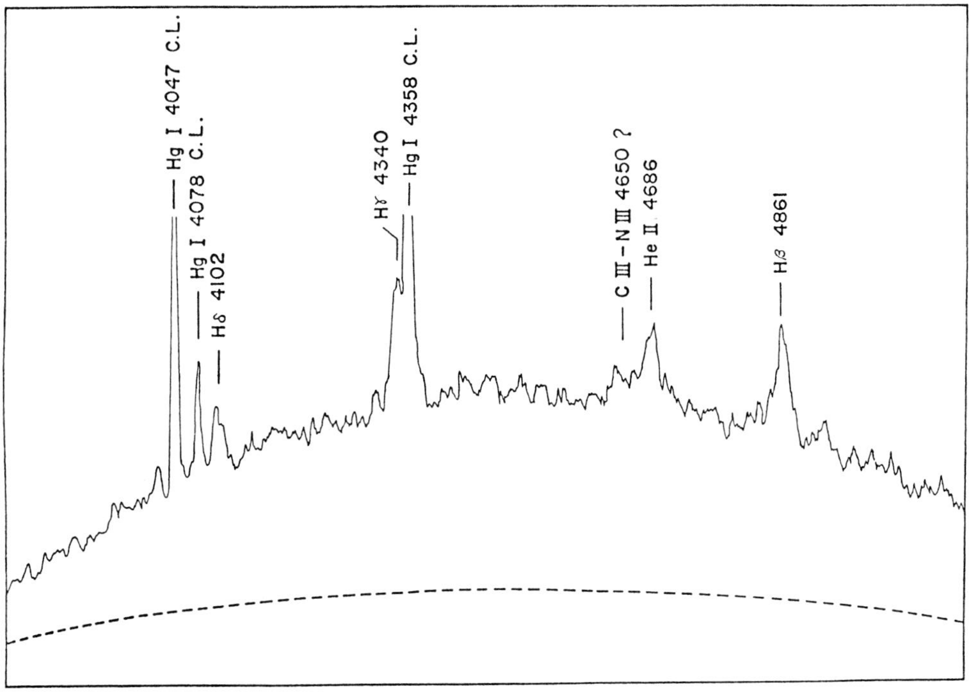

Figure 1. An intensity tracing of the spectrogram of AC Cnc taken on April 30, 1981. The broken line represents the sky background.

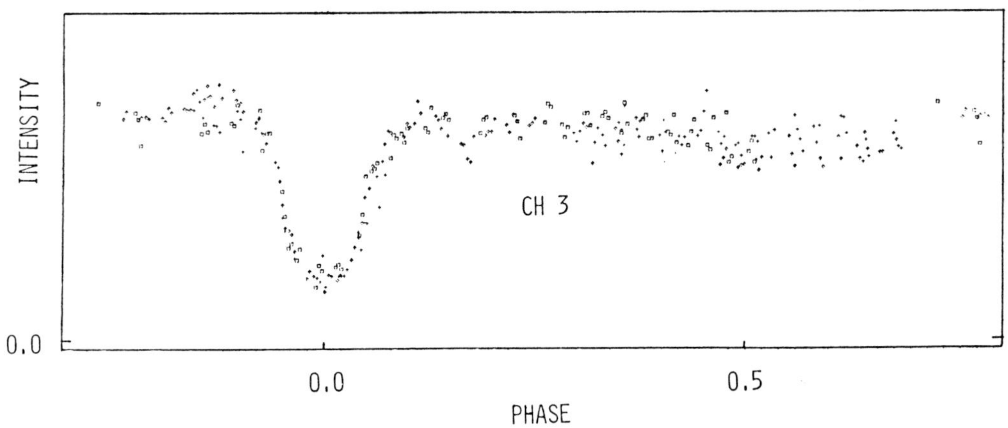

Figure 2. Light curve of AC Cnc in Channel 3 (λλ4300 - 4670).

$i = 75^\circ$
$r(\text{disk}) = 0.43$ (q = 1 assumed)
$r(\text{red dwarf}) = 0.38$
$\beta = 1.25.$

Anyhow it is evident that the size of the disk is fairly large compared with that of a non-viscous disk, and this should be in favour of a viscous disk model. As far as the light curve and the geometry of the system are concerned, it is found that nova-like variables AC Cnc and Lanning 10 (Horne et al., 1982) are very much similar to each other.

REFERENCES

Horne, K., Lanning, H. H. and Gomer, R. H.: 1982, Astrophys. J. 252, 681.
Kurochkin, N. E.: 1960, Astron. Tsirk. No. 212.
Kurochkin, N. E. and Shugarov, S. Yu.: 1980, Astron. Tsirk. No. 1114.
Okazaki, A., Kitamura, M. and Yamasaki, A.: 1982, Publ. Astron. Soc. Pacific, 94, 162.
Shugarov, S. Yu.: 1981, Astron. Zh. 58, 583.
Young, P. and Schneider, D. P.: 1980, Astrophys. J. 238, 955.

RECENT SPECTROSCOPY OF X-RAY SOURCES AND SYSTEMS RELATED TO CATACLYSMIC VARIABLES: I.

A.P. Cowley, D. Crampton, and J.B. Hutchings
Dominion Astrophysical Observatory
Herzberg Institute of Astrophysics
Victoria, B.C.

ABSTRACT: We present results of recent spectroscopic observations of A) some LMC X-ray sources, B) the X-ray burster 1735-44, and C) the peculiar emission-line binary BE UMa. From the short time scale variations and inferred low mass components each shows some similarities to more classical CV's.

A.) In the course of a large scale investigation of the stellar X-ray sources in the Magellanic Clouds being carried out by our group, Thorstensen and Charles, Helfand and others at Columbia, and Pakull some interesting stars have been identified. Of the approximately two dozen LMC sources for which Einstein HRI positions are available, nearly half have been found to be foreground sources, one is a background Seyfert, and several have yet to be identified. Of the half dozen known to be LMC stars, LMC X-4 is the only firmly identified supergiant binary of the type similar to Vela X-1 in the Galaxy. LMC X-1 may be identified with the supergiant R 148 but an equally probable candidate is a fainter star (No. 32 - Cowley <u>et al</u>. 1978) which shows only He II, λ 4686, in emission and lies some 6" away (Pakull, 1980) Single-trailed, 30 minute exposure spectra taken at CTIO show the He II emission to be highly variable in both intensity and velocity. On one plate the emission line was visible for about half the exposure and absent during the last half. On another plate the line was "curved" showing a velocity range of about 100 km s^{-1}. No other stellar features are apparent in the spectrum, although a bright nebula contributes strong H and [O III] emissions which would make it difficult to detect weak stellar hydrogen lines. Pakull has identified 0544-684 with a spectroscopically similar object. Our spectra show variable He II emission with no other obvious stellar features. Both stars are unlike anything we know in the Galaxy. With $M_B \sim -3$ these LMC objects are brighter than Sco X-1 or other bright-disk systems such as cataclysmic type variables even in outburst. The rapid variations suggest a low mass system, but the high disk luminosity is unusual.

Warren and Penfold (1975) identified a very blue star with LMC X-3.

M. Livio and G. Shaviv (eds.), Cataclysmic Variables and Related Objects, 21–24.

The HRI position now confirms this identification. The spectrum shows very broad hydrogen and He I absorptions with occasional Hβ P-Cygni emission. Our scattered observations over a few observing seasons and one by Johnston et al. (1978) show the velocity varies by ∿ 350 km s^{-1} in < 2 days, although as yet we have not determined the period. The low mass implied by these variations suggests we may be seeing a very bright accretion disk, since the spectrum and luminosity do not fit any normal star at the distance of the LMC.

B.) From a photometric investigation of the X-ray burster 1735-44 McClintock and Petro (1981) suggested a possible 4.3 hour (orbital?) period. SIT Vidicon spectra were taken at CTIO with 30 minute time resolution to look for emission line changes. While only 2 hours of observations were obtained on each of two nights, we did find a curious pattern of line strength and velocity variations which repeated on the second night. The ratio of N III (4640 bl)/C III (4650 bl) changed systematically by a factor of 4 while the radial velocity of both blends varied together by ∿ 800 km s^{-1}. The cycle time of these changes was 2.85 hours, but we have insufficient data to know if these variations are periodic. The velocity probably refers to some gas streaming rather than to orbital motion since He II, λ 4686, shows a much smaller velocity range. The changing N III/C III intensity could possibly be due to varying resonance fluorescence of the N III.

C.) BE UMa is a unique emission line binary ($P = 2.29^d$) which has recently been discussed by Margon et al. (1981) and Ferguson et al. (1981). In spite of the strong emission lines causing it to look like a CV, they concluded that the system contains a low mass cool dwarf widely separated and not interacting with the very hot white dwarf. The heating by the white dwarf causes both an ∿ 1 magnitude sinusoidal light variation and the strong emission lines which are apparently produced on the inner heated hemisphere of the cool star. However, more recently Ando et al. (1982) have found deep eclipses at the bottom of the sinusoidal light curve and from the duration of the eclipses and the shape of the partial phases have inferred the primary must be a hot subdwarf.

We have obtained moderately high resolution spectra of the system throughout the orbit in order to determine the masses and to study the line profile and intensity variations. In addition we have modelled the continuous light curve to infer properties of the component stars.

From the phasing of our velocity curve it immediately became obvious that the published period needed some revision. Combining the Ando et al. time of eclipse and the time of minimum from the Harvard plate material, kindly supplied by Tokarz (1982), we derive a new ephemeris:

$$T_{min\ lt} = \text{HJD } 2444998.281 + 2.291171E \text{ days}$$

Further, from the emission line velocities, we find a mean amplitude and systemic velocity from all lines of:

$$K = 102\ \mathrm{km\ s^{-1}},\ V = -60\ \mathrm{km\ s^{-1}}\ (\omega = e = 0)$$

implying $f(M) = 0.253\ M_\odot$

Additional constraints can be put on the geometry, masses and luminosities of the component stars by fitting the sinusoidal light curve using a radiative heating model and by considering the further limits the shape and duration of the eclipse impose. We find that no main-sequence secondary viewed at any angle can account for the observed light curve. Our best model gives:

Hot Star: $R = 0.028 \pm .008\ R_\odot$ $T = 130{,}000 \pm 10\%\ °K$
Cool Star: $R = 1.6 - 2.4\ R_\odot$ $T = 3500 \pm 300\ °K$
$i \sim 74° - 77°$ $q \sim M_{hot}/M_{cool} \sim 1 - 3$

For these ranges of i and q we find from the mass function that $M_{hot} \sim 0.6\ M_\odot$ while $M_{cool} \sim .2$ to $.6\ M_\odot$. Although the mass and temperature derived respectively from the velocities and the light curve of the cool star are in reasonable agreement, it is odd that the star is sufficiently evolved at this low mass to have a size of several solar radii. The system must have previously had a period of considerable mass exchange and loss. Furthermore our modelling shows, contrary to previous ideas, the secondary nearly fills its Roche lobe and some interaction between the components is probably occurring.

From the derived temperature and radius we can derive magnitudes of $M_{V(hot)} \sim +6.5$ and $M_{V(cool)} \sim +7.5$, in agreement with the depth of the visual light eclipse which appears to be total. Thus the hot star is considerably brighter and larger than a white dwarf (as Ando *et al.* found) and the cool star lies well above the main-sequence both in radius and luminosity. Perhaps this system will eventually become a CV like the old nova GK Per which combines a white dwarf/cool giant pair in a two day binary.

REFERENCES

Ando, H., Okazaki, A., and Nishimura, S. 1982, *Pub. Astron. Soc. Japan*, **34**, 141.
Cowley, A.P., Crampton, D., and Hutchings, J.B. 1978, *Astron. J.*, **83**, 1619.
Ferguson, D.H., Liebert, J., Green, R.F., McGraw, J.T., and Spinrad, H. 1981, *Ap.J.*, **251**, 205.
Johnston, M.D., Bradt, H.V., Doxsey, R.E., Gursky, H., Schwartz, D.A., Schwarz, J., and van Paradijs, J. 1978, *Ap.J.*, **225**, 159.
Margon, B., Downes, R.A., and Katz, J.I. 1981, *Nature*, **293**, 200.
McClintock, J.E., and Petro, L.D. 1981, *IAU Circ.* 3615.
Pakull, M. 1980, *IAU Circ.* 3472.
Tokarz, S. 1982, *private communication*; paper in preparation.
Warren, P.R., and Penfold, J.E. 1975, *Mon. Not. Royal Ast. Soc.*, **172** 41P.

DISCUSSION FOLLOWING A. COWLEY'S TALK

KITAMURA: The physical parameters depend upon how accurately the light curve is analyzed, did you analyze the light curve yourself?

COWLEY: Yes. Here is what the temperature distribution on the cool star must look like. You can see that it is enormously heated.

KITAMURA: How about the eclipse light curve?

COWLEY: We did not. Except to look at the duration of partial and total phases because that sets limits on the inclination and the radii, but we did not analyze them completely. The durations tell you that the inclination is not 90° it must be nearly grazing.

TAPIA: I have not seen the recent spectra that you have taken of BE UMa. Could you tell us, are they different from the first ones that you took, you told us that the object was very different now.

COWLEY: No, what I meant by different was that, before, I think that we all had the picture that it was a white dwarf and very far away, not in contact, was a low mass dwarf and the light curve analysis indicates that it must be a much larger star than the dwarf and the stars are nearly a semi-detached system, that is what I meant by different, not that the spectra are different. The spectra look just the same as they looked a year ago and two years ago.

RECENT SPECTROSCOPY OF X-RAY SOURCES AND SYSTEMS RELATED TO CATACLYSMIC VARIABLES: II

D. Crampton, A.P. Cowley and J.B. Hutchings
Dominion Astrophysical Observatory
Herzberg Institute of Astrophysics
Victoria, B.C.

ABSTRACT: Spectroscopic observations of two very different objects are discussed: the first, SS433, is a very massive binary system probably containing an unseen O or WR star and a black hole surrounded by a luminous precessing accretion disk (from which emanate spectacular relativistic jets); the second, GK Per, is a classical old nova consisting of a slightly evolved K2 star and an unseen white dwarf companion surrounded by an accretion disk which is less luminous than the K2 star.

SS433. Although the binary nature of the SS433 system has been extensively studied (Crampton, Cowley and Hutchings 1980, hereafter CCH, Crampton and Hutchings 1981, hereafter CH), and the main parameters of the 13 day orbital cycle are established, the effects of the 164 day precession cycle are not yet known. For example, the 13 day periodicity was first detected through velocity variations of Hβ (CCH) but after the initial observations (carried out over only a few months or $\sim$ 0.4 of the 164 day cycle) Hβ no longer exhibited the same smooth cyclic velocity variation (CH). The discovery that the He II 4686 emission features showed velocity variations that are persistent, have a larger amplitude than Hβ and are phased in such a way that they can easily be reconciled with the light curve (Cherepashchuk 1981, Kemp *et al.* 1981) indicates that the Hβ emission feature must at least partly arise from gas streaming from the early type star to the degenerate companion. It might be expected that this streaming motion would be affected by the precession of the accretion disk and so the behaviour of the velocity curve would change during the 164 day cycle, perhaps elucidating the geometry within the system.

Continuing spectroscopic observations have so far failed to yield any conclusive results. Although there is a suggestion that the gross behaviour of the Hβ velocity curve is as expected, i.e., it has a larger amplitude during phases when the disk is viewed nearly edge-on than when it is face-on, large departures from this pattern sometimes occur. There may be a correlation between these deviations and the

M. Livio and G. Shaviv (eds.), Cataclysmic Variables and Related Objects, 25–28.

sporadic intense flaring activity observed at radio wavelengths by Johnston et al. 1981. At present, it is not clear whether the Hβ velocity variation is more affected by the changing aspect of the accretion disk, or some erratic behaviour which manifests itself as radio flares. Continued monitoring will hopefully yield additional information on the nature of the streaming and structure of the accretion disk.

GK Per. Kraft (1963) demonstrated that Nova GK Per (1901) is a spectroscopic binary with P = 1.904 days. Absorption lines of a K2 IV star are clearly visible in the spectrum in addition to H, He I and He II emission lines arising both from a region near the companion star and from the ejected shell. The orbit which Kraft derived was not well defined and quite eccentric (e = 0.4) and Paczynski (1965) noted that the data could be equally well represented by a circular orbit with P = 0.685 days. Recent observations by Bianchini, Hamzaoglu and Sabbadin (1981) suggest that the longer P is to be preferred, but no substantial improvements of the orbital parameters were deduced. Since GK Per is one of the old novae in which the absorption spectrum is clearly visible making it more amenable to precise analysis than most old novae, and since it was discovered to be an intermittent hard X-Ray source (King, Ricketts, and Warwick 1979), a new spectroscopic study was undertaken. Only a preliminary analysis of the data, 30 A mm^{-1} image tube spectra, is reported here.

The emission lines display large profile changes throughout the orbital cycle and hence the radial velocities derived from them are subject to large random and systematic errors. The absorption lines yield quite accurate (± 10 km s^{-1}) radial velocities. It was quickly discovered that the velocities could not be fitted by either Kraft's or Paczynski's periods but determining the exact period was difficult since it is very close to two days so that during an observing season observations on alternate nights are virtually identical. A combination of the current observations with a few observations from the past five years and Kraft's data yield P = 1.99679 days. Kraft's data are equally well fitted by this period as by his original one, and, surprisingly perhaps, the orbit is circular, or not formally significantly different from circular. Preliminary orbital elements are: V_o = 28 km s^{-1}, K_{abs} = 124 ± 2 km s^{-1}, K_{em} = 26 ± 4 km s^{-1} (Hβ wings), T_0 (maximum positive velocity) = JD 2444881.016 ± 0.01. The mass function f(m) = 0.395 $M_⊙$ but the mass ratio is quite uncertain since the emission velocities may not represent orbital motion. The masses are constrained to relatively narrow limits in any case. Assuming that the unseen companion is a white dwarf with a maximum mass of 1.2 $M_⊙$, the mass function implies that either the K2 star is much less massive than normal (i.e. it has lost a substantial fraction of its mass) or that i must be $\sim$ 90°, or both. Eclipses have never been observed but due to the fact that the period is nearly an integral number of days, they could easily have been missed.

It seems probable that GK Per consists of a white dwarf of nearly maximum mass and an undermassive slightly evolved K2 star viewed at a high angle of inclination. Photometric observers are encouraged to look for eclipses at the appropriate phase to enable further constraints to be placed on the physical parameters of the system.

REFERENCES

Bianchini, A., Hamzaoglu, E., and Sabbadin, F. 1981, A. and A., **99**, 392.
Cherepashchuk, A.M. 1981, M.N.R.A.S., **194**, 761.
Crampton, D., Cowley, A.P., and Hutchings, J.B. 1980, Ap.J.(Letters), **235**, L131 (CCH).
Crampton, D., and Hutchings, J.B. 1981, Ap.J., **251**, 604 (CH).
Johnston, K.J. et al. 1981, A.J., **86**, 1377.
Kemp, J.C., Barhous, M.S., Kemp, G.N., and Hagood, D.M. 1981, Proc. of Rome Meeting on SS433, Vistas in Astronomy, **25**, 31.
King, A.R., Ricketts, M.J., and Warwick, R.S. 1979, M.N.R.A.S., **197**, 77p.
Kraft, R.P. 1964, Ap.J., **139**, 457.
Paczynski, B. 1965, Acta Astr., **15**, 197.

DISCUSSION FOLLOWING D. CRAMPTON'S TALK

KING: Just a comment about the X-rays from SS 433, I don't think you would expect them to eclipse, because almost certainly they can't be coming from directly within the binary system. The reason for that is that the stellar wind or the great cloud around the system is simply too dense. That is actually confirmed by the timescale of activity of the X-rays, there is variability on all the timescales, until you get to quite short ones and there is no variability beyond that and that is simply because you do not see them coming from within the system. They are also very faint, so it is presumably some interaction of this cloud with whatever else is outside.

CRAMPTON: Yes. We actually had thought about that possibility, but did not know of any confirming evidence of it, we knew that there were no eclipses.

TAPIA: Considering that you have these large variations in the radial velocity, why do you still insist in believing that the emission lines come from a disk, couldn't we have some contamination from the blobs that we know are being ejected from SS 433?

CRAMPTON: In the line profiles, the lines change in a very ordered fashion during the cycle, so that I don't think that we actually see great blobs moving out, except that ocassionally when there are very large flares in the light curve, the velocities do change. We are not measuring individual peaks, we are measuring further down the profile, so possibly you are right.

BIANCHINI: Did you try also noncircular orbits with HeII emission lines, or did you base your observations on the absorption which normally are better defined lines. I suggest to try with the emission also, because in my previous work on the H_α emission I found an eccentric orbit with the same eccentricity as given by Kraft, this does not mean necessarily that there is an eccentricity because, as I have already pointed out, you can have a spurious eccentricity because of the presence of the hot spot.

CRAMPTON: It is interesting that when Kraft observed this star he also got exactly the same part of the velocity curve that we did, just by chance. I am not aware that anyone has actually observed the light curve at minimum. The H_α velocities that you published do not fit our period and in fact, whether or not they have an eccentricity, they just produce a scatter. The formal eccentricity, if you decided this could be an eccentric orbit comes out to be 0.09, but it is not terribly well defined because you need more points.

BIANCHINI: I would like to ask Ritter if he can confirm the fact that we can have a very small mass, evolved secondary of 0.2 $M_\odot$?

RITTER: The data is insufficient so far to pin down the nature of the object.

CRAMPTON: The luminosity that we get, is more or less what you expect for a slightly evolved K star.

WARNER: Surely, there is no alternative, if you have got a K2 star in a two day orbital period, it cannot be an ordinary main sequence, it has to be evolved to fill its Roche lobe. So there is no alternative but to have an evolved star there.

EGGLETON; I just wonder, does that not conflict with the supposed magnitude of GK Per from Gallagher and Oinas, they give a magnitude of +5 I think, a K 2 giant in that sort of a period ought to have a magnitude getting to zero.

CRAMPTON: Yes, it cannot be a giant it has to be just a class IV star.

EGGLETON: Well, I think it presents something of an evolutionary problem then, but maybe not an insuperable one.

ORBITAL SOLUTIONS FOR WZ SAGITTAE DURING QUIESCENCE

Ronald L. Gilliland
High Altitude Observatory, National Center for Atmospheric Research*

1. INTRODUCTION

WZ Sagittae was one of the first cataclysmic variables shown to be a binary system (Kraft 1961) from spectroscopic observations. Krzeminski (1962) found that WZ Sge was an eclipsing binary with an 81 min 38 s orbital period. Krzeminski and Kraft (1964, hereafter KK) in the first of many comprehensive analyses of the WZ Sge system to appear in the literature stated "a lower limit to certain detection of orbital velocity changes for the double emission feature is 2K = 75 km s^{-1}." Interpreted as an upper limit to K_1 of 37.5 km s^{-1} this fundamental result has remained untested for two decades, although interpretations and observations of WZ Sge have led to a vast literature. Despite the lack of a well determined orbital velocity, WZ Sge is one of the most thoroughly studied cataclysmic variables. The photometric properties during quiescence have been thoroughly studied and discussed (Robinson, et al.1978; Fabian, et al.1978; Ritter and Schroder 1979), although a consensus on model parameters has remained elusive. Results of spectroscopic observations are presented here which were obtained seven months after the 1978 outburst of WZ Sge. The primary result is a substantial upward revision for the orbital velocity amplitude of the primary component. The higher velocity allows solutions for the component masses in which the secondary ($M_2 \sim 0.1\ M_\odot$) is resident on the zero-age main sequence while the white dwarf mass remains under the Chandrasekhar limit ($M_1 \sim 1.1\ M_\odot$).

2. OBSERVATIONS

On the night of 1979 July 16 we obtained time-resolved spectroscopy of WZ Sge over 4 orbital cycles with the ITS Cassegrain spectrograph on the 3 m Shane telescope of Lick Observatory. The 1200 line grating and red-sensitive image tube were used to provide optimal sensitivity and resolution at Hα. Continuum source scans of incandescent lamps were

*The National Center for Atmospheric Research is sponsored by the National Science Foundation.

M. Livio and G. Shaviv (eds.), Cataclysmic Variables and Related Objects, 29–34.

taken before and after the observing sequence in order to remove effects of tube irregularities. Scans of a flux standard star (HD 182487) were obtained to allow reduction of the scans to an absolute flux calibration. Where trade-offs were necessary to obtain precise radial velocity information versus accurate flux calibration all efforts were directed toward optimization for radial velocities. A narrow entrance slit was used to obtain better resolution. Scans of comparison lamps were obtained after every third observation of the object to maintain precise wavelength calibrations. In the red (~ 5750-7100 Å) we obtained 21 8 min scans at a dispersion of 0.66 Å per channel resulting in 2-3 Å resolution. These were obtained as two (12 and 9 respectively) long time series with different slits separated by 2 hours. In the blue (~3800-5150 Å) we obtained 14 8 min scans at a dispersion of 0.67 Å per channel resulting in ~ 3 Å resolution.

3. RADIAL VELOCITIES

All scans were reduced with the following procedure: 1) Object and sky scans taken with the two slits were divided by the respective continuum scan to remove image tube response irregularities. 2) The sky background is subtracted from the object. 3) Using the procedures described in Stover, et al. (1980) and Gilliland (1982) the sky-subtracted scans were transformed to a uniform log (λ) scale. 4) Continuum was flattened by division with a 4th order polynomial for the red scans and a cubic for the blue scans. Care was taken to insure that the polynomial division neither obscure, nor introduce any large scale features into the resulting spectra. Sums based upon 168 min for the red spectral region and 112 min for the blue are shown in Figure 1. The wavelength calibrations have relative precisions of 5 km s^{-1} for the red scans near Hα and ~ 10 km s^{-1} for the blue scans from $\lambda\lambda$ 4100-4900 Å. The calibration precisions are smaller than the precision with which the position of any spectral feature in WZ Sge can be measured, therefore final radial velocity errors are dominated by the ability to measure given features.

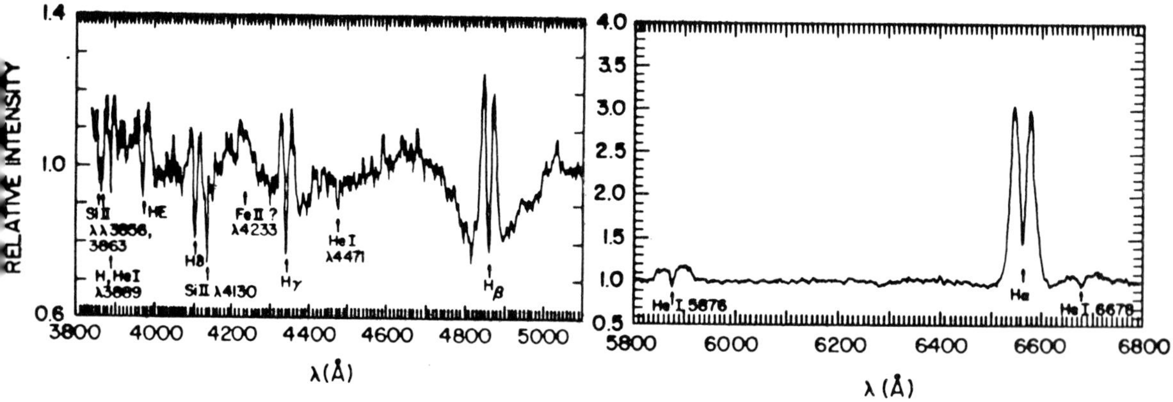

Figure 1 -- The right panel shows the sum of 21 8 min red scans of WZ Sge after shifting to a uniform wavelength scale and flattening the continuum with polynomial division. The left panel shows the same for 14 8 min blue scans.

The centroid of the doubled Hα feature was determined by fitting a 3 parameter (height, position and width) Gaussian profile to the outer wings with varying amounts of the central core weighted to zero. Radial velocity curves evaluated for different amounts of excluded core show systematic trends in the derived velocity amplitude and mean scatter. Fitting only the outermost parts of the wings results in slightly smaller derived orbital amplitude with increased scatter and resultant errors. As points successively closer to the line core are included in the fit the scatter about an orbital solution as shown in Fig. 2 reaches a minimum at an excluded core width of 1700 km s^{-1} (the peaks have a mean separation of ~1400 km s^{-1}). The velocity amplitude is K_1 = 48.2 ± 4.1 km s^{-1} with an rms scatter of 12.9 km s^{-1}. The optimal fit is thus obtained by fitting a Gaussian profile from 850 to 2000 km s^{-1} on each side of the line centroid. Reducing the excluded core to 1450 km s^{-1} resulted in K_1 = 56.7 km s^{-1} with a scatter of 15.8 km s^{-1}. The opposite extreme of excluding 2000 km s^{-1} from the core (starting fit ~40% below peaks) yields K_1 = 4.1 km s^{-1} with a scatter of 19.2 km s^{-1}. The latter trend is expected, fitting farther out in the wings (and thus discarding most of the velocity information) should lead to smaller derived amplitudes with increased scatter. The analyses with different core definitions all gave consistent phases for maximum recessional velocity (Φ = 0.83 ± 0.01) and were consistent with circular orbits. The optimal case discussed above, reduced without assuming a circular orbit produced K_1 = 48.7 km s^{-1}, ε = 0.10 ± 0.09. We have also analyzed the wings of the Hβ emission. The result is of much lower significance due to: 1) lower velocity resolution at Hβ than Hα, 2) fewer comparison lines, 3) lower signal to noise, 4) weaker emission at Hβ, 5) strong, perhaps variable, underlying absorption at Hβ, and 6) fewer scans. The resulting velocity is K_1 = 63 ± 16 km s^{-1}. Table 1 summarizes the radial velocity results. We believe the adopted velocity amplitude of 49 ± 6 km s^{-1} reasonably reflects the best solution and possible associated errors, but that possible systematic errors are more likely to have resulted in too small an estimate, than too large. Thus it is believed there is a significant discrepancy between the KK upper limit of 37.5 km s^{-1} and these results.

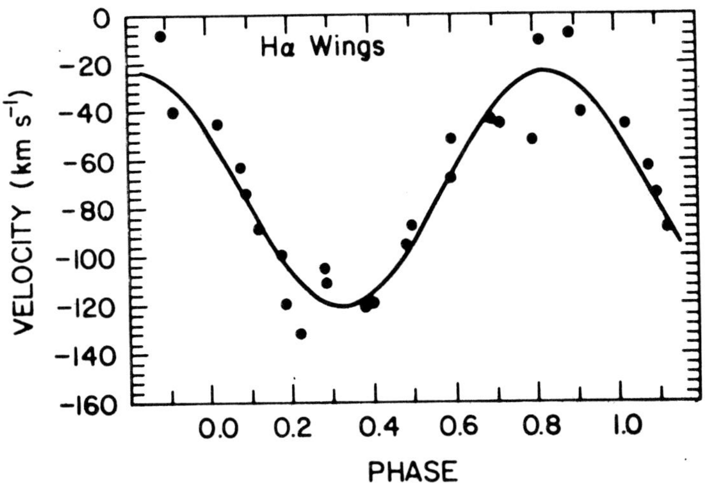

Figure 2 -- Radial velocity curve for WZ Sge. ε = 0 is assumed, as well as the Robinson, et al. (1978) ephemeris. Velocities are for the wings of H α. Continuous curve is the first solution from Table 1.

4. INTERPRETATION

The wings of Hα can be traced out to a velocity of 2400 km s^{-1} from line center. Interpreted as Keplerian velocities in the inner disk this places a lower limit to the primary mass of $M_{WD} > 0.44\ M_\odot$. This is a nontrivial limit, since many recent studies have considered masses smaller than this. A canonical approach to estimating masses in cataclysmic variables is to use the assumption that the secondary is precisely filling its Roche lobe, thus allowing a calculation of its mean density. Then with knowledge of a mass-radius relation for an assumed secondary star type (usually zero-age main sequence) one directly obtains a secondary mass. Use of the mass function from the radial velocity analysis, with constraints on the orbital inclination then allow one to solve for the primary mass. This technique has been avoided with WZ Sge, since its application for $K_1 < 38$ km s^{-1} (KK) invariably leads to white dwarf masses well above the Chandrasekhar limit. The higher value for K_1 of 49 km s^{-1} implies that the estimated primary mass will be smaller for a given, assumed secondary mass. In particular, using the mass-radius relation for a lower main sequence unevolved secondary, the orbital period implies $M_2 = 0.125\ M_\odot$. Assuming an orbital inclination of 80° and the adopted orbital solution ($f(M_\odot) = 6.67 \times 10^{-4}$) we find $M_{WD} = 1.55 \pm 0.32\ M_\odot$ -- still uncomfortably large. If, however, we take the secondary mass as $0.1\ M_\odot$, consistent in magnitude with claims of distended radii due to mass loss on shorter than thermal time scales (Rappaport, et al. 1982), then with the other assumptions the same $M_{WD} = 1.1\ M_\odot$. The 'canonical' approach therefore yields a large, but physically reasonable value for the primary mass.

Are the results $M_2 = 0.1\ M_\odot$, $M_{WD} = 1.1\ M_\odot$ consistent with the recent analyses of eclipse profiles for WZ Sge (Fabian, et al. 1978; Ritter and Schroder 1979)? Interpolation in Fig. 2 of Fabian, et al. (1978) for $K_1 = 49$ km s^{-1} suggests possible solutions with q,i = (11, 69°) to q,i = (7, 78°). If we use constraint number 1 of Ritter and Schroder (1979) and choose $i = 75^o$, then $q = 10.6$ is required with $K_1 = 49$ km s^{-1} -- this is satisfied with $M_2 = 0.1\ M_\odot$ and $M_{WD} = 1.06\ M_\odot$, essentially the result obtained above with the 'canonical' assumptions. The full range of allowed system parameters are: q from 8-11, i ~75-77°, $M_2 \sim 0.06$-0.11, $M_{WD} \sim 0.5$-$1.2\ M_\odot$. We note that, if $M_2 \sim 0.06\ M_\odot$, $M_{WD} \sim 0.5\ M_\odot$ the secondary is degenerate and the orbital period should be increasing, if $M_2 \sim 0.1\ M_\odot$, $M_{WD} \sim 1.1\ M_\odot$ the secondary is still nondegenerate with orbital evolution toward decreasing periods. Since Robinson, et al. (1978) were unable to detect a period change based on data over nearly half an outburst period, observational discrimination based on period changes between the high and low mass solutions seems unlikely.

The observations were obtained in collaboration with John Faulkner, Edward Kemper and Nicholas Suntzeff; a more complete discussion is in preparation for publication elsewhere.

TABLE 1

ORBIT SOLUTIONS FOR WZ SAGITTAE

Line	K (km s^{-1})	γ (km s^{-1})	φ (km s^{-1})	σ SD
Hα wings	48. ± 4.	-72. ± 3.	0.83 ± 0.01	12.9
Hβ wings	63. ±16.	-149 ± 12.	0.86 ± 0.04	40.2
Adopted Solution	49. ± 6.	-72 ± 5.	0.83 ± 0.02	

Using the photometric elements (Robinson, *et al.* 1978): Min = JD_0 2437547.72845 + $0.^{d}$ 0566878455E, and assuming $\varepsilon = 0$.

REFERENCES

Fabian, A.C., Lin, D.N.C., Papaloizou, J., Pringle, J.E., and Whelan, J.A.J.:1978, M.N.R.A.S., **184**, p. 835.
Gilliland, R.L.,:1982, Ap. J., **261**, p. 000.
Kraft, R.P.: 1961, Science, **134**, p. 1433.
Krzeminski, W.: 1962, Pub. A.S.P., **74**, p. 66.
Krzeminski, W., and Kraft, R.P.: 1964, Ap. J., **140**, p. 921 (KK).
Rappaport, S., Joss, P.C., and Webbink, R.F.: 1982, Ap. J., **254**, p. 616.
Ritter, H., and Schroder, R.: 1979, Astro. Ap., **76**, p. 168.
Robinson, E.L., Nather, R.E., and Patterson, J.: 1978, Ap. J., **219**, p. 168.
Stover, R.J., Robinson, E.L., Nather, R.E., and Montemayor, T.J.: 1980, Ap. J., **240**, p. 597.

DISCUSSION FOLLOWING R. GILLILAND'S TALK

RITTER: In the workshop in Leicester, this Spring, Charles talked about measurements of WZ Sagittae and he found a substantially smaller radial velocity, so can you comment on this method of deriving this?

GILLILAND: I have not seen his analysis, so I have no idea how he went about it, I can say that if he measured the peaks, the same analysis technique that Kraft and Krzeminski used, he would have found a much smaller result, in a sense the Kraft and Krzeminski upper limit is correct if you analyse the peaks, but the canonical best way to do this is analyze the wings and I am confident that this is the right result in that case. I am not confident really that it means anything.

RITTER: I would like to remark that the first conclusion you drew that it might be an eclipse feature from the opposite of the disk, one cannot rule this out by the fact that you don't see eclipses from the center of the disk, because if you have a feature on the opposite edge of the disk which undergoes eclipses, you would also expect that the light source in the center of the disk undergoes an eclipse.

ROBINSON: Obviously, the disk has reformed itself at that point, but yet you see no signs of the S wave.

GILLILAND: No obvious signs but the V/R ratio does change slightly, so there is something going on, of an asymmetric nature in the disk.

ROBINSON: Why is the S wave missing?

GILLILAND: Perhaps, the disk is still brighter at this point intrinsically, it could be that the S wave is there just as strong and the system brightness is up, so that it is not such a strong relative contribution.

SPECTRA OF SYMBIOTIC STARS

Hagai Netzer and Elia M. Leibowitz
Department of Physics and Astronomy and the Wise Observatory, Tel Aviv University

and Gary Ferland
Department of Physics, University of Kentucky, Lexington

ABSTRACT

We have observed Z And and other symbiotic stars at the Wise Observatory and have obtained absolute flux measurements of many emission lines. Preliminary analysis of the data shows that the line emitting gas is probably photoionized by a hot, $T_* \gtrsim 200,000$ °K continuum, which can explain the extremely strong HeII lines. The hydrogen Balmer line intensities are different from what is predicted by recombination theory and the optical depths in Hα and Hβ must be significant. Our photoionization model can be used to deduce the geometry, density and temperature of the line emitting gas, and the nature of the hot companion.

INTRODUCTION

Symbiotic stars, unlike other objects discussed at this meeting, have long orbital periods, of the order of 1000 days, and may not be associated with a compact object. Spectroscopic study of them is the most important tool for understanding the physical processes in their line emitting zone. In this paper we present preliminary results from an ongoing project at the Wise Observatory, Israel, to monitor such objects.

OBSERVATIONS

Spectrophotometric observations of several symbiotic stars were carried out with the Wise Observatory 1-m telescope during 1981 and 1982. The detector used is a 2x936 diode self-scanned Digicon. It enables absolute flux measurements with low to moderate (1-6Å) spectral resolution over a 3500Å wavelength range. The spectrum of Z And, shown in Fig. 1, is a typical example. This spectrum was obtained with a large entrance aperture, and the flux put on absolute scale by observations of spectrophotometric standards. The linearity and stability of the Digicon enables us to measure many weak emission lines, up to a level of about

M. Livio and G. Shaviv (eds.), Cataclysmic Variables and Related Objects, 35–40.

2 percent the intensity of Hβ. The real limitation in measuring such weak features is not the detector but the underlying continuum, with the strong molecular bands. We have therefore subtracted an M giant continuum prior to the measurement of line intensities.
The present work is concerned mainly with Z And and we defer discussion of other objects to a later date.

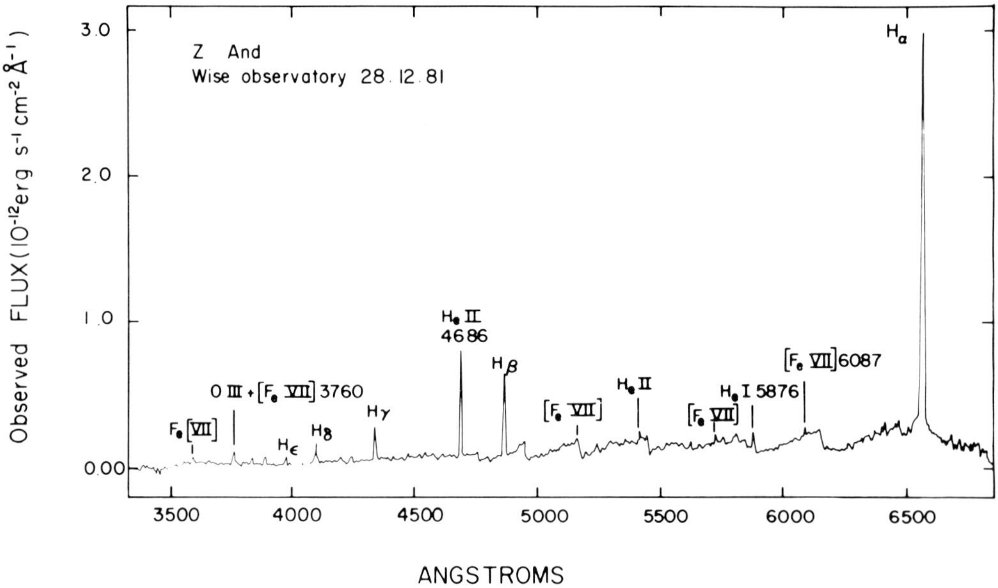

Figure 1. Digicon 110 Å/mm spectrum of Z And, obtained at the Wise Observatory in December 1981.

DISCUSSION

Our spectrum of Z And and the recent IUE data of Altamore et al. (1981) show lines of low (e.g. OI 1304), moderated (e.g. CIII] 1909) and very high excitation (e.g. [FeVII] lines). Such a large range in excitation is usually a result of photoionization by a hot continuum, and we shall assume therefore that this process is indeed the main source for ionizing and exciting the line emitting gas. We investigate the consequences of a simple model where a hot central source (hot blue companion? disc around a compact source?) illuminates some gas around it, causing heating and ionization. The line emitting gas can be anything from a skin of an accretion disc to expanding gas shell or even the atmosphere of the M star. The model is spherically symmetric so all the gas is at the same distance from the ionizing source. We assume a black body ionizing continuum and attempt to deduce its temperature from comparison of the model calculations with the observations. This simplified picture does not intend to represent a full realistic model for the system, but has some general consequences regarding some of the more important physical processes.

The spectral analysis of Z And shows:

a. [FeVII] (Nussbaumer and Storey 1982), CIII] and NIII] (Altamore et al. 1981) lines indicating $10^8 \leq N_e \leq 10^{11}$ cm^{-3} and $T_e \sim 2$ -$3x10^4$ ^oK in the line emitting gas

b. extremely strong HeII lines, in particular, HeII 4686 which is stronger than Hβ

c. very small reddening, $E_{B-V} < 0.3$

d. Balmer decrement very different from that predicted by radiative recombination theory

Point b above is most interesting, and typical also for other symbiotic stars. Possible explanations for it include:

(i) Very Hot Ionizing Source: The HeII 4686/Hβ line ratio reflects mainly the relative number of ionizing photons below and above 4 Ryd. (most photons with $h\nu > 4$ Ryd. will be absorbed by He^+). Large relative abundances of helium cannot therefore be the explanation for the abnormal line ratio (see for example Williams 1971). A high ionization temperature is needed, of at least 150,000 ^oK and possibly even 200,000 ^oK or more.

(ii) Density Bounded Emission Line Clouds: Photons with energy larger than 4 Ryd. are absorbed by He^+ ions close to the inner illuminated surface of the cloud. Softer photons, that ionized hydrogen, are absorbed deeper inside. Density bounded situations, in comparison with radiation bounded cases, would favour the excitation of HeII 4686 over that of Hβ.

(iii) Optically Thick Hydrogen Balmer Lines: The strong HeII 4686/Hβ may in fact be the result of a relatively weak Hβ, rather than a very strong HeII 4686. The observed Hα/Hβ ratio in Z And is about 6, and is much larger than the recombination case B value of 2.8. The reddening deduced from other lines is too small to explain this and also there is no single E_{B-V} that can explain both Hα/Hβ and Hγ/Hβ. One must look therefore for other processes that affect the hydrogen spectrum.

Study of Balmer decrement in Quasars and Seyfert galaxies shows that self-absorption in the Balmer lines must be common and important (Netzer 1975). The result of this is a steep Balmer decrement, quite similar to the one seen in Z And. The optical depth in the Balmer lines depends on the incidence flux of ionizing photons, and can be calculated once this is known (Ferland and Netzer 1979). We have calculated $\tau(H\alpha)$ for our best fit photoionization model and found it to be much larger than 1. Strengthening this conclusion are the observed strong OI lines at 8446 and 1304Å. These lines are excited by OI fluorescence with $L\beta$ and are further enhanced if $\tau(H\alpha)$ is larger than 1. Their unusual strength in the spectrum of Z And is another indication of considerable optical depth in the Balmer lines.

A common property of nebulae with $\tau(H\alpha) > 1$ is an $H\beta$ weaker by a factor of ~ 2 than predicted by recombination theory. This helps to explain the anomalous HeII 4686/$H\beta$ in Z And. Another prediction, yet to be checked observationally, are changes of $H\alpha/H\beta$ and $H\gamma/H\beta$ intensity ratios, following variations of the ionizing continuum.

To summarize, our model and analysis indicate physical conditions in the emission line gas around Z And that are very different from those in low density nebulae. Line transfer effects are probably important and the gas may be density bounded. A very high ionization temperature continuum, possibly of $T_* > 200{,}000\ ^{\circ}K$ is also indicated. This puts severe limitations on models of symbiotic stars involving discs around main-sequence stars. Obviously much more work is still needed, especially if discs are present, and disc-type continuum, rather than a simple black body is causing the ionization. We are attempting to incorporate these modifications into our model, so that a better fit with the observation is achieved.

ACKNOWLEDGEMENT

Research at the Wise Observatory is supported by "The Fund for Basic Research Administered by The Israel Academy of Sciences and Humanities"

REFERENCES

Altamore, A., Baratta, G.B., Cassatella, A., Friedjung, M., Giangrande, A., Ricciardi, O. and Viotti, R., 1981, Ap. J. 245, 630

Ferland, G. and Netzer, H., 1979, Ap. J. 229, 274

Netzer, H., 1975, M.N.R.A.S. 171, 395

Nussbaumer, H. and Storey, P.J., 1982 (preprint)

Williams, R.E., 1971, Ap. J. Lett. 167, 27

DISCUSSION FOLLOWING H. NETZER'S TALK

BATH: I would just like to make the comment that by taking a single temperature source you have immediately taken a source which is not the expected source in symbiotic stars, in the sense that, most of us now believe that accretion is important and if disks are formed, then you have a temperature distribution over them which in main sequence accretion goes from 3000°K to 10^5°K and with white dwarf accretion it can go to 10^6°K, before it becomes super Eddington. Therefore you've got to worry in your photoionization balance calculations about a variation in the temperature of the exciting source and I wonder how much that will effect the conclusions that you make about the properties of the emission line region.

NETZER: I certainly agree with you. We are definitely going to put there something which is similar to the models that you have predicted. The main thing is that you need a very hot component there, you must have something there which is as hot or maybe hotter than $2x10^5$°K.

FRIEDJUNG: Just two or three little points: firstly there are some problems about the energy distribution of symbiotics that are not properly understood, we need high resolution IUE observation. Secondly, there is good reason to suppose in many symbiotic stars that a lot of the emission lines are formed in the winds from the red component, there are some indications for a radial velocity shift in the UV. Finally, I think that the physics of line excitation is more complicated, photoionization obviously does occur, there may be some regions with high electron temperature.

WILLIAMS: Why couldn't all of this be happening in the chromosphere of the secondary star?

NETZER: That is fine with me, I can tell you roughly what is the distance between the photoionizing source and the cloud and that is about the right distance, also I can tell you that the covering factor of the system is much smaller than one.

WILLIAMS: So you don't have to postulate the existence of clouds.

NETZER: No, this was just for illustration.

ROBINSON: Is it your impression looking at published spectra by other people that they will need higher temperatures for their central sources or lower than Z And, in particular, can you see if you will ever need temperature more like 10^6 rather than a few times 10^5?

NETZER: I have seen at least two or three others with very strong He II lines, so for them it will be just the same, there may be some others which have weaker lines, but $2x10^5$°K is a lower limit for some of them.

LIVIO: You have actually stated a lower limit of $2x10^5$°K, if you could say that you can fit the observations better with 10^6°K this may mean that you need a compact object in there, while if you can do it with 10^5°K you may not need it.

NETZER: My impression is that you cannot do it with 10^6°K black body, you will need a disk component, which will contribute at softer energies, so if you take just a black body at 10^6°K, the answer is no, but that is not a good approximation anyhow.

KING: Since you have such large line opacities wouldn't it be the case that your ratios will be affected by the deviation from spherical symmetry, so that if it is the secondary, as it moves around, your predicted ratios should change.

NETZER: Yes, definitely, if you talk about Balmer decrement, my prediction is that it should change.

SHAVIV: Is the continuum between the lines real continuum or noise?

NETZER: First, you have seen the M giant which was very strong and that was subtracted by another M giant.

SHAVIV: Do you then have an estimate for the relative energy in all the lines and the continuum.

NETZER: Yes, I do, this comes from other people's observations with the IUE, this relates to my answer about the covering factor (how much of the source is covered by clouds) and this is a very small number of the order of 10%.

KQ MON AND THE NATURE OF THE UX URSA MAJORIS NOVA-LIKE VARIABLES

Edward M. Sion and Edward F. Guinan
Astronomy Department, Villanova University

ABSTRACT

KQ Mon, a new UX UMa type nova-like variable discovered by H.E. Bond, exhibits optical spectra, UBV and high speed photometric characteristics strongly similar to other members of the UX Ursa Majoris subset. However its ultraviolet spectra when compared with other UX UMa stars shed considerable light on the nature of these objects. Its spectrum is dominated by strong broad highly ionized absorption lines including the NV, SiIV and CIV UV resonance doublets. No emission lines or P Cygni type features are evident and the velocity displacements are smaller suggesting the absence of a hot high velocity wind characterizing other UX UMa stars. A physical interpretation of the UV absorption spectra and optical spectra and their observed temporal variations is presented. Based upon (1) high mass accretion rates derived by fitting the observations with theoretical continuum fluxes from steady accretion disk models by R.E. Williams, (2) the presence of outflowing winds with the determination of corresponding mass loss rates and properties of the theoretical wind models by Cassinelli, Olson and Stalio, and (3) new time independent steady state disk structure computations, a model is suggested to explain the observed properties of the UX UMa stars and their lack of major outbursts.

1. INTRODUCTION

The UX Ursa Majoris stars form a group of nova-like variables with common photometric and spectroscopic properties. The optical characteristics of these objects are thoroughly described by Warner (1976). They appear to be related to the cataclysmic variables but are not known to have undergone major outbursts such as those observed for dwarf novae, recurrent novae, classical novae, or symbiotic variables. In order to understand the still uncertain evolutionary connection between the UX UMa stars, the old novae, the dwarf novae in the outburst or extended high states and other cataclysmic variables as well as the physics of steady state disks, it is fundamentally

M. Livio and G. Shaviv (eds.), Cataclysmic Variables and Related Objects, 41–49.

important to know why any outburst behavior is absent in these stars. Critical new information about the UX UMa stars has recently become available from x-ray observations with the Einstein satellite (cf. Cordova, Mason and Nelson 1981) and ultraviolet spectroscopy with the International Ultraviolet Explorer satellite (cf. Greenstein and Oke 1982, Holm, Danek and Schiffer 1982, Guinan and Sion 1982a, 1982b, Sion and Guinan 1982, Wargau et al 1982). A new object, KQ Mon, has been added to our program of study with the IUE satellite. In this paper we discuss its relationship to the other UX UMa stars and suggest a model to explain their observed properties and absence of major outbursts.

2. OBSERVATIONS OF KQ MON

Bond's (1981) optical spectra of KQ Mon revealed very shallow Balmer absorption lines and HeI (λ4471) absorption, while his UBV and high speed photometric observations in 1978 and again in early 1981 revealed no light variations due to possible orbital phase modulation nor was an appreciable brightness change expected given the absence of radial velocity variations in the optical spectra. However, Bond's high speed photometry revealed the presence of low amplitude flickering that when coupled with the appearance of the optical spectrum and KQ Mon's location near the old novae and white dwarfs in the two color diagram, suggested a strong similarity to the UX Ursa Majoris subset of the nova-like variables (Warner 1976).

KQ Mon was observed at low dispersion with the IUE satellite on December 5, 1981 and February 5, 1982. The instrumentation and space-craft characteristics are described by Boggess et al.(1978). Four spectra with the short wavelength prime (SWP) camera, and four spectra with the long wavelength redundant (LWR) camera were obtained. The exposure times were 35 minutes and 40 minutes respectively, with the large and small aperture and two of the spectra were trailed.

Using the Fine Error Sensor (FES) photometer, visual magnitudes of +13.06 and +12.97 were obtained in December 1981 and February 1982 respectively while Bond obtained V=+13.06 in 1978, indicating that the light level appears reasonably constant. Broad band color indices for KQ Mon were measured by Bond (1982) to be B-V=+0.08, and U-B=-0.72. The U, B and V magnitudes were converted to absolute flux units using the absolute calibration of Hayes (1980).

In figure 1 a single SWP spectrum typical of the others is shown together with labeled identification of the strongest line features. The spectrum is dominated by strong broad absorption lines due to NV, OI + SiIII, SiIV, CIV, HeII and NIV. Possible Lyα absorption is obscured by the strong geocoronal feature. The long wavelength (LWR) spectrum appears essentially featureless. The data was de-reddened using the interstellar extinction law of Nandy et al.(1975) and a value of E(B-V)=0.08 was adopted. In figure 2 we plot the de-reddened satellite continuum fluxes on a log F_λ vs. log λ(Å) scale along with the broad band U B and V fluxes. On the same figure we show, for comparison, theoretical continuum fluxes from model steady state accretion disks calculated by Williams (1981b) assuming a disk inclination angle i=0 (pole-on). The models shown in figure 2 are normalized to the optical V band flux.

3. COMPARISON TO OTHER UX UMA STARS

The UV line spectrum of KQ Mon shows strong broad absorption lines of the resonance doublets and excited species but unlike other members of the UX UMa class (e.g. TT Ari in its high state, LS I 55 8, V 3885 Sgr, RW Sex, UX UMa) they are not appreciably violet-displaced and P Cygni-type emission is entirely absent. Thus a hot wind from the disk driven by X-radiation from the inner disk may be weak or absent in KQ Mon. Moreover, the absorption lines of KQ Mon appear to be marginally broader than those of LS I 55 8, V 3885 Sgr and RW Sex. The strengths and shapes of the CIV, NV and SiIV absorption features show significant variations as a function of time over the longest time baseline of our observations, 130 minutes. Similar line variations were found in other low inclination UX UMa stars (e.g. LS I 55 8, TT Ari, RW Sex) and may arise due to the changing aspect of gas observed against the disk at different orbital phases or possibly may result from disk asymmetry.

The estimates of mass loss rates due to wind outflow in the UX UMa stars give $\dot{M}_{wind} \sim 10^{-2}$ to 10^{-4} $\dot{M}_{acc}$ when $\dot{M}_{acc}$ is the mass accretion rate (cf. Guinan and Sion 1982a, 1982b, Greenstein and Oke 1982). The weakness or absence of an outflowing wind in KQ Mon may be due to (1) a somewhat lower (sub-critical?) $\dot{M}_{acc}$ than other UX UMa stars (2) accreting gas flowing above and/or below the disk plane (3) a weak white dwarf magnetic field preventing disk gas from forming a high pressure zone at the equator which could drive the type of polar wind described by Greenstein and Oke (1982) or (4) a nearly pole-on disk with conical outflow of hot gas in a narrow cone angle.

Like other UX UMa stars, the UV continuum is essentially flat from the UV to the optical and can be fitted quite well with accretion disk fluxes. The disk model fits in figure 2 clearly indicate an accretion rate near $10^{-8} M_{\odot} yr^{-1}$ for KQ Mon.

4. NATURE OF THE UX URSA MAJORIS STARS

Given the high ionization states of the UV absorption lines and the evidence for mass outflow in most of the UX UMa stars, the following tentative interpretation of the UV line and continuum spectra is suggested based in part on the wind models of Cassinelli Olson and Stalio (1978 cf. section IV, Guinan and Sion 1982). If we assume the X-radiation in these systems emerges from the hot inner disk boundary layer region rather than say a "hot spot" at the disk edge (because the disk is too luminous relative to a hot spot) it is likely that the UV resonance doublets such as N V (λ1240), C IV (λ1550) and Si IV (λ1396) arise in an outflowing wind. These high ionization features, seen in combination with a relatively "cool" integrated disk continuum, are similar to the UV spectra of O and B stars with outflowing winds (Cassinelli, Olson and Stalio 1978). A thin hot corona ($\sim 10^6$K) at the base of the flow in the boundary layer region provides the X UV photon source needed for the high ion states. The high ionization stages can be produced by the Auger mechanism, whereby two electrons are removed from C, N and O following K shell absorption of x-rays. Greenstein and Oke (1982) have independently proposed for RW Sex than an outflowing wind from the disk is driven by X-radiation, with the gas in conical geometry at the poles of the degenerate dwarf. They propose a broad cone while Guinan and Sion (1982) did not specify the geometry for V 3885 Sgr. It should be noted that for values of the viscosity parameter between 0.0 and 1 (Bath and Pringle 1981) derived for dwarf novae in outburst, a wind outflow geometry other than spherical would constrain the UX UMa systems to have magnetic fields (Wiita 1982). Indeed one of the UX UMa stars, TT Ari, appears to be an "intermediate polar" (Warner 1982) with a magnetic field strength lower than that associated with the AM Her objects and magnetic white dwarfs but strong enough to funnel a radial accretion flow. In either case the outflowing gas is seen silhouetted against a luminous disk for a large range of orbital inclinations which explains the appearance of the blue-shifted absorption lines and P Cygni-type features observed in a number of UX UMa stars. It is likely that when the accretion rate $\dot{M}$ is high, X UV quanta do not penetrate to the disk edge because the accretion flow is optically thick to that flux region. When the disk luminosity declines (i.e. $\dot{M}$ decreases), it is predicted that mass loss should cease, the absorption lines formed in the wind should disappear, the continuum should fade and the UV line spectrum should go into emission with the high excitation (barely displaced) UV emission lines originating in a "chromosphere" surrounding the inner disk region (cf. TT Ari in its low state; Wargan _et al_.1982). The site of this "chromosphere" is predicted to be the upper parts of the disk atmosphere more than one scale height from the central plane of the disk. When $\dot{M}$ again increases, the boundary layers' luminosity goes up and the UV spectrum again goes into absorption and an outflowing wind may again be driven.

That UX UMa stars may have higher accretion rates than other types of cataclysmic variables is indicated directly or indirectly by (1) derived accretion rates from continuum fits (e.g. RW Sex, Greenstein and Oke 1982; V 3885 Sgr, Guinan and Sion 1982a; LS I 55 8, Guinan and Sion 1982b; KQ Mon, Sion and Guinan 1982; Wargau _et al_.1982), (2) the flat continuum of a luminous, thick disk dominating the light from the UV through the optical and IR, (3) the low ratio of L_X/Lopt for those UX UMa stars observed with the Einstein satellite (Cordova Mason and Nelson 1981) and (4) the outflowing winds from accretion disks which may manifest a higher boundary layer luminosity or _local_ super Eddington accretion.

We contend here that the accretion rates of the UX UMa stars are higher than, for example, the dwarf novae during quiescence and that the higher accretion rates of the UX UMa stars are responsible for their lack of major outbursts. Based upon the excellent agreement between our observed fluxes and the theoretical disk continuum fluxes of Williams (1981b), steady accretion rates $\gtrsim 10^{-8} M_\odot yr^{-1}$ are implied for the UX Ursa Majoris stars. A higher mass accretion rate implies more massive thicker disks with some mechanism or viscosity source which allows continuous accretion onto the white dwarf thus keeping the system in continuous "outburst" (or extended high state). Time independent steady state disk models with the central plane convection occurring for the high accretion rates associated with the UX UMa stars, have been calculated by Vila (1982). Here central plane convection would be an effective viscosity source allowing accretion onto the white dwarf. However, recent steady disk models by Meyer and Meyer-Hofmeister (1981) have stable radiative structure for $\dot{M}_{acc} \gtrsim 10^{-9.5} M_\odot yr^{-1}$. Since the accretion rates of the UX UMa stars seem to be well above this value, steady state radiative disk structure can be maintained as long as $\dot{M}$ remains high enough for the disk to be in a stable radiative state. Thus these models lead us to suggest that the UX UMa stars possess steady state equilibrium disks due to their high mass accretion rates which would explain the absence of "outbursts." A stable steady state disk may also provide a natural explanation for the observed winds in RW Sex, LS I 55 8, V 3885 Sgr and UX UMa. If the accretion flow is high enough to be super-critical, the wind outflow may result from the response of the disk to re-establish a steady state accretion flow. In this case, KQ Mon may lack an outflowing wind because the accretion flow is just sub-critical. On the other hand, the accretion rate derived for RW Sex by Greenstein and Oke (1982) is super-Eddington for a $1 M_\odot$ white dwarf.

Finally further work is needed on the driving mechanism and structure of the winds in UX UMa systems. Since systemic mass loss appears to be occurring via these winds, the effects of angular momentum loss on the evolution of UX UMa stars should be investigated.

ACKNOWLEDGEMENTS

It is a pleasure to thank Dr. Robert E. Williams for kindly making available unpublished model accretion disk continuum fluxes for normal and helium-rich compositions. We are very grateful to Dr. Howard Bond for calling our attention to KQ Mon as a possible UX UMa star and providing a finding chart as well as unpublished optical photometry. Useful discussions with Drs. S.G. Starrfield, S.C. Vila and P. Wiita are gratefully acknowledged. This work was supported by the National Aeronautics and Space Administration under contracts NSG 5375 and NSG 7385.

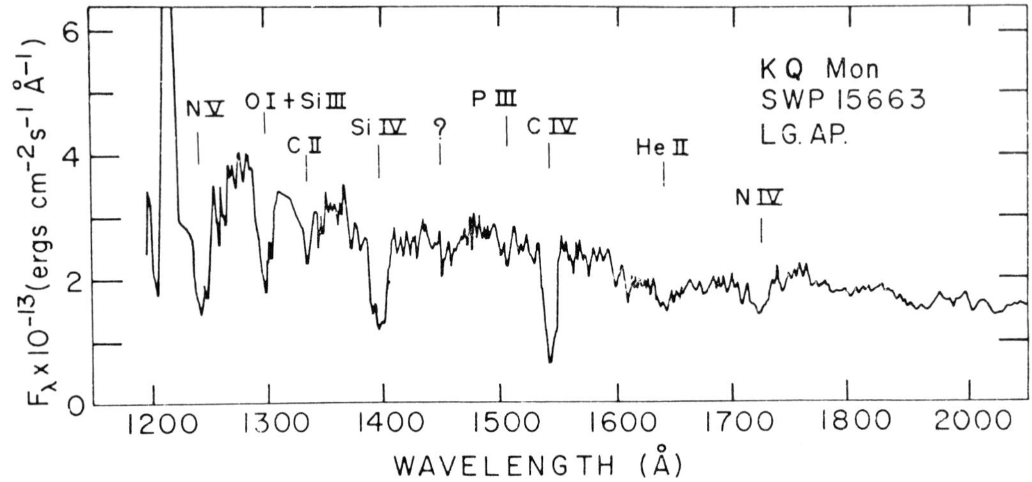

Figure 1: Short Wavelength Prime (SWP) low dispersion IUE spectrum of KQ Mon. The strong off-scale emission line at λ1216 is geocornal Lyman Alpha. The stronger absorption features are identified

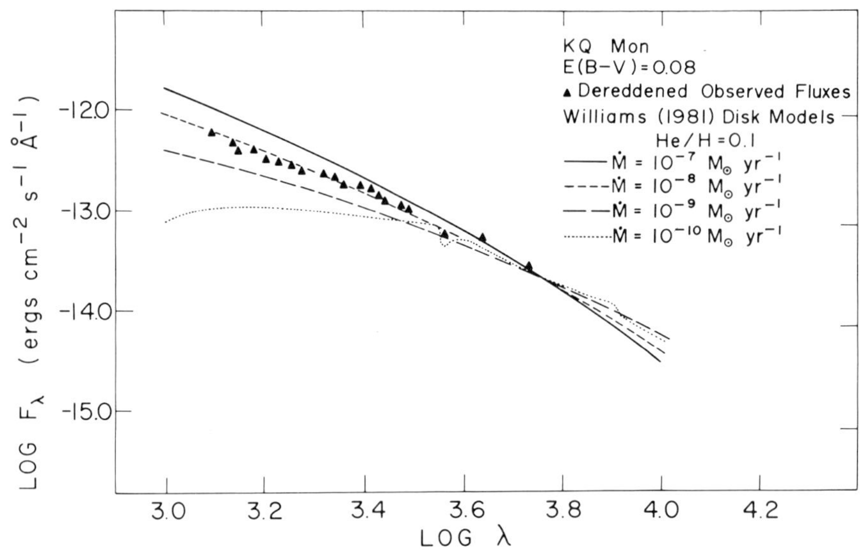

Figure 2: The de-reddened continuum fluxes of KQ Mon, log F_λ versus logλ, compared with three solar composition accretion disk models of Williams (1981b) normalized at the optical $\underline{V}$ band flux. Optical fluxes are labeled with corresponding bandpass designations.

REFERENCES

Bath, G.T., and Pringle, J.E.: 1981, Monthly Notices Roy. Astron. Soc. 194, p. 967.

Bond, H.E.: 1978, Publ. Astron. Soc. Pacific 90, p. 216.

Bond, H.E.: 1981 (private communication).

Bond, H.E.: 1982 (unpublished observations).

Boggess, A., et al.: 1978, Nature 275, p. 372.

Cassinelli, J.P., Olson, G.L., and Stalio, R.: 1978, Astrophys. J. 220, p. 573.

Cordova, F.A., Mason, K.O., and Nelson, J.E.: 1981, Astrophys. J. 245, p. 609.

Greenstein, J.B., and Oke, J.B.: 1982, Astrophys. J. 258, p. 209.

Guinan, E.F., and Sion, E.M.: 1982a, Astrophys. J. 258, p. 217.

Guinan, E.F., and Sion, E.M.: 1982b, "Advances in Ultraviolet Astronomy: Four Years of IUE Research," NASA Conf. Publ. 2238, p. 465.

Hayes, D.S.: 1979, "Dudley Obs. Report" No. 14, p. 297.

Holm, A.V., Panek, R.J., and Schiffer, F.H.: 1982, Astrophys. J. 252, p. L35.

Meyer, F., and Meyer-Hofmeister, E.: 1981, Astron. Astrophys. 104, p. L10.

Nandy, K., Thompson, G.I., Jamar, C., Monfils, A., and Wilson, R.: 1975, Astron. Astrophys. 44, p. 195.

Sion, E.M., and Guinan, E.F.: 1982, "Advances in Ultraviolet Astronomy: Four Years of IUE Research," NASA Conf. Publ. 2238, p. 460.

Wargau, W., Drechael, H., Rahe, J., and Vogt, N.: 1982, Astron. Astrophys. 110, p. 281.

Warner, B.: 1976, Observatory 96, p. 49.

Warner, B.: 1982 (preprint).

Williams, R.E.: 1981a, Astrophys. J. 235, p. 939.

Williams, R.E.: 1981b (unpublished models).

Vila, S.C.: 1982 (unpublished models).

Wiita, P.: 1982 (private communication).

DISCUSSION FOLLOWING E. SION'S TALK

WARNER: The wind that you see, is the velocity greater than the escape velocity from the system?

SION: The wind velocities in all but one case are less than the escape velocity from the surface of the white dwarf. In the case of BD-7307 terminal velocities that exceed the escape velocity from a white dwarf are indicated.

WARNER: I am interested in escape from the system, I just wonder whether it is a method of losing angular momentum from the system, from the binary, that is.

SION: We haven't checked that yet.

WARNER: A bit of historical accuracy, the idea that UX Ursa Majoris systems, as they are now called, are systems in permanent outbursts, really goes back to Bob Kraft in about 1965 or so.

ROBINSON: Is there any correlation of the strength of the P Cygni profile with orbital inclination? That is to say, have you looked at UX Ursa Majoris, which has a very high inclination and does it show the P Cygni profiles, evidence of a wind and so forth?

SION: Yes, there is a correlation, the deeper the absorption trough the more nearly pole-on seems to be the system and in UX Ursa Majoris those same absorption features are gone, it is mostly emission in the UV, because you are looking at the gas above, you are seeing the disk edge on.

FRIEDJUNG: First, a reply to Robinson, there is a recent preprint by Cordova and Mason where they say there is a correlation between the intensity of the P Cygni component and the inclination . A second thing, I would like to say about the wind, these winds from accretion disks may not be quite as unusual as they first appear, they may resemble very much winds from normal hot stars and may be something completely different from the winds you get from novae during outburst. An approximate calculation I did, using the most recent form for radiation pressure in the lines, by Abbot, and applying it to the old nova HR Del predicted the mass loss rate to within a factor of two. My impression is that these winds may be driven just by the radiation pressure in the lines as in certain theories for hot star winds.

SION: This is precisely my interpretation.

MEYER: I want just to clarify what continous outburst means, in view of the model that has been suggested by different people and also by us, it would just be the stationary disk, which is not modulated by the nonstationary flows in convective regions. So you don't need a special process to dump matter on the star, but rather you get rid, with your high mass flow rates, of the modulation process.

SION: In TT Ari the disk goes away or seems to shrink and then the luminosity goes down and then goes back up again.

MEYER: As long as you are above the critical mass accretion rate, of course you can have such variations. I would like to add that the velocities in your P Cygni profiles would probably give you information about the distance from the white dwarf from which the main bulk of this flow originates.

WILLIAMS: I was going to say that the fact that the velocities of outflow deduced from the resonance lines, are less than the escape velocity from the surface, requires that the mass loss is coming from further out, from the accretion disk. This is consistent with Friedjung's point that you would in fact expect mass loss from accretion disks, particularly in systems where you have heating of the disk by X-rays. One further point, N IV absorption that you identified, I think in KQ Mon that line is forbidden, it is very difficult to get it in absorption, I think you should recheck the identification of that, I think it may be a silicon line.

KING: I would like to support Williams' point. We had some UV spectra of UX UMa and we found that the C IV line was not eclipsed, when the system was eclipsed, which means this region of gas is very large, it must be as big as the shadow the secondary casts.

ON THE VARIATION OF THE ORBITAL PERIOD AND THE 67-MIN OSCILLATION IN EX HYDRAE *

C. Sterken
Astrophysical Institute, Vrije Universiteit Brussel,
Pleinlaan 2, B-1050 Brussels, Belgium

N. Vogt
Universitätssternwarte, Scheinerstrasse 1,
D-8000 München 80, Germany

R.F. Freeth
Department of Physics, The University of Auckland,
Private Bag, Auckland, New Zealand

H.D. Kennedy
Midway Observatory, Kelvin Grove College of Advanced
Education, Victoria Park road, Brisbane, Qld 4059

B.F. Marino
Queen Street 156, Northcote, Auckland 9, New Zealand

A.A. Page
Mount Tamborine Observatory,
17 Hillhouse street, Aspley, Qld 4034

W.S.G. Walker
Appleyard Crescent 14, Auckland, New Zealand

We report the first results of an observational campaign consisting of coordinated photoelectric photometry of Ex Hya from four observatories at different geographical longitudes for a total of 146 hours in April 1979.

Ex Hya was monitored differentially for a total of 109 hours, partly simultaneously at different telescopes. The largest time-interval of uninterrupted observation is of nearly 18 hours. 68 times of eclipse, and 89 times of maxima in the 67-minute cycle have been obtained. Figure 1 is an O-C diagram representing all residuals listed by Vogt et al. (1980), and those as obtained during this run. It is clear that the trend of a decreasing orbital period as present till 1976 does not

* Based on observations obtained at the European Southern Observatory, La Silla, Chile and at the South African Astronomical Observatory, Sutherland, South Africa.

M. Livio and G. Shaviv (eds.), Cataclysmic Variables and Related Objects, 51–54.

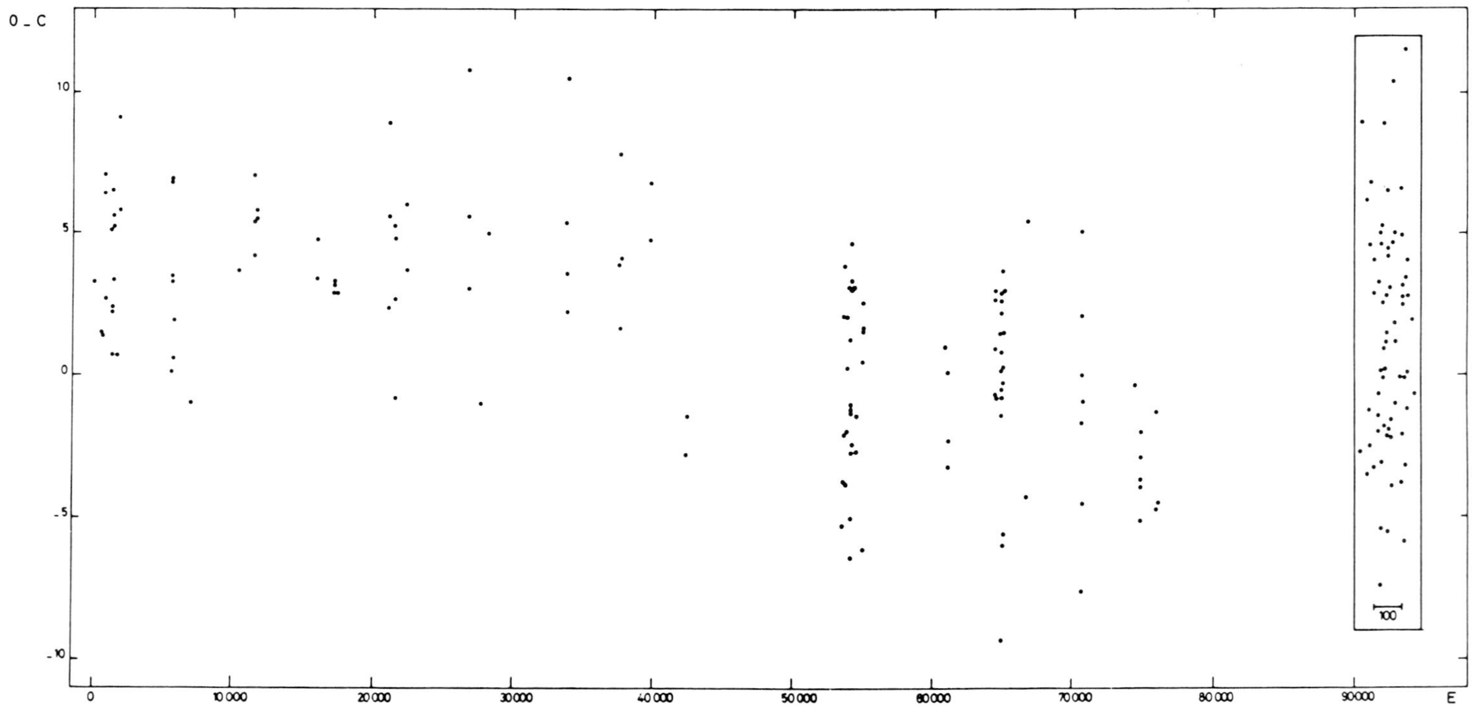

Fig.1. O-C residuals (in units of 0.0001 day) based on the elements of Mumford (1967), vs cycle number E for 206 eclipse minima of Ex Hydrae. The cycle number scale of the insert is a 20-fold (length of 100-cycle fragment is indicated; the center of the rectangle is correctly positioned).

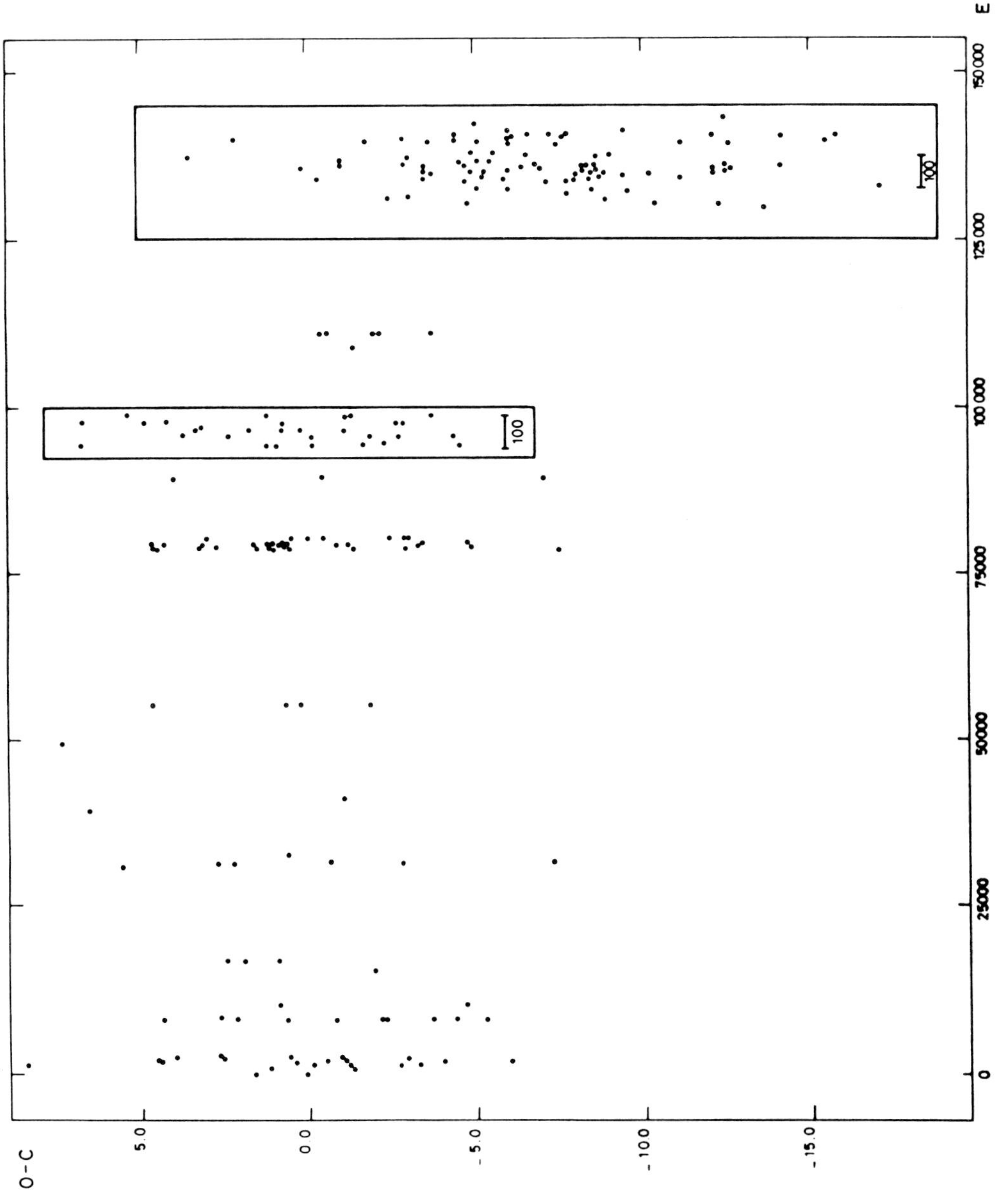

Fig.2. O-C residuals (in units of 0.001 day) based on the ephemeris of Vogt et al. (1980), vs cycle number E for 213 maxima in the 67-min cycle for Ex Hya. The scale of the insert is a 50-fold (length of 100 cycle fragments are indicated; the center of the rectangle is correctly positioned).

continue, and that instead of a secular period decrease, more erratic changes of the period are present.
Figure 2 shows O-C deviations based on the elements given by Vogt et al. (1980) for all times of maxima listed by their paper, and for the 89 times as obtained during the present campaign. It is clear that Ex Hya exhibits a significant decrease of the 67-min period with a time scale of 10^6 years. The best fit of all moments of maxima is :

$$\mathrm{HJD(max)} = \underset{\pm 6}{2437699.8895} + \underset{\pm 2}{.04654656}E - \underset{\pm 1}{1.2}\ 10^{-12}E^2$$

with a standard deviation of .0036 day.
Our ephemeris agrees with those derived by Gilliland (1982) and by Kruszewski et al. (1982), both based on independent new data, combined with the earlier moments of maxima from Vogt et al. (1980).
A more extensive account of our results will be published separately.

REFERENCES

Gilliland, R.L. 1982, Astrophys.J. 258, 576

Kruszewski, A., Mewe, R., Heise, J., Chlebowski, T., van Dyk, W., Bakker, R. 1982, in "Binary and Multiple Stars as Tracers of Stellar Evolution", IAU Coll. 69, Z. Kopal and J. Rahe (eds.), Bamberg, D. Reidel Publishing Co., Holland, 457.

Mumford, G.S. 1967, Astrophys.J.Suppl. 15, 1

Vogt, N., Krzeminski, W., Sterken, C. 1980, Astron.Astrophys. 85, 106

DISCUSSION FOLLOWING C. STERKEN'S TALK

ROBINSON: What fraction of the 67 min cycle has the O-C diagram displaced itself from the linear Ephemeris, is it only out of synchronization by a tenth of a cycle of something like that?

GILLILAND: About two tenths of a cycle.

ALPHA BETA DISCS

G.T. Bath, A.C. Edwards & V.J. Mantle
Department of Astrophysics, Oxford.

1) Introduction

Following earlier work of Lynden-Bell & Pringle (1974) and Lightman (1974a, 1974b), Bath & Pringle (1981) have presented a simple method for studying the time-dependent evolution of viscous accretion discs. These models are axisymmetric, with the vertical structure reduced to integrated averages of local physical conditions. Published work examines models of dwarf nova eruptions driven by mass transfer bursts (Bath & Pringle 1981 - Paper I), eruptions produced by global viscous changes within the disc (Bath & Pringle 1982a - Paper II), and the time-dependent properties of giant discs in symbiotic binaries (Bath & Pringle 1982b - Paper III).

In the next section we use these models to examine the theoretical basis of the observed correlation between the eruption decay-time and binary-period of dwarf novae (Bailey 1975, Mattei and Klavetter 1982). The main section of the paper describes an improved method for treating the interaction between the mass-transfer stream and disc matter, and presents results for bursting mass-transfer dwarf-nova models.

2) Theoretical eruption decay-rates

Bailey (1975) first pointed to the existence of a correlation between the outburst decay-time and the binary period for dwarf novae. More recently Mattei and Klavetter (1982) have conducted an extensive analysis of AAVSO data and confirm that the decay-rate, t_d, defined as the slope of the most linear portion of the declining light curve, in day per magnitude, depends on the binary period, with the best fit straight line being $t_d = 11.1P + 0.11$ (see Fig.1).

We have examined the predicted decay-rate/binary-period relation with time-dependent α -disc evolution models and

M. Livio and G. Shaviv (eds.), Cataclysmic Variables and Related Objects, 55–67.

conclude that outbursts caused by sudden bursts in the flow rate of the mass-transfer stream obey the observed decay-rate/binary-period relation subject to the following conditions being satisfied: Firstly, the mass-transfer burst must decay on a faster timescale than the viscous evolution timescale of the disc. Secondly, to fit the observed decay times, the viscosity in the disc must have a value corresponding to $\alpha \simeq 1.5 \pm^{1.5}_{0.5}$. The observed relation is then simply a consequence of the increase in the time needed for matter to diffuse through larger discs in longer binary-period systems.

We construct models according to Paper I, with minor and unimportant changes to the opacity. The dominant parameters which determine the eruption decay-time in any system are, in order of decreasing importance, the size of viscosity (i.e. α), the maximum outer disc radius, R_{OUT} (assumed to be equal to the tidal radius, $R_{OUT} = 0.88R_{RL}$, where R_{RL} is the average Roche lobe radius (Papaloizou and Pringle 1977, Paczynski 1977)), and the mass of the accreting white dwarf, which determines the inner disc radius, R_{IN}, and the gravitational potential. Note $R_{OUT} = R_N$ and $R_{IN} = R_1$ in the notation of paper I.

To determine the variation of R_{OUT} with period we specify the white dwarf mass, M_1. The condition that Kepler's law is satisfied, the secondary obeys a main-sequence mass-radius relation, and fills its Roche lobe then determines the size of the primaries Roche lobe, R_{RL}, and hence R_{OUT}. The circularization radius, R_K, of the mass transfer stream is obtained from the binary separation by interpolation from tables given in Lubow and Shu (1975). The values of R_{OUT} and R_K are given in Table 1 for periods in the range $0.05 < P(\text{days}) < 0.5$ and $1M_\odot$ white dwarf primary.

Table 1

q	M_2 ($M_\odot$)	Period (days)	R_{OUT} (cm)	R_K (cm)
0.59	1.68	0.50	7.52×10^{10}	1.85×10^{10}
0.67	1.49	0.45	7.05×10^{10}	1.75×10^{10}
0.77	1.30	0.40	6.56×10^{10}	1.65×10^{10}
0.89	1.12	0.35	6.05×10^{10}	1.55×10^{10}
1.06	0.94	0.30	5.53×10^{10}	1.45×10^{10}
1.29	0.77	0.25	4.97×10^{10}	1.34×10^{10}
1.65	0.61	0.20	4.36×10^{10}	1.23×10^{10}
2.23	0.45	0.15	3.71×10^{10}	1.12×10^{10}
3.38	0.30	0.10	2.96×10^{10}	9.90×10^{9}
6.85	0.15	0.05	2.00×10^{10}	8.13×10^{9}

Theoretical visual light-curves of dwarf nova eruptions have been computed for the case of a mass transfer burst, with the same burst profile independent of binary period, and the decay time of the most linear portion of the light curve determined. We find that the light curve decay-time, t_d, is not sensitive to the form of the mass-transfer changes so long as the burst shuts off on a timescale shorter than the viscous disc diffusion-time. Variations of total burst mass, the maximum transfer rate in the burst, and the quiescent mass transfer rate have insignificant effect on t_d.

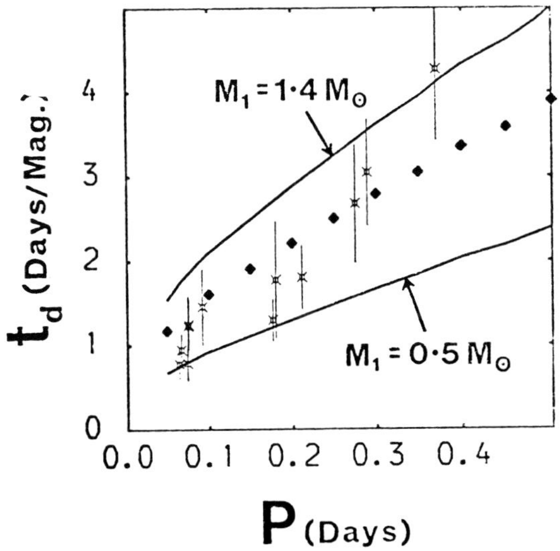

Fig 1. Decay time, t_d, versus binary period for observed systems (stars with error bars), and computed models with $\alpha = 1.5$ and $M_1 = 1M_\odot$ (solid diamonds). The model relation for $M_1 = 1.4\ M_\odot$ and $M_1 = 0.5\ M_\odot$ is also shown.

In Fig.1 the theoretical values of t_d are plotted (filled diamonds) together with the observed values of individual systems from Mattei and Klavetter (1982) (stars with error bars). For the assumed value of $1M_\odot$ for M_1, the best fit is obtained with a value of $\alpha = 1.5$. Also shown are the limits on t_d for $1.4\ M_\odot > M_1 > 0.5\ M_\odot$. The observed scatter can clearly be accounted for by variations in white dwarf mass from system to system. Equivalently, variations in α in the range $3.0 > \alpha > 0.5$ would account for the observed scatter, with a $1\ M_\odot$ white dwarf. It is evident from this study that the allowed variation of α is small, and that the decay-time/binary-period relation is a natural consequence of increasing disc size in longer period binaries.

Note furthermore that the decay-time is independent of the duration of the mass-transfer burst. Supermaximum outbursts can be obtained with long duration, larger mass flux bursts. These decay on the same timescale as normal bursts, and satisfy Bailey's relation, as is observed (Mattei and Klavetter 1982).

3) Mass input and stream/disc mixing

The method described in Bath and Pringle (1981) for computing disc evolution assumes instantaneous mixing of the mass-transfer stream with disc matter. Stream material is fed into the disc with the appropriate specific angular momentum, the change in disc radius computed, and the input mass mixed in an appropriate ratio in the two outer zones at the disc edge, R_{EDGE}, where angular momentum balance is achieved. Whether the disc edge remains at its maximum radius, R_{OUT}, or shrinks within this radius is determined by the balance between angular momentum transport within the disc (which tends to expand the disc towards R_{OUT} or keep it in contact with R_{OUT}), and the angular momentum of the mass transfer stream (which tends to shrink the disc towards R_K).

At R_{OUT} it is assumed that all excess angular momentum can be efficiently removed by tidal interaction with the companion. That is, the effective couple in the disc is zero at R_{OUT}, and $\partial(\Sigma\nu R^{\frac{1}{2}})/\partial R = 0$ at that point (note that the outer boundary condition is stated incorrectly in papers I and II. The computations in both papers use the correct boundary condition, ensuring mass conservation).

The instantaneous mixing hypothesis is employed largely for computational convenience. It conserves mass and angular momentum appropriately, but the extent of the region in which mixing takes place is arbitrarily determined by the zone size.

In practice the mass transfer stream will penetrate the disc on a "quasi-elliptical" orbit. How deeply the stream penetrates before all the stream material has been sheared off by shocks driven by disc matter in circular orbits (as discussed for the case of low viscosity discs by Lubow & Shu 1975) is unknown. However there is evidence (e.g. Stover 1981) that extensive penetration of the stream occurs in some systems. We have therefore developed an alternative approach to the mass input problem which is fully conservative, and includes mass and angular momentum sources implicitly in the conservation equations for disc structure.

If the mass flux fed from the stream into the disc at radius R and time t is $\partial(\dot{m}(R,t))/\partial R)$ then, in the thin disc approximation, conservation of mass gives,

$$\frac{\partial \Sigma}{\partial t} = \frac{-1}{R}\frac{\partial(R\Sigma v_R)}{\partial R} + \frac{1}{2\pi R}\frac{\partial \dot{m}}{\partial R} \qquad (1)$$

where Σ is the surface density and V_R the radial velocity. If the stream has specific angular momentum which would place it in an orbit at radius R_K, then conservation of angular momentum implies,

$$\frac{\partial(\Sigma R^2 \Omega)}{\partial t} = \frac{-1}{R}\frac{\partial(R^3 v_R \Sigma\Omega)}{\partial R} + \frac{1}{R}\frac{\partial}{\partial R}\left(R^3 \nu\Sigma\frac{\partial\Omega}{\partial R}\right) + \frac{R_K^2 \Omega_K}{2\pi R}\frac{\partial \dot{m}}{\partial R} \qquad (2)$$

where ν is the effective kinematic viscosity. Combining (1) and (2), eliminating V_R, and assuming Keplerian angular velocities, leads to an evolution equation for the surface density as a function of time of the form,

$$\frac{\partial \Sigma}{\partial t} = \frac{3}{R}\frac{\partial}{\partial R}\left(R^{\frac{1}{2}}\frac{\partial(\nu\Sigma R^{\frac{1}{2}})}{\partial R}\right) + \frac{1}{2\pi R}\frac{\partial \dot{m}}{\partial R} + \frac{1}{\pi R}\frac{\partial}{\partial R}\left(R\left(1-\frac{R_K^{\frac{1}{2}}}{R^{\frac{1}{2}}}\right)\frac{\partial \dot{m}}{\partial R}\right) \qquad (3)$$

This is the revised evolution equation including stream material as a source implicitly. The first term is the basic diffusion term describing matter and angular momentum redistribution due to viscous stress, and is the fundamental expression controlling disc evolution (see e.g. Pringle 1981). The second term is the stream source term from equation (1). The third term describes the influence on the evolution of Σ of the angular momentum of the mass transfer stream. The first, diffusion term drives material out (and in) into a disc by viscous stress, whilst the third term describes the tendency of new input material with angular momentum appropriate to an orbit at R_K to squeeze the disc towards an annulus at radius R_K.

To solve equation (3) the rate at which matter is stripped off the stream and redirected into circular orbits, $\partial\dot{m}/\partial R$, must be specified. This is not a well defined problem. In order to make progress, and in particular to examine the effect on disc evolution of deep disc penetration as opposed to instantaneous mixing, we have adopted the following parametrization. We assume that the stream is stripped at a rate which is a fraction β times the circular mass flux in the disc at that radius, i.e.

$$\frac{\partial \dot{m}}{\partial R} = \beta\, \Sigma \left(\frac{GM_1}{R} \right)^{\frac{1}{2}} \tag{4}$$

If $\beta = 1$ the momentum imparted to stream material by "collision" of disc matter moving in the azimuthal direction is that of the disc material itself (though note that the form of equation (3) ensures that angular momentum is conserved). It is clear that $\beta \simeq 1$ is an upper limit, and in general $\beta < 1$.

We have explored the effects of different values of β with numerical models. Low values of β allow deep stream penetration, high values of β ($\beta = 1$) approach instantaneous mixing. To obtain numerical solutions equation (3) is transformed to variables $X = 2R^{\frac{1}{2}}$ and $S = X\Sigma$ (see paper I).

$$\frac{\partial S}{\partial t} = \frac{12}{X^2} \frac{\partial^2 (\nu S)}{\partial X^2} + \frac{4(GM_1)^{\frac{1}{2}}}{\pi X^2} \frac{\partial}{\partial X}\left(\left(1 - \frac{X_K}{X}\right) \beta S \right) + \frac{4(GM_1)^{\frac{1}{2}}}{\pi X^3} \beta S \tag{5}$$

Equation (5) is solved with a centred finite difference scheme with appropriate boundary conditions which ensure conservation of mass within the grid, allow mass and angular momentum loss at R_{IN} and angular momentum loss at R_{OUT}. The vertical structure is solved as in Paper I, with values of $\alpha = 1.5$, $R_{OUT} = 5 \times 10^{10}$ cm, $R_K = 1.5 \times 10^{10}$ cm, $R_{IN} = 5 \times 10^8$ cm, and $M_1 = 1\ M_\odot$.

In Figs 2 and 3 the change in structure of the disc following a burst of mass transfer is shown for values of $\beta = 1.0$ (a), $\beta = 0.1$ (b), and $\beta = 0.01$ (c). In Fig 2 the radius of the disc, R_{EDGE}, together with the point at which complete mixing of the stream into the disc is reached, R_{STRIP} is shown as a function of time. Material is being progressively stripped off the stream between R_{EDGE} and R_{STRIP} according to the prescription of equation (4). In Fig 3 the change in surface density as a function of radius during the rise in the mass transfer burst is shown. The surface densities increase with time as matter is injected into the disc from the enhanced flux in the stream. In Fig 3 each curve is separated by 2×10^4 sec evolution time, starting at 1×10^5 sec with a steady-state disc (curve A) and finishing at 2×10^5 sec (curve B) when the burst reaches its maximum rate of 10^{19} g s^{-1}. The form of the mass transfer burst is given in Fig 4.

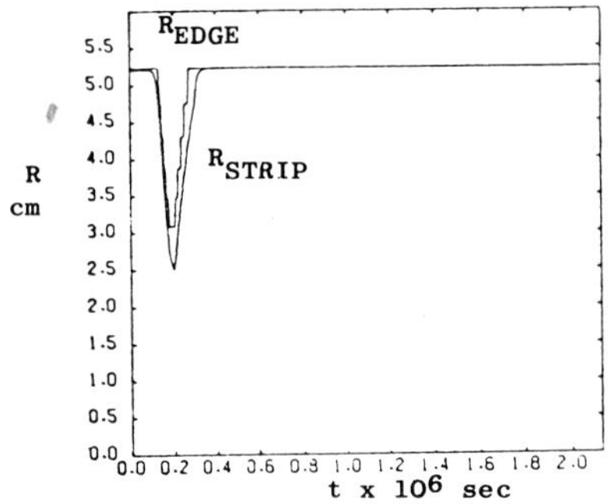

Fig 2a β = 1.0

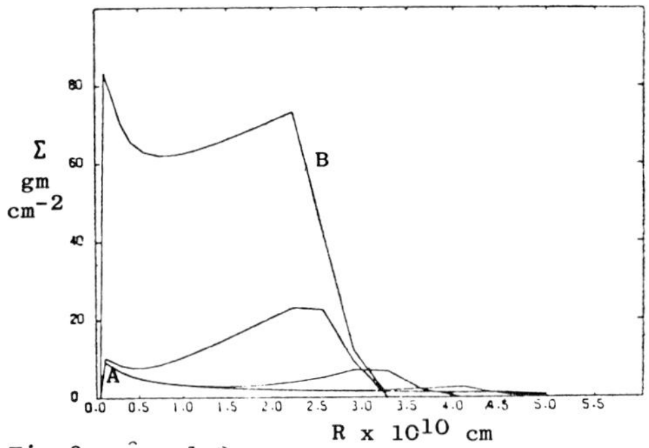

Fig 3a $\hat{p}$ = 1.0

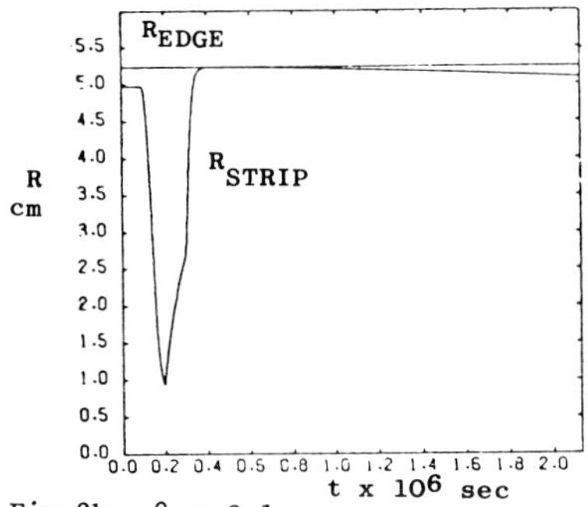

Fig 2b β = 0.1

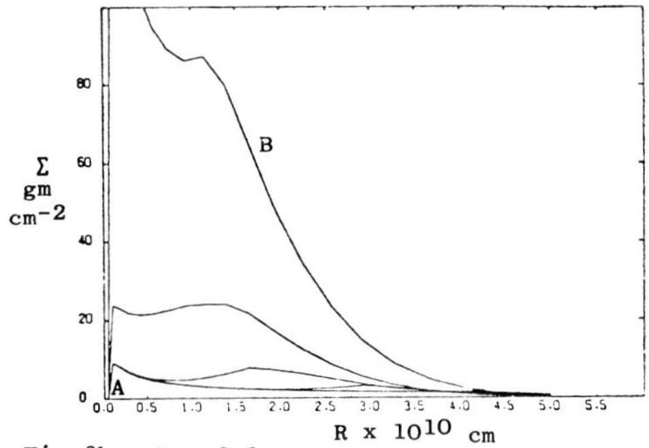

Fig 3b β = 0.1

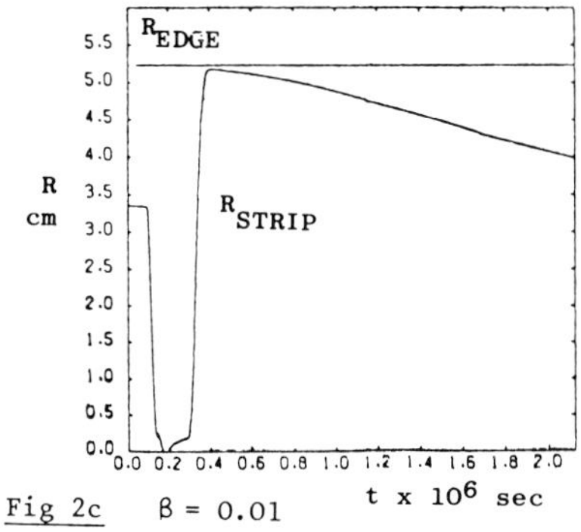

Fig 2c β = 0.01

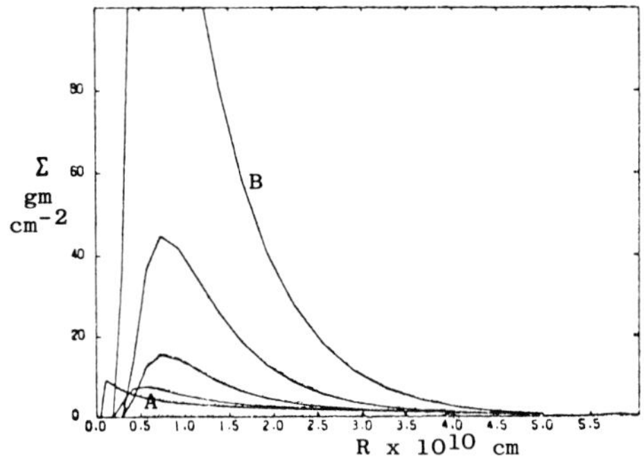

Fig 3c β = 0.01

Fig 2. Variation of disc edge, R_{EDGE}, and stream penetration radius R_{STRIP}, with time.

Fig 3. Evolution of surface density during the rise in the mass transfer rate. Curves labelled A correspond to steady state conditions at t = 1 x 10^5 sec and B to the state at t = 2 x 10^5 sec when the maximum mass transfer rate is achieved. Each curve is separated by T = 2 x 10^4 sec.

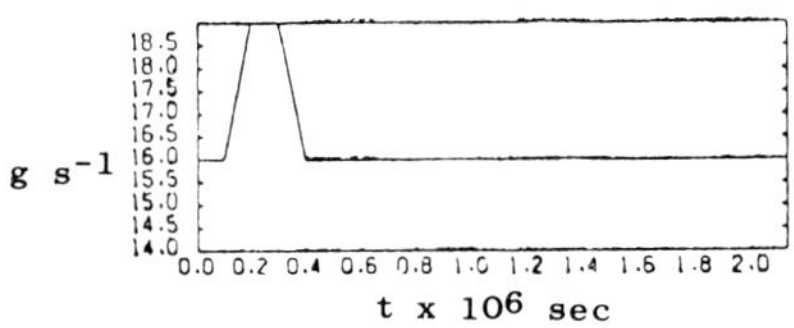

Fig 4. Mass transfer variation with time.

At a value of $\beta = 1.0$ the stream is stripped in the outermost region of the disc (Fig 2a). The behaviour is close to the instantaneous mixing of Paper I. During the mass transfer burst the disc edge shrinks due to deposition of low angular momentum material in the outer disc rim, and then expands back to R_{OUT} at 5×10^{10} cm. The form of the surface density distribution (Fig 3a) shows how the stream builds up the outer edge of the disc with new material which subsequently diffuses inward.

At a value of $\beta = 0.1$ the stream penetrates deep into the disc in the burst (Fig 2b), down to a radius of 1.0×10^{10} cm (R_K). The disc edge no longer shrinks, but, as can be seen in Fig 3b, R_{EDGE} remains at R_{OUT} and material is stripped progressively off the stream as it penetrates all the way through the disc down to R_{STRIP}.

More extreme disc penetration occurs at the lowest value of $\beta = 0.01$. Even in the steady-state before outburst the stream penetrates well into the disc (as must be occurring in RU Peg, see Stover 1981). During the burst most of the material is deposited close to R_K, and stripping extends all the way down to the minimum distance of approach to the white dwarf by the stream. Because this material has angular momentum which tries to force it into an orbit at radius R_K, and viscosity has not yet had time to operate effectively, the surface density profile peaks at a point close to R_K. As expected, with low values of β most of the stream material goes into an orbiting annulus near R_K. This persists until viscosity acts to spread the new burst of matter out into a disc.

Although the value of β has a major influence on the surface density evolution of the disc, the effect on the form of the eruption light curve is much less significant. The only difference between the light curves with $\beta = 0.01$ compared to $\beta = 1.0$ is enhanced ultraviolet flux at maximum light (Fig 5). This enhanced ultraviolet emission is due to the increased fraction of matter deposited directly in high temperature inner disc regions. After the transient effects associated with the period of enhanced mass flux are over, the subsequent viscous evolution is closely similar and independent of β. The value of β affects the rise and maximum light behaviour, but the decay is unaffected. Thus the decay-time/period relation discussed in Section 2 is insensitive to the processes discussed in this Section. One way in which an estimate of β in individual systems may eventually be obtained is from the changes in the continuum spectral distribution during the rise and at maximum light. In Fig 6 the continuum distribution during the rise (at the same time intervals as Fig 3) is shown. Major differences between the three cases are apparent, and in particular during the rise, and at

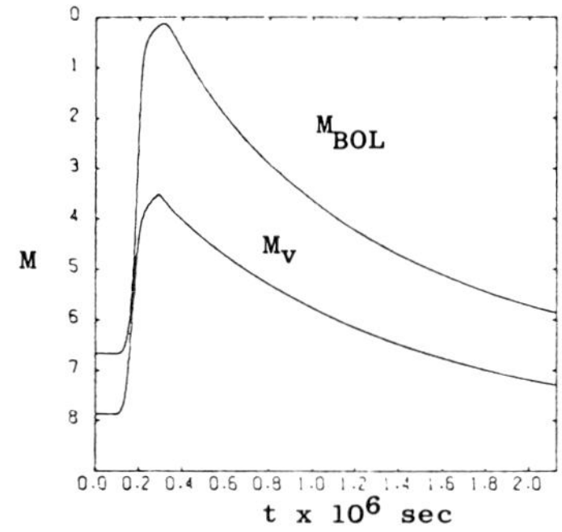

Fig 5a   β = 1.0

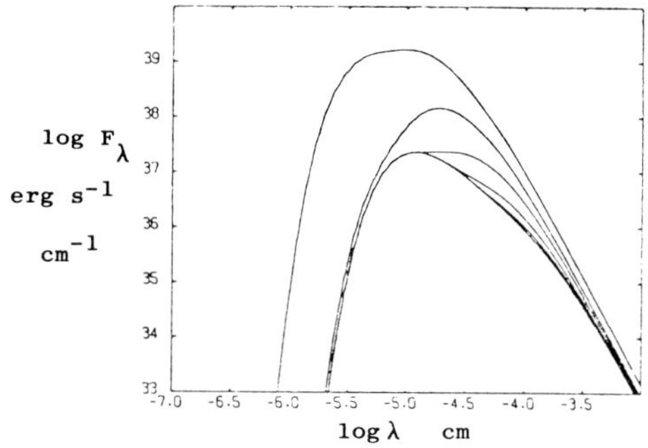

Fig 6a   β = 1.0

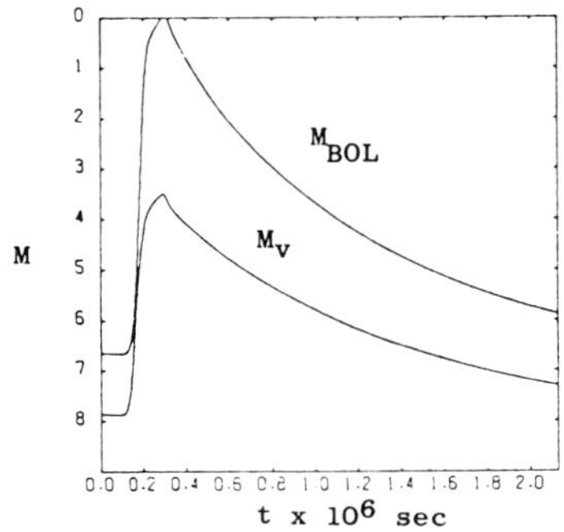

Fig 5b   β = 0.1

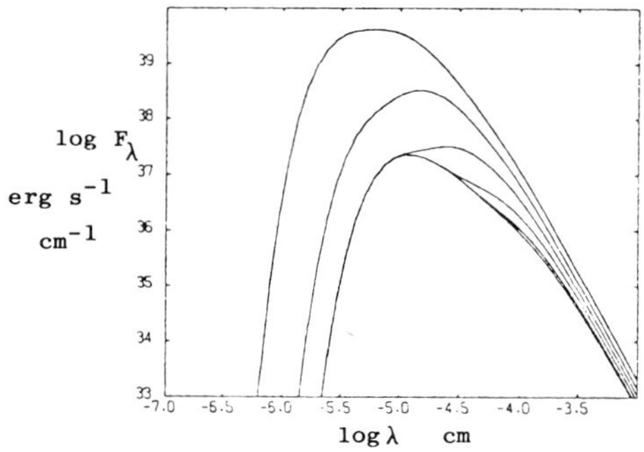

Fig 6b   β = 0.1

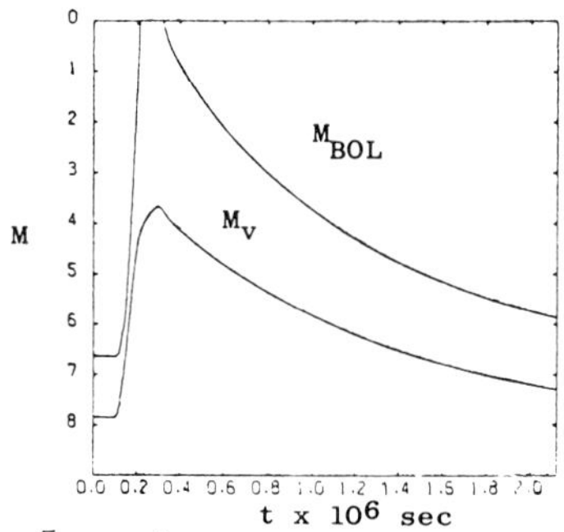

Fig 5c   β = 0.01

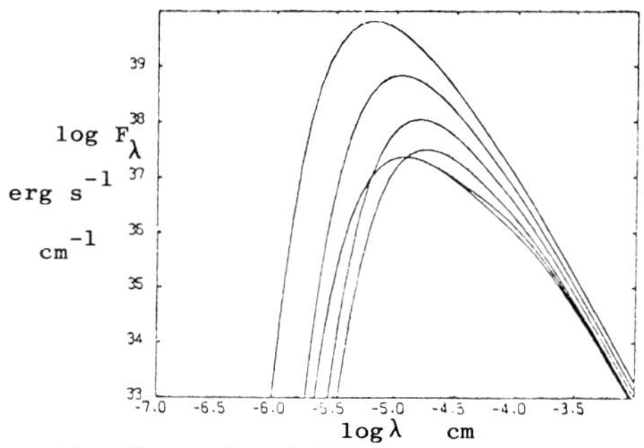

Fig 6c   β = 0.01

Fig 5. Bolometric and visual light curves.

Fig 6. Spectral evolution during the rise in the mass transfer rate for the same times as Fig 3.

maximum light, the slope is not a steady-state $\lambda^{-2.33}$ power law. During these phases, the effects of mass input and time-dependent evolution introduce significant deviations from the predictions of steady models.

Hassall et al. (1982) find that on the rise to outburst the UV flux in VW Hydri lags at least one day behind the optical, with the continuum spectral distribution rising initially only in the visual, as we find for models with $0.1 < \beta < 1.0$. Comparing their results with models in Fig.6 we find that $\beta > 0.1$. There is no need at this stage to invoke optically thick winds in dwarf novae to account for deviations of the continuum from steady state predictions as Klare et. al (1982) have claimed.

An upper limit on β can be set from the observed variation of the "hot-spot" luminosity during the outburst rise. It is well established that during the rise the hump luminosity does not change in proportion to the overall increase in luminosity from the disc. Indeed it is often claimed that this indicates no change in the mass flux in the stream, and that the outbursts must be produced by disk instabilities. However this conclusion is based on the naive assumption that all the available stream kinetic energy is released within a distance of order the height of the disc, H at the disc edge. If stream penetration occurs this will not be the case. In Fig.7 the light curve of VW Hyi in outburst (Warner 1974) is shown together with the hump luminosity. The delay of the rise in the hump behind the overall outburst is evident.

In Fig.8 the changes in the disc, spot and stream luminosity are shown for a $\beta = 0.1$ disc. The spot luminosity, L_{spot} is computed as that fraction of the stream-disc impact energy liberated within a region of thickness H of the disc edge (i.e. that flux which could produce an asymmetric radiation pattern with the properties of the so-called "hot-spot"). The stream luminosity, L_{stream}, is the stream kinetic energy liberated deeper within the disc (which may be radiated anisotropically but not with the expected distribution in binary phase of the "hot-spot").

It is evident that the spot luminosity does not rise simultaneously with the outburst rise, but is delayed. This is simply due to the fact that until the surface density at the outer edge has grown by viscous transport of material deposited interior to the disc edge, the stream continues to penetrate, and the spot flux is unaffected by the outburst. A simultaneous increase of L_{disc} and L_{spot} occurs only if $\beta = 1.0$, i.e. with instantaneous mixing. We conclude that there is no observational support for the view that mass transfer variations cannot be driving the outbursts. With $\beta = 0.1$ all the major observed features presently known to be associated with the disc component of dwarf nova outbursts are realised. Further evidence in support of this interpretation is given in a subsequent paper in this volume.

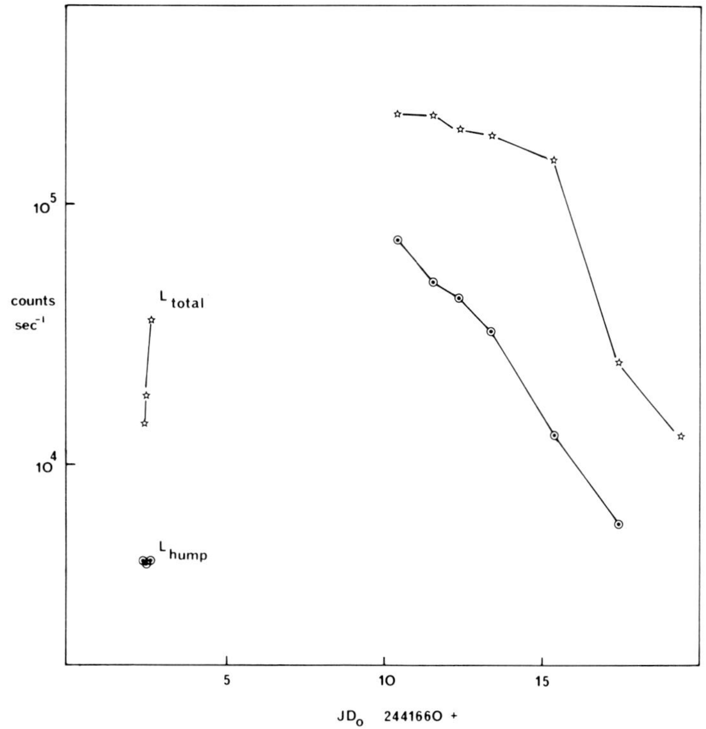

Fig.7. Observed total flux variation and hump luminosity through outburst in VW Hydri.

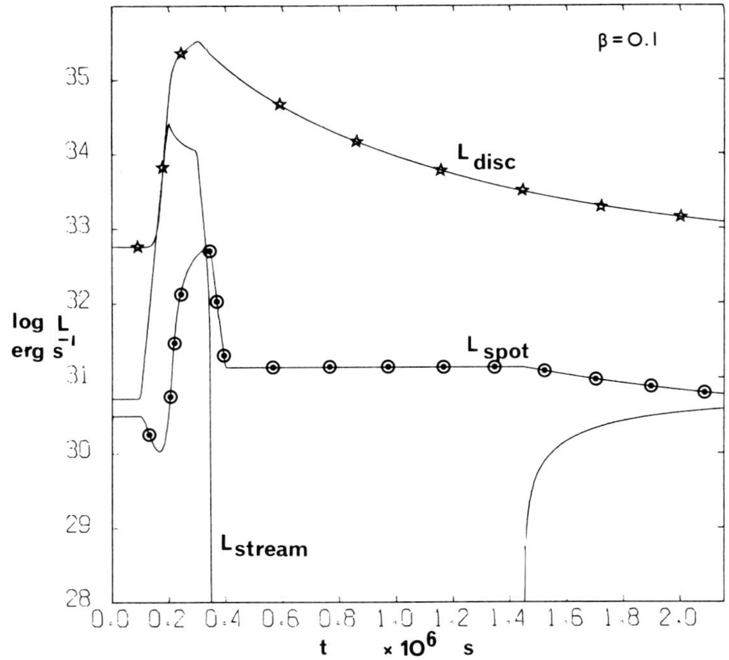

Fig.8. Changes in disc, spot, and stream luminosity in a $\beta = 0.1$ disc.

References

Bailey, J., 1975, J.Br.astr.Ass., 86, 30.
Bath, G.T. & Pringle, J.E., 1981, Mon.Not.R.astr.Soc., 194, 967.
Bath, G.T. & Pringle, J.E., 1982a, Mon.Not.R.astr.Soc., 199, 267.
Bath, G.T. & Pringle, J.E., 1982b, Mon.Not.R.astr.Soc. 200, (In press).
Hassell, B.J.M., Pringle, J.E., Schwarzenberg-Czerny, A., Wade, R.A., Whelan, J.A.J., 1982, Mon.Not.R.astr.Soc. (In press).
Klare, G., Krautter, J., Wolf, B., Stahl, O., Vogt, N., Wargau, W., & Rahe, J., 1982, Astr. & Astrophys, In press.
Lightman, A.P., 1974a, Astrophys.J., 194, 419.
Lightman, A.P., 1974b, Astrophys.J., 194, 429.
Lubow, S.H. & Shu, F.H., 1975, Astrophys.J., 198, 383.
Lynden-Bell, D. & Pringle, J.E., 1974, Mon.Not.R.astr.Soc., 168, 603.
Mattei, J.A. & Klavetter, J.K., 1982, In press.
Paczynski, B., 1977, Astrophys.J., 216, 822.
Papaloizou, J. & Pringle, J.E., 1977, Mon.Not.R.astr.Soc. 181,
Pringle, J.E., 1981, Ann.Rev.Astr. & Astrophys., 19,
Stover, J.R., 1981, Astrophys.J., 249, 673.
Warner, R., 1974, Mon.Not.R.astr.Soc., 170, 219.

DISCUSSION FOLLOWING G. BATH'S TALK

LAMB: It seems to me that the distinction you are making between the luminosity of the spot and the luminosity of the whole stream is a little shaky.

BATH: I agree. Let me just clarify, this is just an approach to study a problem which at the moment is being pushed under the carpet and upon which an awful lot of arguments depend. All I am saying is that this is the first attempt to study it and when I do, it turns out that many of the arguments that have been made in the past are not valid, they are only valid if $\beta=1$, which is an extreme case.

ROBINSON: Even if you accept that there is an observational distinction between L_{stream} and L_{spot} it looks to me like L_{spot} does achieve a new bright equilibrium value long before the eruption has decayed completely so in this model, the spot should reappear at some point in the eruption.

BATH: I am going to talk about details of the anisotropic pattern of this in my second talk.

SCHATZMAN: My question concerns the turbulent viscosity which you assumed in order to find values of α and β close to 1, do you know of any work which gives more justification than just the parametrization of the scales?

BATH: No. I know that people have done work, but I don't know of any work that has been accepted.

RITTER: You have shown how the stream penetrates into the disk if you release a blob of matter at L_1. What about disk penetration in the quiescent state?

BATH: These were evolved in a steady state, constant $\dot{M}$. No stream penetration with $\beta=1$ in the steady state, slight penetration with $\beta=0.1$ and extensive penetration with $\beta = 0.01$.

RITTER: So this means that if you make observations of dwarf novae in the minimum state, then the instantaneous mixing assumption is not too bad.

BATH: Right. In most cases, except I am worried about Stover's observations of U Peg.

RITTER: I have another question. When you have shown the radiation coming out, you gave the argument that if you have a very low β, you have in the bolometric flux a huge peak in the radiation, but there you have obviously assumed that all the matter which is streaming in, is circularized at the corresponding orbit, because if you have $\beta=0.01$ then the stream behaves as a more or less non dissipative.

BATH: It goes around ten times in fact, before it gets completely stripped, with $\beta=0.01$.

LIVIO: Concerning the Bailey relation, it seemd to me that if the last observational point should be taken seriously that the theoretical graphs that you have obtained, always did not have the same slope as the slope of the observations.

BATH: I don't think that this is important. I think it is amazing that with an "α viscosity" here, which is not real physics, we get as good a correlation as we have. Also, α may be changing as a function of time. I would like to mention here that it was John Whelan who in the conference in Cambridge in 1975 suggested that maybe mass transfer bursts generate disk instabilities.

MATTEI: The points on left side of the graph of the Bailey relation are all for SU UMa stars and the ones after the gap of course are for U Gem and Z Cam type stars and even though we can fit a line like the one in the figure, there appears to be a better fit it we fit one line through Z Cam and U Gem stars and another one through SU UMa stars. Is there any way of explaining the different values of α, in order to have two relations?

BATH: If there is a distinction, I don't understand what it is at the moment. I would like to make the comment that it is very difficult to fit this relationship with disk instability models of a simple type. This is because if you vary the viscosity globally over the disk, the decay of the light curve is determined initially by the diffusion time through the disk, but at later stages of the outburst it is determined by the rate of change of viscosity itself, and that has to scale with binary period if it is going to fit the Bailey relationship and I don't see any reason why that should be the case.

HUMPS AND SUPERHUMPS

G.T. Bath, A.C. Edwards & V.J. Mantle
Department of Astrophysics, Oxford.

1) Stream-disc accretion shocks

It is now well established, following the classic work of Smak, Warner and Nather, that discs in dwarf novae possess an anisotropic radiation pattern responsible for the hump, or shoulder, that occurs prior to eclipse in the quiescent state, and, in systems with suitable inclination such as U Gem, responsible also for the primary eclipse itself. In systems with higher inclination such as Z Cha, the primary eclipse is composed of both this anisotropic disc component and the inner-disc/white-dwarf primary component.

In the quiescent state the anisotropic component generates a radiation pattern and an eclipse profile which place it in the region where the mass transfer stream impacts the disc edge, while radiating away from the white dwarf in the form of a spot of dimension $\simeq 10^9$ cm. This component is commonly called the "hot-spot", though it is now clear that it is <u>not</u> hot compared to the inner disc region. It is hotter than disc regions at different azimuth, but the same radius, but cooler than regions further in and is, relatively speaking, optically bright. It is probably better termed a "bright-spot" as some authors have previously indicated.

It is normally assumed that this quiescent bright-spot results from shock-heating of the stream/disc gas at the point of impact of the stream with the disc. However, in situations in which deep penetration of the stream is occurring it is not at all clear that such a spot will be formed. It is well established that during eruption the phase of maximum hump light is not stable. It may undergo somewhat irregular phase variations (humps), or behave as a periodic wave with a period longer than the quiescent hump period by a few percent (superhumps) (Vogt 1974, Warner 1975). We describe here two independent mechanisms which could account for these two phenomena.

M. Livio and G. Shaviv (eds.), Cataclysmic Variables and Related Objects, 69–76.

2) Humps and peapod diagrams

In circumstances in which the main anisotropic component is generated by the disc, two extreme disc/stream shock heating patterns can be distinguished. These depend on the density of the stream and of the outer disc regions. In one case the stream hits a higher density disc with negligible penetration and formation of a standard bright-spot. In the second case the disc hits the stream as the stream penetrates the disc, a shock-heated "wall" is produced at the point where low density disc material impacts the stream, and some fraction of stream material is stripped into the disc.

The sort of conditions we envisage here are illustrated in Fig.1.

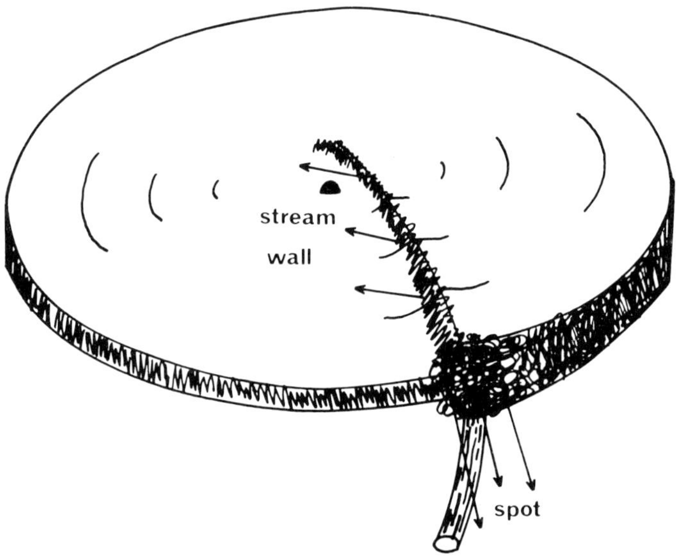

Fig.1

At its most extreme this stream-stripping process can be envisaged as producing a wall, tilted at an angle, ψ above and below the disc plane (we assume below a tilt angle of 45°) and radiating in the direction indicated in Fig.1. In practice the wall will not be a stable feature, and downstream disc emission will doubtless be significant. Nonetheless we anticipate that the radiation pattern will tend towards this extreme case in circumstances of deep stream penetration.

In order to study the anisotropic pattern generated by stream/disc impact we plot the anisotropic component generated by these two processes as a function of phase in a polar diagram. In Fig.2 we show the radiation generated by a quiescent state alpha beta disc (α = 1.5, β = 0.1) in which the bright-spot component dominates. Within region (2) we plot the total anisotropic radiation component above the disc on a logarithmic scale, relative to the isotropic disc component. In region (1) we plot the anisotropic component alone. The continuous line is the visual luminosity, and the thin broken line the bolometric luminosity.

The thick dotted line shows the polar distribution of anisotropic radiation from Z Cha during the second orbit on Dec 1972 (Warner 1974). It is evident that all the characteristics of the radiation pattern are represented by the standard bright-spot model in this "peapod" diagram (peripheral excess anisotropy polar diagram). The diagram also suggests that stream penetration and wall radiation may be responsible for a small degree of excess flux at phases 0.25 to 0.50 with respect to the spot.

At outburst in Z Cha it is well established that the hump may move in binary phase to an apparent position almost opposite the spot. In Fig.3 we show the radiation pattern produced by an outburst in Z Cha during the second orbit on Jan 8 1973 (Warner 1974). We compare this with the radiation pattern generated by deep stream penetration. The movement of the hump is clearly reflected by a shocked wall model. In practice we find that such a wall always produces excess flux between phase $\simeq$ 0.25 and $\simeq$ 0.50 with respect to the spot, with small changes in phase in response to increased penetration. However, with increased penetration and disc evolution the contribution of the wall relative to the spot can change significantly. During the decline of the same outburst in Z Cha the hump pattern returned to a standard spot pattern, but with some excess flux in the region 0.50 to 0.75 in phase. Clearly the division of anisotropy effects into a bright-spot and a penetration wall is an over-simplification. We consider the general principles outlined here incorporate the fundamental features of stream/disc penetration and provide a basis for further discussion of anisotropy effects resulting from stream/disc impact.

Superhumps

Superhumps have probably led to more sleepless nights than any other phenomena so far discovered in cataclysmic variables. The increase of hump period by a few percent commonly found in SU UMa super maxima outbursts has proved resistant to any acceptable explanation when closely examined. We point out that Warner's (1975) original suggestion that the superhumps are due to anisotropic radiation from a non-synchronously rotating red component is still not excluded observationally, and we present here both observational and theoretical arguments in support of such a model.

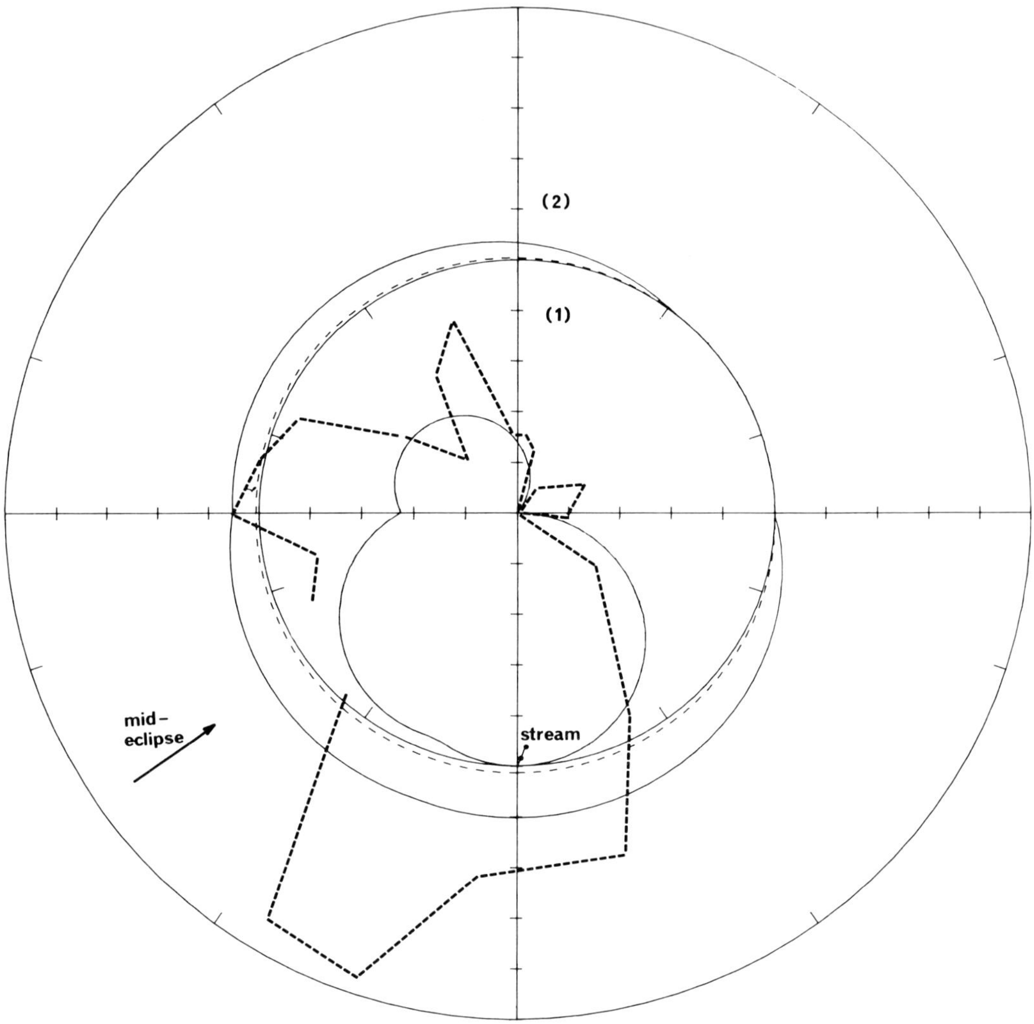

Fig.2 Distribution of anisotropic radiation with binary phase in the quiescent state (peapod diagram). In region (1) the anisotropic component only is shown, in region (2) the total relative to the background isotropic component. The model has β = 0.1 (thin line) and is compared with the observed hump distribution of Z Cha (thick dotted line) on Dec 7 1972.

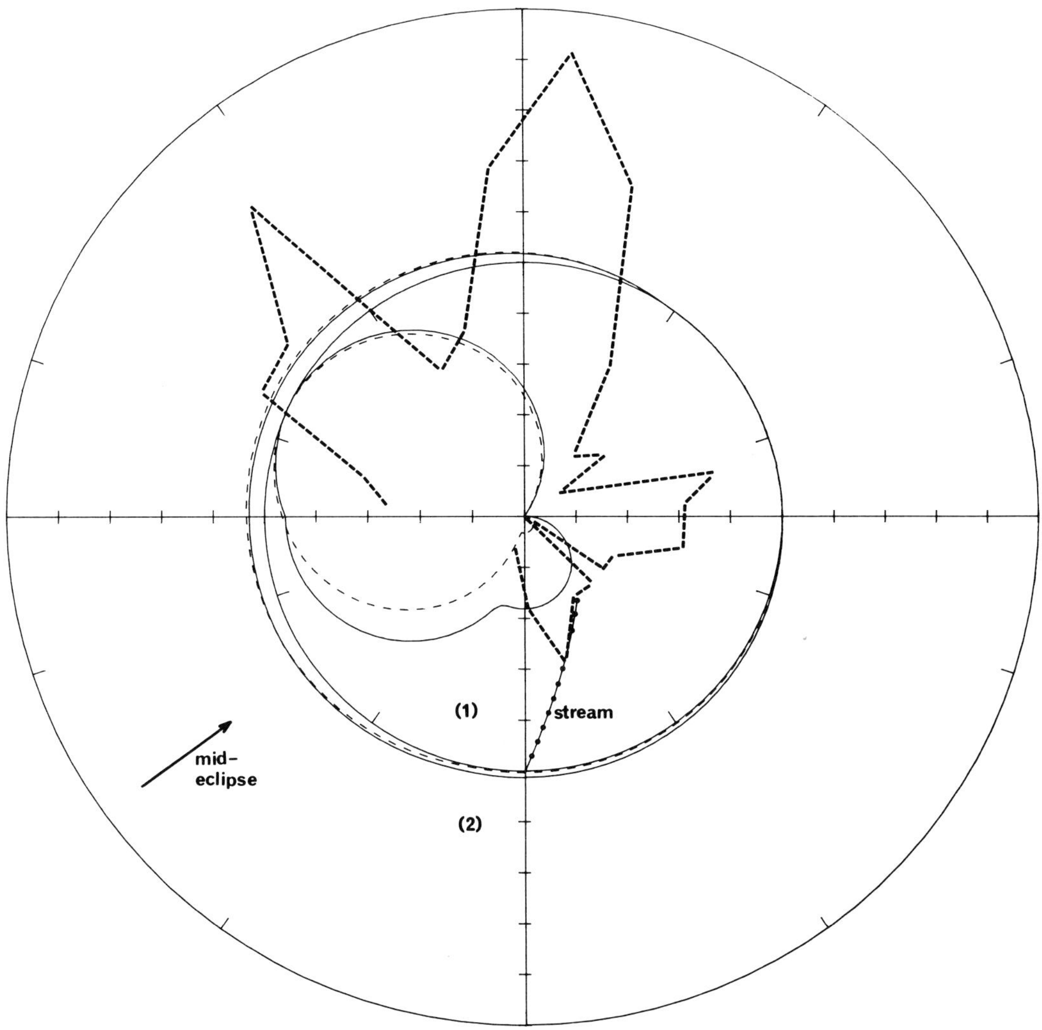

Fig.3 Peapod diagram in the outburst state. The model exhibits deep stream penetration and radiation from the wall now dominates. The distribution in phase is similar to that seen at outburst in Z Cha on Jan 8 1973 (thick dotted line).

In VW Hydri the quiescent hump period is well established to be stable with P = .07427107 (±5) day (Vogt 1974, Warner 1975, Bailey 1976). At supermaxima the typical quoted superhump period is P = .074676 day. If we interpret the quiescent period as the orbital period, then the superhump period corresponds to a beat period of 2.2 day, which would be produced by nonsynchronous rotation of an asymmetric radiation component with a backward rotation period of 2.2 day in the rotating frame of the binary. Evidently a slowly rotating red component with a large scale asymmetric radiation pattern due to a surface "bright-feature" radiating sufficient luminosity in the optical would produce the effect observed.

We argue here that the shortest period cataclysmic systems (i.e. the SU UMa systems exhibiting supermaxima) contain slowly rotating red components, which have lost angular momentum through previous evolutionary mass transfer and have yet to achieve synchronism through tidal dissipation. Normal outbursts are produced by a discrete local dynamical instability in the neighbourhood of the inner Lagrangian point (Bath 1975, Papaloizou and Bath 1975). Supermaxima outbursts are produced by continued dynamical mass transfer as material slides past the inner Lagrangian point through non-synchronism. The superhumps result from asymmetry in the radiation field on the surface of the red component through variation in the strength of the instability around the equator and variations with longitude in the subsequent thermal relaxation time.

It is a notable feature of many supermaxima outbursts that they show structure, or repetitive mini-maxima following the rise. In VW Hydri we find that all four supermaxima outbursts published by Bateson (1974) show evidence for modulation with a period 2.0-2.5 day. The average of all four light curves is shown in Fig.4, with the outbursts aligned to rise simultaneously. There is clear evidence for periodic modulation at about the beat period. Cross-correlating the data shows the signal disappears when shifts of an odd number of days are introduced. Could this be evidence that the red component is transferring mass into the accretion disc at a variable rate as the superhump region on the red component is brought past the inner Lagrangian point? It is certainly the case that the superhump becomes more isotropic and is progressively smeared out as the outburst progresses, in the same way as the 2.2 day modulation of the overall superoutburst dies away.

A decisive test which would locate the asymmetric radiating region on the red component (or possibly outside the system) is the depth of the primary eclipse. We predict that the depth of the primary eclipse below the background should be constant, that is, the bottom of the primary eclipse rise simultaneously with the march of the superhump through it (apart from any independent variations in disc luminosity which may occur). We do not anticipate deep secondary eclipses, since the primary eclipsing object is only a thin, flat, bright strip.

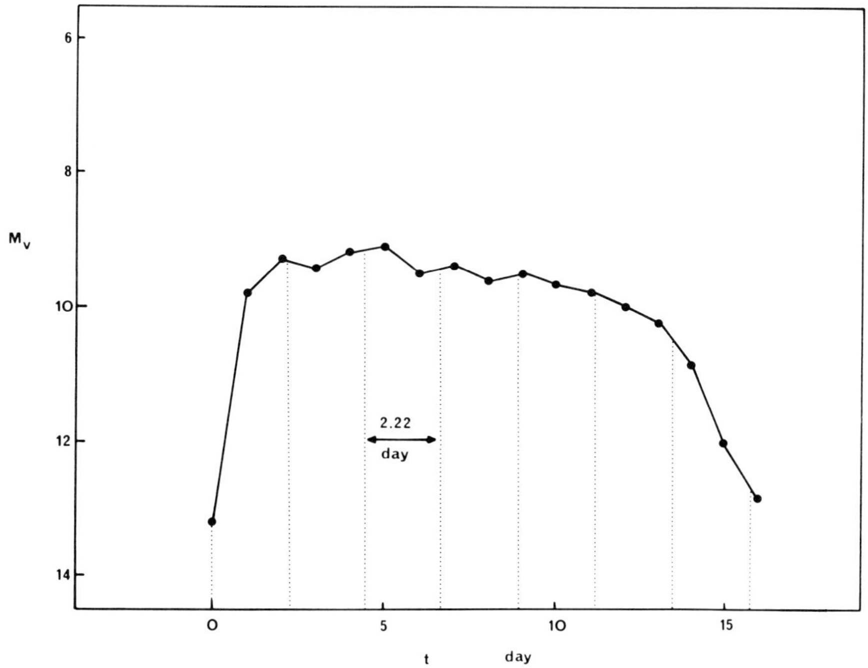

Fig.4 Average outburst light curve of four successive supermaxima outbursts of VW Hydri. Each individual superoutburst shows evidence of a 2.0-2.5 day oscillation, as does this average light curve.

References

Bailey, J., 1976, V.S.S. R.A.S. N.Z. 4 (C76), 1.
Bateson, F.M., 1974, V.S.S. R.A.S. N.Z. 1 (C73), 2.
Bath, G.T., 1975, Mon.Not.R.astr.Soc., 171, 311.
Papaloizou, J.C.B. & Bath, G.T., 1975, Mon.Not.R.astr.Soc., 172, 339.
Vogt, N., 1974, Astr. & Astrophys., 36, 369.
Warner, B., 1974, Mon.Not.R.astr.Soc., 168, 235.
Warner, B., 1975, Mon.Not.R.astr.Soc., 170, 219.

DISCUSSION FOLLOWING G. BATH'S TALK

ROBINSON: If your model is right one might expect a correlation between the amplitude of the hump and the inclination of the orbit,do you know of any such correlation?

BATH: No, I don't, I haven't thought about that.

RITTER: If this model is right, your bright spot on the seconary star must be bright indeed, because the overall luminosity of the system is very large and we know that in these systems we cannot see the secondary in quiescence in the visible region, so if you now create a spot which you can see in visible light, when the system is at least one hundred times brighter than in quiescence, you must have a fantastically hot region on the surface and so, you could immediately predict something about the colour behavior of your superoutbursts, so have you looked into that?

BATH: No, I have not looked into this. Let me just say about the luminosity of the hump or the superhump. Whatever model you propose for the superhump, it has to have that luminosity, that is the first point. The second point is that in the spherical models of mass transfer instabilities, I found that the luminosity did increase by a factor of the order of a hundred, over the whole surface of the star and the question is how much of this is covered.

WARNER: You are concentrating on VW Hyi here, if the mechanism applies, in say, OY Car, than you would expect deep secondary eclipses, secondary minima, which are not observed during a supermaximum of OY Car. So, the absence of deep secondary minima during superoutbursts, I think, excludes the secondary as the source of the extra luminosity .

MATTEI: I will study some more supermaxima light curves to see if there is a real structure, but I want to ask how do you explain the appearance of superhumps two or three days later than the start of the supermaximum, according to the model? Why don't they appear right away?

BATH: I don't know.

CNO ABUNDANCES AND NOVA OUTBURST STRENGTHS

Jim MacDonald
University of Illinois at Urbana-Champaign

ABSTRACT

Simple theories of hydrogen shell flashes on accreting white dwarfs are used to find the requisite conditions for classical novae. It is shown that enhancement of CNO abundances, relative to the Sun, is necessary for fast novae yet slow novae, such as DQ Her, can still occur with large CNO abundances. An analysis of observational data indicates that the white dwarfs in nova binaries have masses typically 1.1 $M_{\odot}$. For some old novae, the accretion rate, deduced from the accretion disk luminosity, is too high to permit a strong enough hydrogen flash to give the observed nova strength. Some possible resolutions to this paradox are suggested.

Recurrent novae are probably not thermonuclear runaways on accreting white dwarfs.

1. INTRODUCTION

It is now generally accepted that classical novae are the results of thermonuclear runaways in hydrogen-rich white dwarf envelopes accreted from a binary companion (Gallagher & Starrfield 1978 and references therein). Hydrogen shell flashes on accreting white dwarfs may be relevant to symbiotic stars (Paczynski & Zytkow 1978) and also play a role in the evolution of type I supernova progenitors. For these reasons it is of interest to know how the hydrogen-flash strength depends on the binary system properties, i.e. mass transfer rate, white dwarf mass and luminosity etc. In the following section we investigate this dependence by use of simple models of the "accretion" and "expansion" phases of the flash evolution. Particular attention is paid to the possibility that the CNO abundances in the accreted envelope are greatly enhanced relative to the Sun. Such CNO enhancements have been observed in the ejecta of a number of classical novae (Ferland & Shields 1978, Williams et al 1978, Gallagher et al 1980, Stickland et al 1981, Williams 1982) and may also be present in the accretion disk (Williams, this volume). Further it has been suggested that CNO enhancement is necessary to give fast

M. Livio and G. Shaviv (eds.), Cataclysmic Variables and Related Objects, 77–95.

novae (Starrfield, Truran & Sparks 1978).

In addition to outlining the conditions necessary for the occurrence of classical novae, our simple models can be used to interpret observational data and give a handle on white dwarf masses, accretion rates, etc. in nova systems. Brief results of such an investigation are given in the final section along with discussion as to why the kinetic energy of nova ejecta is small when compared to the binding energy and nuclear energy of the accreted envelope and whether or not recurrent novae are thermonuclear powered.

2. DEPENDENCE OF FLASH STRENGTH ON BINARY SYSTEM PROPERTIES

To survey how the hydrogen flash strength depends on the binary system properties we need to consider two stages of the flash: i) an "accretion" phase, in which the mass of hydrogen-rich envelope, M_{env}, is determined and ii) an "expansion" phase, in which the flash strength is related to white dwarf mass, M_{env}, and envelope composition.

2.1. The Accretion Phase

We consider a pure carbon white dwarf of mass M_{WD}, radius R_{WD} and luminosity, L_{WD}, accreting hydrogen-rich material at rate F. Spherical symmetry is assumed throughout.

The white dwarf envelope remains in thermal equilibrium if

$$F \frac{\mathcal{R}}{\mu} T_c \ll L_{WD} \tag{2.1}$$

(T_c is the white dwarf central temperature) i.e. the thermal timescale must be less than the accretion timescale (MacDonald 1980). Otherwise, gravitational energy released in the envelope increases the surface luminosity above L_{WD}. We consider first the low accretion rate regime for which condition (2.1) is satisfied.

To find M_{env}, the envelope mass at the onset of the thermonuclear runaway, the usual stellar structure equations are integrated inwards from the photosphere at $r = R_{WD}$ down to the envelope-core boundary. M_{env} is taken to be the minimum envelope mass such that the following three criteria are simultaneously satisfied:

A) The total nuclear energy production in the accreted material, L_{nuc}, is greater than L_{WD} (note L_{nuc} is not included in L in the envelope integration). Physically, this means that the white dwarf cannot radiate away the energy produced by thermonuclear reactions.

B) The thermonuclear timescale at the base of the accreted material is shorter than the timescale on which T_a, the temperature at the base of the accreted material, is changing due to accretion alone, i.e.

$$C_P \dot{T}_a \leq \varepsilon_{nuc} \; . \tag{2.2}$$

C) Conduction into the core is unimportant. We quantify this last condition by calculating conduction and nuclear timescales, t_{con} and t_{nuc} respectively, defined by

$$t_{con} = 2 \frac{\ell^2}{C} \tag{2.3}$$

$$t_{nuc} = \frac{C_P T}{\eta \; \varepsilon_{nuc}} \tag{2.4}$$

where 2ℓ is the thickness of the energy generation zone

$$2\ell = \frac{L_{nuc}}{4\pi R^2 \rho \; \varepsilon_{nuc}} \tag{2.5}$$

C is the conduction coefficient (evaluated for core composition)

$$C = \frac{4acT^3}{3\kappa C_P \rho^2} \tag{2.6}$$

and

$$\eta = \left. \frac{\partial \ln \varepsilon_{nuc}}{\partial \ln T} \right|_P \; . \tag{2.7}$$

In the above all quantities are evaluated in the burning zone at the base of the accreted material. Conduction is taken to be negligible when

$$t_{con} > t_{nuc} \tag{2.8}$$

We now consider the high accretion rate regime for which condition (2.1) is severely violated. As discussed in MacDonald (1980), the accreted material attains a quasi-equilibrium with thermal energy generation rate

$$\varepsilon_{th} \simeq FT \frac{\partial S}{\partial m} \tag{2.9}$$

The structure equations are again solved but with an energy generation term given by equation (2.9). The procedure used to find M_{env} is to guess the photospheric luminosity and integrate from the surface inwards to the point where L = 0. (This is a justified approximation because the accretion luminosity is much greater than L_{WD}). L_{nuc} and $\dot{T}_a$ are then evaluated. A second guess for the photospheric luminosity is then used and the procedure repeated. From these two guesses we estimate a third photospheric luminosity such that criteria A and B are approximately satisfied. (Conduction into the core has not been considered for this case). We then iterate until the minimum accreted mass for which criteria A and B are simultaneously satisfied is found.

For intermediate values of F we find the accreted mass by interpolating between the low and high F results. We take

$$M_{env} = \frac{M_{LO} + x\, M_{HI}}{1 + x} \tag{2.10}$$

where M_{LO} and M_{HI} are the low and high F accreted masses and

$$x = \left(\frac{\mathcal{R} F T_c}{\mu L_{WD}}\right)^2 \tag{2.11}$$

As a test of this simple model we give in table 1 a comparison with the results of a number of detailed numerical calculations. In all cases the composition of accreted material is essentially solar. The mean error in our approximate values is 24% and, in general, our approximate value is an underestimate. We feel that this accuracy is good

Table 1. Comparison of accretion model with detailed calculations.

$M_{WD}/M_\odot$	$L_{WD}/L_\odot$	$F(M_\odot y^{-1})$	$M_{det}/M_\odot$	$M_{sim}/M_\odot$	ref.
0.456	7×10^{-4}	10^{-13}	6.89×10^{-4}	4.00×10^{-4}	2
0.5	5.7×10^{-2}	10^{-10}	5.01×10^{-4}	4.16×10^{-4}	4
1.0	5×10^{-4}	10^{-10}	2×10^{-4}	1.37×10^{-4}	1
1.0	10^{-3}	10^{-13}	2.4×10^{-4}	1.64×10^{-4}	2
1.0	10^{-3}	10^{-10}	1.38×10^{-4}	1.35×10^{-4}	4
1.0	10^{-3}	10^{-8}	5.85×10^{-5}	6.11×10^{-5}	3
1.0	5×10^{-3}	10^{-10}	1×10^{-4}	9.47×10^{-5}	1
1.0	9.8×10^{-2}	10^{-10}	5.25×10^{-5}	6.49×10^{-5}	4
1.0	9.8×10^{-2}	10^{-7}	3.63×10^{-5}	2.91×10^{-5}	4

Refs: 1) Taam & Faulkner 1975, 2) Taam 1977, 3) Kutter & Sparks 1979, 4) MacDonald 1979, 1980

enough for a parameter space survey. M_{env} is found to depend most sensitively on M_{WD}. This is simply because surface gravity and hence pressure gradient increase steeply with M_{WD} so that less mass has to be accreted onto higher mass white dwarfs before thermonuclear temperatures and densities are reached. In figures 1 and 2 we show how M_{env} depends on F and L_{WD} for two envelope compositions, X = 0.7, Z_{CNO} = 0.014 and X = 0.35, Z_{CNO} = 0.507 respectively. M_{WD} = 1.0 $M_\odot$ in both cases. We see that the higher mass envelopes occur for low F and low L_{WD}. In this case the onset of nuclear burning is due to the pp chains rather than the CNO cycle and hence there is only a weak dependence on composition via opacity and molecular weight. However for high L_{WD} or high F the thermonuclear runaway is initiated by the CNO cycle and a higher CNO abundance leads to a significantly smaller accreted mass.

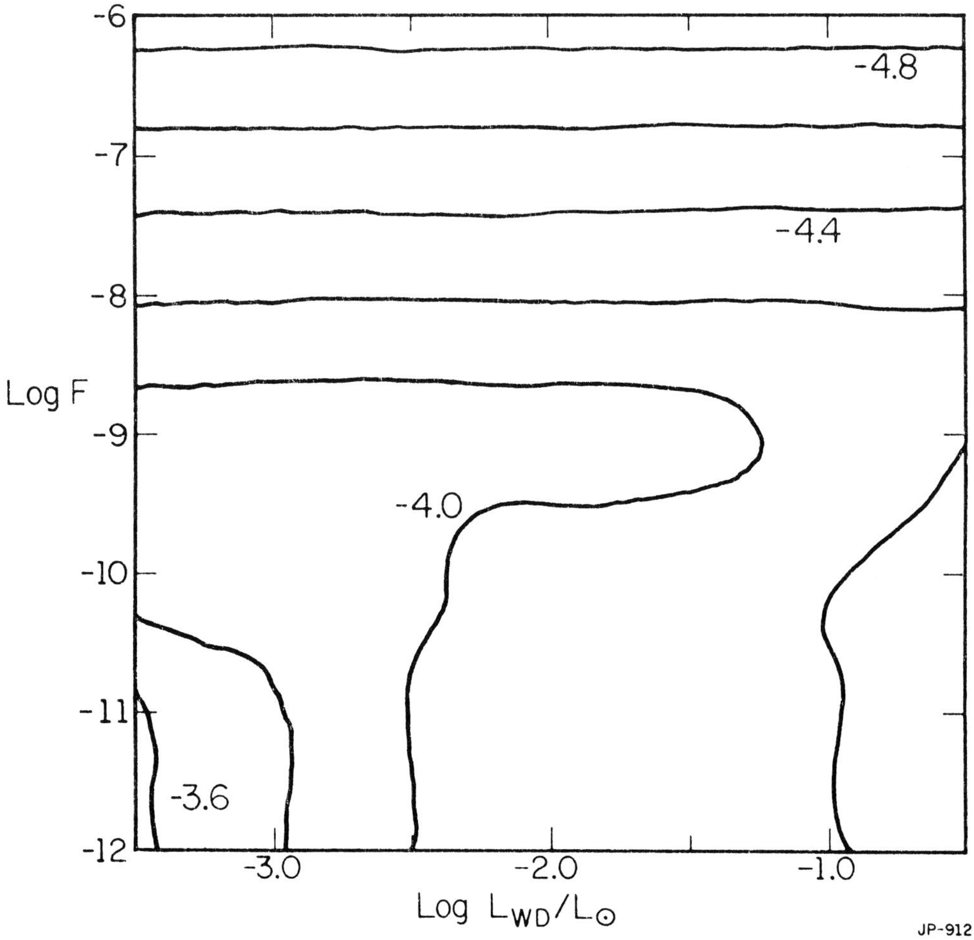

Figure 1. Contours of accreted mass in L_{WD} - F space for X = 0.7, Z_{CNO} = 0.014 and M_{WD} = 1.0 $M_\odot$. Contours are labelled with log $M_{env}/M_\odot$.

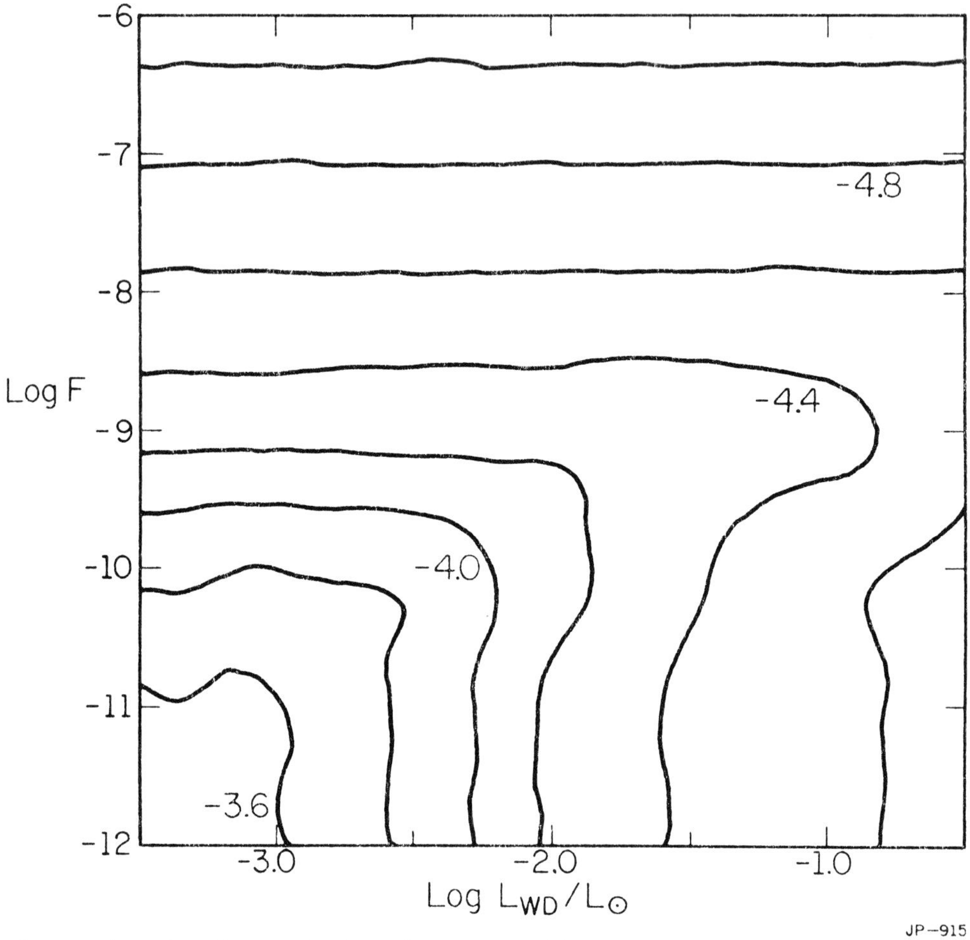

Figure 2. As figure 1, but for X = 0.35, Z_{CNO} = 0.507.

2.2. The Expansion Phase

The thermonuclear runaway proceeds to increase the shell source temperature, T_{ss}, until radiation pressure causes envelope expansion and subsequent cooling. Just after maximum T_{ss} is reached the envelope is convective throughout. Also the bulk of the envelope remains convective during the expansion towards maximum photospheric radius. The convective energy transport is efficient so we can approximate the envelope structure by an adiabat. In this phase only radiation and gas pressure are important. This allows us to find the envelope structure analytically in terms of β, the ratio of gas pressure to total pressure:

$$P = P_o \frac{(1-\beta)^{5/3}}{\beta^{8/3}} \exp(32/3\beta) \qquad (2.12)$$

$$\rho = \rho_o \frac{1-\beta}{\beta} \exp(8/\beta) \tag{2.13}$$

$$T = T_o \left(\frac{1-\beta}{\beta}\right)^{2/3} \exp(8/3\beta) \tag{2.14}$$

where $$P_o = \frac{\mathcal{R}}{\mu} \rho_o T_o = \frac{1}{3} a T_o^4 . \tag{2.15}$$

Further, at maximum T_{ss} and for some time after, the envelope is in hydrostatic equilibrium. If we take P = 0 at r = R, solving the hydrostatic support equation gives

$$r = \frac{R}{1 + q\, I(\beta)} \tag{2.16}$$

where $$I(\beta) = \frac{(1-\beta)^{2/3}}{\beta^{5/3}} \frac{8-3\beta}{2} \exp(8/3\beta) \tag{2.17}$$

and $$q = \frac{\mathcal{R} T_o}{\mu} \frac{R}{GM_{WD}} \tag{2.18}$$

The mass of the accreted envelope is then given by

$$M_{env}/M_s = q^4 \int_{\beta_{in}}^{1} \frac{J(\beta)}{[1+qI(\beta)]^4} d\beta \tag{2.19}$$

where $$M_s = 4\pi \left(\frac{\mu}{\mathcal{R}}\right)^4 \frac{a}{3} (GM_{WD})^3 \tag{2.20}$$

$$J(\beta) = \frac{\beta-1}{\beta} I'(\beta) \exp(8/\beta) \tag{2.21}$$

and β_{in} is the value of β at the base of the accreted envelope. β_{in} is found from

$$R = R_{WD} [1 + q\, I(\beta_{in})] \tag{2.22}$$

For a given value of R, the envelope structure is now completely known. The envelope evolution is then determined from the energy conservation equation,

$$\frac{d(U+V)}{dt} = L_{nuc} - L_* \tag{2.23}$$

where U and V, the internal and gravitational energies of the accreted envelope respectively, are given by integrals over the envelope. L_{nuc} is the total nuclear energy generation in the envelope and L_* is the energy radiated at the photosphere,

$$L_* = L_{ed}\ (1-\beta_p)\ \frac{(32-24\beta_p)}{(32-24\beta_p-3\beta_p^2)} \tag{2.24}$$

where L_{ed} is the Eddington luminosity and β_p is the value of β at the photosphere, taken to be at optical depth $\tau = 2/3$. Electron scattering is the dominant opacity source and hence

$$L_{ed} = \frac{4\pi c G M_{WD}}{\kappa_{es}} = \frac{6.5 \times 10^4}{1+X}\ \frac{M_{WD}}{M_\Theta}\ L_\Theta \tag{2.25}$$

For given M_{WD}, M_{env} and envelope composition, we find the early evolution of the nova by integrating equation (2.23) from the time of maximum T_{SS} until either i) the expansion velocity, $v_{exp} = \dot{R}$, exceeds the escape velocity, v_{esc}, so that some mass is ejected or ii) $v_{exp} < 0$ i.e. the envelope has reached an equilibrium radius without any ejection of matter.

We quantify the strength of the nova outburst by the value of v_{exp} when escape velocity is reached. For M_{WD} close to the Chandrasekhar limit, M_{ch}, we also allow for the change in X and μ due to nuclear conversion of H into He.

Fujimoto (1982) has shown that, for solar CNO abundances, the flash strength depends mainly on the "proper" pressure in the shell source.

$$P_* = \frac{G M_{WD}}{4\pi R_{WD}^4}\ M_{env} \tag{2.26}$$

The present calculations are in exact agreement with this conclusion. We find, as did Fujimoto, that $P_* \gtrsim 10^{20}$ dynes cm^{-2} is required for ejection with solar CNO abundances. With Z = 0.51, the critical value of P_* is reduced to $\sim 2 \times 10^{19}$ dynes cm^{-2}, indicating that substantially smaller envelope masses are sufficient for nova outbursts to occur, when the CNO nuclei are enhanced.

Figures 3 and 4 are contour diagrams of v_{exp} at the time when $v_{exp} = v_{esc}$ for the two compositions considered above. Also shown are the regions of (M_{WD}-M_{env}) space which our simple accretion model predict to be realisable. We see that no ejection of material is possible for solar CNO unless $M_{WD} \gtrsim 1.1\ M_\Theta$ and the ejection velocity will be low ($\lesssim$ 250 km s^{-1}) even for optimal conditions. For $Z_{CNO} = 0.51$ ejection occurs if $M_{WD} \gtrsim 0.8\ M_\Theta$. Further the highest ejection velocity, $\sim$ 1600 km s^{-1}, is comparable to the principal absorption velocities observed in the fastest known nova, V1500 Cyg (Duerbeck & Wolf 1977). This clearly shows that strongly enhanced CNO abundances are necessary for fast novae. However slow novae such as DQ Her can also occur with CNO greatly enhanced if the envelope mass is sufficiently small.

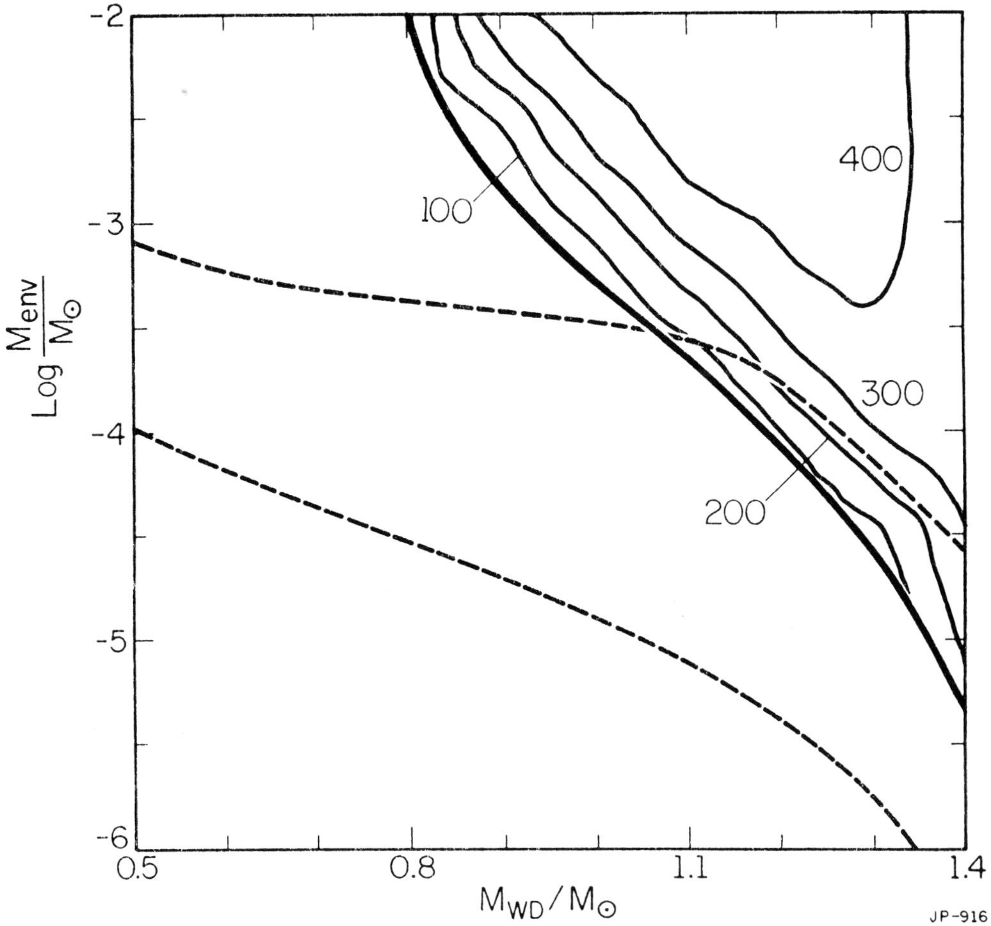

Figure 3. Contours of expansion velocity, v_{exp}, when it equals escape velocity in M_{WD}-M_{env} space for X = 0.7, Z_{CNO} = 0.014. Contours are labelled with v_{exp} in km s^{-1}. Also shown (dashed lines) is the domain of M_{WD}-M_{env} space accessible by accreting white dwarfs.

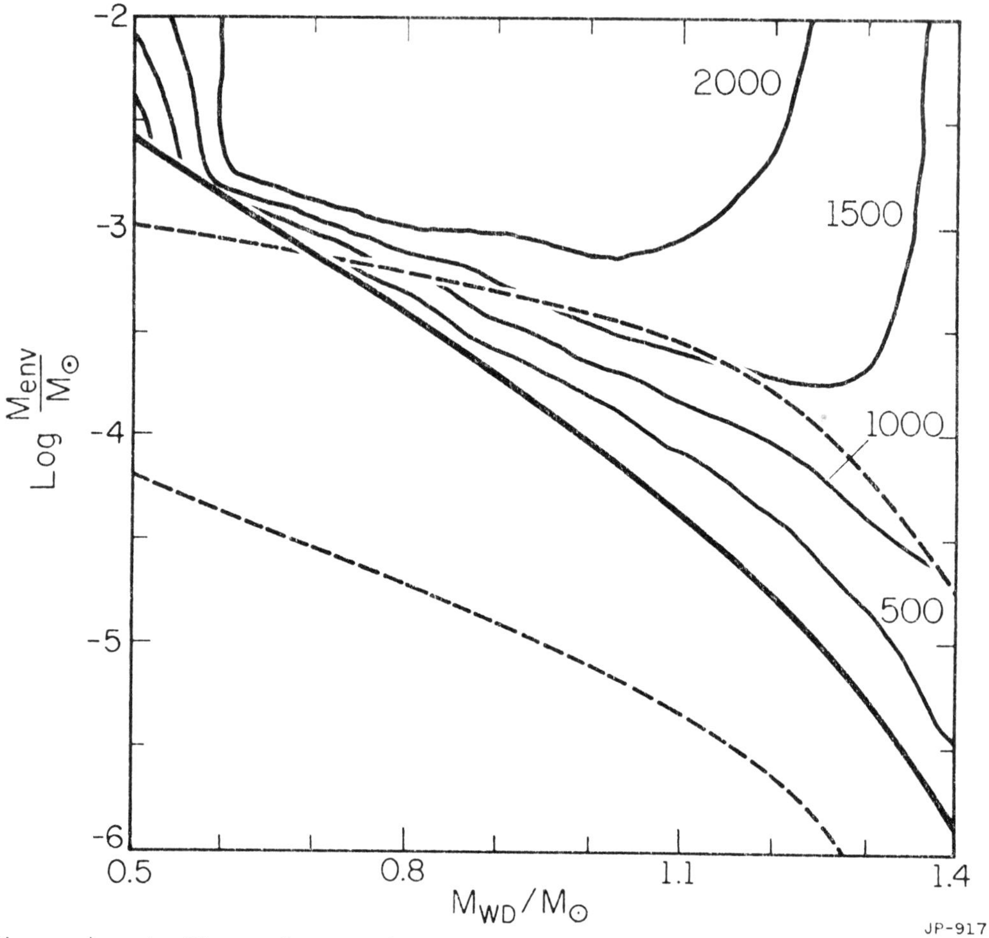

Figure 4. As figure 3, but for X = 0.35, Z_{CNO} = 0.507.

3. DISCUSSION AND CONCLUSIONS

In table 2 we have collected together abundance estimates for the ejecta of a number of recent novae and translated them into mass fractions. Also given are decline time, t_3, principal absorption velocity, v_{princ}, and ejecta mass estimate, M_{ej}, adjusted to the given distance, D. The most uncertain of these quantities are the ejecta masses which are usually determined from emission line strengths with the assumption of a uniform nebula. Clumping in the ejecta will lead to an overestimate of M_{ej}. Conversely only the mass of ionized hydrogen is measured and M_{ej} may be underestimated if neutral hydrogen is present. Taking this data at face value we can, in principle, find the masses of the white dwarfs on which these particular novae occurred. We need to make one further assumption: that all the material above the shell source is ejected so that $M_{env} = M_{ej}$. Starrfield (1979) has considered the timescale on which novae turnoff and concluded that the bulk of the envelope must be ejected by some means on a timescale short compared to the nu-

Table 2. Abundances and ejecta masses.

Object	D(kpc)	t_3(d)	v_{princ}(kms^{-1})	$M_{ej}/M_\odot$	X	Y	Z_{CNO}	ref.
V1500 Cyg 1975	1.6	4	1600	$3x10^{-4}$	0.49	0.21	0.28	1,2
CP Lac 1936	1.3	10	1300	$3x10^{-5}$	0.60	0.26	0.13	3,4
V1668 Cyg 1978	3.9	24	740	$2x10^{-4}$	0.47	0.22	0.31	5
IV Cep 1971	3.6	42	1000	$8x10^{-5}$	0.40	0.21	0.39	4,6
DQ Her 1934	0.42	100	310	$5x10^{-5}$	0.30	0.17	0.53	7
RR Pic 1925	0.48	150	300	$3x10^{-4}$	0.49	0.45	0.04	8
HR Del 1967	0.86	220	350	$1x10^{-4}$	0.43	0.47	0.073	9

Refs: 1) Ferland 1978, 2) Ferland & Shields 1978, 3) Pottasch 1959, 4) Ferland 1979, 5) Stickland et al 1981, 6) Pacheco 1977, 7) Williams et al 1978, 8) Williams & Gallagher 1979, 9) Tylenda 1979

clear burning timescale in order to give the observed rapid decline in visual luminosity. Thus the assumption of $M_{ej} \simeq M_{env}$ appears justified at the present time. We then find that the white dwarfs in classical novae have mean mass, $\overline{M}_{WD} = 1.10 \pm 0.08\ M_{\odot}$.

DQ Her, a slow nova with a very high CNO abundance (Williams et al 1978) has always been a problematic object for thermonuclear runaway theories of novae. We see, however, that a low accreted mass consistent with the observed ejecta mass can give a slow nova outburst even when the CNO abundances is as high as it is in DQ Her. We suggest that the differences in rates of light curve development for DQ Her and V1500 Cyg are primarily due to differences in accreted mass rather than a difference in white dwarf mass, as had been suggested by Truran (1982).

In figure 5 we show the relationship between speed class, CNO abundance and envelope mass. $M_{WD} = 1.1\ M_{\odot}$ has been adopted. $v_{princ} = 500$ km s^{-1} and $v_{princ} = 1000$ km s^{-1} have been taken to be the dividing lines

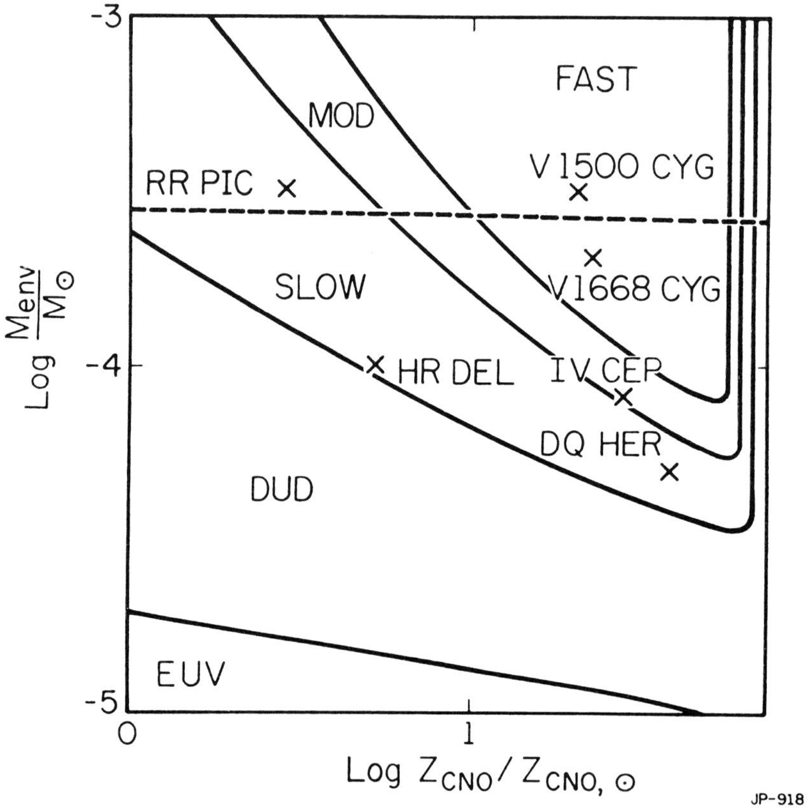

Figure 5. Nova speed class against CNO abundance and envelope mass. $M_{WD} = 1.1\ M_{\odot}$ has been assumed. The crosses mark the positions of classical novae. Also shown is the maximum envelope mass that can be accreted by a 1.1 $M_{\odot}$ white dwarf before thermonuclear runaway. See text for definition of speed classes.

between slow, moderate and fast novae. Duds are defined to be objects that do not give ejection. EUV objects are a subset of the duds which expand only slightly but brighten to near Eddington luminosity so that the bulk of their radiation is emitted in the EUV. Both novae and duds will also pass through an EUV stage as their photospheres shrink due to mass loss or consumption of nuclear fuel.

Our expansion model also gives a clue to why the deduced kinetic energy of the ejecta (see table 2) is so small when compared to the binding and nuclear energies of the accreted envelope. There are two timescales of interest: t_{nuc}, the time for the total nuclear luminosity to decrease from its maximum value to e^{-1} its maximum value and t_{therm} the timescale on which nuclear energy is being used to unbind the accreted envelope. More exactly,

$$t_{therm} = \frac{|E_{bin}|}{L_{nuc}} \tag{3.1}$$

where E_{bin} is the binding energy of the accreted envelope at the time of maximum L_{nuc}. If $t_{nuc} \gg t_{therm}$ the envelope would become unbound before L_{nuc} decreased significantly and a large amount of kinetic energy would be generated. If $t_{nuc} \ll t_{therm}$ L_{nuc} decreases too rapidly to unbind the envelope and no ejection occurs. From our expansion model we find, in general, $t_{nuc} \sim t_{therm}$ and hence a small kinetic energy results. $t_{nuc} \sim t_{therm}$ because it is the envelope expansion, which occurs on timescale $\sim t_{therm}$, that shuts off the nuclear burning.

We now put together the results of the accretion and expansion phase models to relate the conditions under which the hydrogen flash gives envelope ejection to the properties of the binary system. We find that, for given M_{WD} and composition, critical values of F and L_{WD} can be defined, F_c and L_c say, such that if $F \leq F_c$ and $L_{WD} \leq L_c$ then envelope ejection occurs and results in a nova. Simple analytic fits to F_c and L_c as functions of M_{WD} are

$$\log F_c = -8.775 - 15.088 \left(\frac{M_{WD}}{M_\odot} - 1.459\right)^2 \tag{3.2}$$

$$\log L_c = -0.629 - 5.923 \left(\frac{M_{WD}}{M_\odot} - 1.766\right)^2 \tag{3.3}$$

for Z = 0.02 and

$$\log F_c = -8.632 - 4.596 \left(\frac{M_{WD}}{M_\odot} - 1.334\right)^2 \tag{3.4}$$

$$\log L_c = -1.375 - 7.027 \left(\frac{M_{WD}}{M_\odot} - 1.308\right)^2 \tag{3.5}$$

for $Z = 0.51$.

Similar results can be derived for each nova speed class. We content ourselves with discussion of some of the better observed novae. The most energetic classical nova observed to date is V1500 Cygni which has $v_{princ} \simeq 1600$ km s^{-1} and $t_3 = 4$ days. For the ejecta mass and composition given in table 2, we find $M_{WD} = 1.04\ M_\odot$ is required. The large ejecta mass, $M_{ej} \simeq 3 \times 10^{-4}\ M_\odot$, can be accreted if the accretion rate is low, $F \lesssim 10^{-11}\ M_\odot\ y^{-1}$ and the underlying white dwarf was initially cool, $L_{WD} \lesssim 10^{-3}\ L_\odot$. The absolute visual magnitude of the accretion disk can then be estimated to be $M_V \simeq 9.8$ which, for distance 1.6 kpc and E(B-V) = 0.5 (Ferland 1977), corresponds to apparent magnitude $m_V \simeq 22.3$. This is consistent with V1500 Cygni being too faint to appear on the Palomar Sky Survey (Beardsley et al 1975).

In quiescence, DQ Her has $m_V = 14.6$ which corresponds to $M_V = 6.1$ for distance 420 pc and 0.38 magnitudes of reddening (Ferland 1980). For our mass estimate, $M_{WD} = 1.10\ M_\odot$, we find $F \simeq 8 \times 10^{-10}\ M_\odot\ y^{-1}$ from the accretion disk luminosity, in close agreement with $F = 1.1 \times 10^{-9}\ M_\odot\ y^{-1}$ found from analysis of period changes (Nelson 1976). The luminosity is then $L_{WD} \lesssim 5 \times 10^{-3}\ L_\odot$. We concluded earlier that V1500 Cygni and DQ Her essentially differed only in accreted mass. We now see that this difference is probably mainly due to a two orders of magnitude difference in accretion rate.

For a number of novae, including RR Pic, CP Lac and probably V603 Aql, we find that the accretion rates deduced from visual light in quiescence are too high, typically $\sim 10^{-8}\ M_\odot\ y^{-1}$, for the estimated ejecta mass to be accreted. It has also been pointed out that the soft X-ray emission from classical novae is 2 to 4 orders of magnitude less than predicted by boundary layer theory (Ferland et al 1982). One possible resolution to both these problems is that part of the optical luminosity is due to reprocessing by the accretion disk of UV radiation from a white dwarf that is still hot from the outburst. Alternatively the accretion rate deduced from the optical flux is correct but the secondary is in a perturbed thermal state due to the recent nova outburst and is presently transferring matter at a rate much greater than in equilibrium.

Similarities in the spectra and light curves of classical and recurrent novae have naturally led to the suggestion that recurrent novae are also thermonuclear runaways on white dwarfs. The smallest accretion times occur for white dwarf masses close to the Chandrasekhar limit, $M_{ch} = 1.454\ M_\odot$ in our model. For these high masses conversion of hydrogen to helium must be taken into account. When this is included in our expansion model we find the shortest accretion times for models that give ejection are 400 yr for $Z = 0.51$ and 450 yr for $Z = 0.02$ and occur for accretion onto 1.44 and 1.45 $M_\odot$ white dwarfs respectively. In each case the change in hydrogen mass fraction is $\Delta X \simeq 0.3 - 0.4$. These minimum accretion times are much greater than the interval, typically ~ 35y, between recurrent nova outbursts. We conclude that recurrent novae are not thermonuclear runaways on accreting white dwarfs.

In addition to this theoretical argument against a thermonuclear runaway model for recurrent novae there are a number of observations which seem to rule out a thermonuclear model. The large mass estimate for the blue component of T CrB (Paczynski 1965) is inconsistent with it being a white dwarf but is consistent with the accretion model proposed by Webbink (1976). Also the luminosity of the unusual recurrent nova, WZ Sge, at maximum is much less than the Eddington limit (Fabian et al 1980) in contrast to the classical novae which always attain or exceed the Eddington limit, in agreement with the thermonuclear models. Abundance analyses of recurrent nova ejecta indicate essentially solar abundances and no significant CNO enhancement (Williams et al 1981, Williams 1982). One of these novae is U Sco which has the fastest decline rate of all novae (Payne-Gaposchkin 1957) and a very high ejection velocity $v_{ej} \sim 5000$ km s^{-1} (Barlow et al 1981). If U Sco were a thermonuclear runaway it should have a large CNO enhancement. Further the ejecta mass estimate for U Sco (Williams et al 1981) is roughly 3 orders of magnitude less than is typical of classical novae.

We have shown how the energetics of nova outbursts depend on white dwarf mass, accreted envelope mass and envelope composition. In particular, it has been shown that enhancement of CNO is necessary for fast novae. The source of this enhancement is still unclear but there are two promising enhancement mechanisms. Kippenhahn & Thomas (1978) have suggested that shear instabilities mix chemical composition and angular momentum between the accretion disk and the surface layers of the white dwarf. If the white dwarf has no outer helium layer significant carbon and/or oxygen enrichment can occur. A second possibility for CNO enrichment is a helium shell flash, initiated after a series of weak hydrogen flashes that eject little or no material. Since the nuclear energy content of helium is significantly less than hydrogen, a large helium fraction, $\Delta Y \gtrsim 0.2$, will be converted into carbon and/or oxygen and convectively mixed throughout the envelope. Shear instabilities can then mix freshly accreted solar composition material with this CO enriched material. Alternatively, since the white dwarf envelope expands to engulf the secondary during the helium flash, the red dwarf can accrete CO rich material and later transfer it back to the white dwarf. Abundance analyses of the accretion disks in old novae would show whether this pollution mechanism is a viable possibility. A first step in this direction has been taken by Williams (this volume) who finds $Z_{CNO} \sim 0.1$-0.2 for old novae disks.

To summarize, our main conclusions are:

1) Enhanced CNO abundances are necessary for fast novae. Slow novae, e.g. DQ Her, can still occur if Z_{CNO} is large.

2) The mean white dwarf mass in classical nova systems is predicted to be ~ 1.1 $M_{\odot}$.

3) Ejection of accreted material occurs only if the proper pressure at the base of the accreted envelope is greater than a critical value, P_c, which depends strongly on composition but only weakly on white dwarf mass. We find $P_c \simeq 10^{20}$ dyne cm^{-2} for solar abundances and $P_c \simeq 2 \times 10^{19}$ dyne cm^{-2} for $Z_{CNO} = 0.5$.

4) There exist critical accretion rates and white dwarf luminosities. F_c and L_c, which both depend strongly on M_{WD} and Z_{CNO} such that both $F < F_c$ and $L_{WD} < L_c$ are necessary for a nova outburst. Accretion rates, deduced from the optical luminosity of the old nova, are found to be appreciably higher than F_c for some novae including RR Pic, CP Lac and V603 Aql. Ferland et al (1982) have suggested that some of the optical luminosity of post-novae is reprocessed UV radiation from a hot white dwarf. Alternatively, the post-outburst accretion rate may be significantly higher than average.

5) Recurrent novae are not thermonuclear runaways.

A more detailed version of this paper will appear in the Astrophysical Journal. Support for this work is from NSF grant AST 80-18198.

REFERENCES

Barlow, M.J., Brodie, J.P., Brunt, C.C., Hanes, D.A., Hill, P.W., Mayo, S.K., Pringle, J.E., Ward, M.J., Watson, M.G., Whelan, J.A.J., & Willis, A.J., 1981. M.N.R.A.S., 195, 61.

Beardsley, W.R., King, M.W., Russell, J.L., & Stein, J.W., 1975. P.A.S.P., 87, 943.

Duerbeck, H.W., & Wolf, B., 1977. A. & Ap. Suppl., 29, 297.

Fabian, A.C., Pringle, J.E., Stickland, D.J., & Whelan, J.A.J., 1980. M.N.R.A.S., 191, 457.

Ferland, G.J., 1977. Ap. J., 215, 873.

Ferland, G.J., 1978. Ap. J., 219, 589.

Ferland, G.J., 1979. Ap. J., 231, 781.

Ferland, G.J., 1980. Observatory, 100, 166.

Ferland, G.J., Langer, S.H., MacDonald, J., Pepper, G.H., Shaviv, G., & Truran, J.W., 1982. to appear in Ap. J. Lett.

Ferland, G.J., & Shields, G.A., 1978. Ap. J., 226, 172.

Fujimoto, M.Y., 1982. Ap. J., 257, 752.

Gallagher, J.S., Hege, E.K., Kopriva, D.A., Williams, R.E., & Butcher, H.R., 1980. Ap. J., 237, 55.

Gallagher, J.S., & Starrfield, S., 1978. Ann. Rev. A. & Ap., 16, 171.

Kippenhahn, R., & Thomas, H.C., 1978. A. & Ap., 63, 265.

Kutter, G.S., & Sparks, W.M., 1979. White Dwarfs & Variable Degenerate Stars, ed. H.M. Van Horn & V. Weidemann (University of Rochester), p. 290.

MacDonald, J., 1979. Unpublished Ph.D. Thesis, Cambridge University.

MacDonald, J., 1980. M.N.R.A.S., 191, 933.

Nelson, M.R., 1976. Ap. J., 209, 168.

Pacheco, J.A.F., 1977. M.N.R.A.S., 181, 421.

Paczynski, B., 1965. Acta Astr., **15**, 197.
Paczynski, B., & Zytkow, A.N., 1978. Ap. J., **222**, 604.
Payne-Gaposchkin, C., 1957. The Galactic Novae (North Holland: Amsterdam).
Pottasch, S., 1959. Ann. d'Ap., **22**, 412.
Starrfield, S.G., 1979. White Dwarfs & Variable Degenerate Stars, ed. H.M. Van Horn & V. Weidemann (University of Rochester), p. 274.
Starrfield, S., Truran, J.W., & Sparks, W.M., 1978. Ap. J., **226**, 186.
Stickland, D.J., Penn, C.J., Seaton, M.J., Snijders, M.A.J., & Storey, P.J., 1981. M.N.R.A.S., **197**, 107.
Taam, R.E., 1977. Ap. Lett., **19**, 47.
Taam, R.E., & Faulkner, J. 1975. Ap. J., **198**, 435.
Truran, J.W., 1982. Essays in Nuclear Astrophysics, ed. Barnes et al, (Cambridge University Press).
Tylenda, R., 1979. Acta Astr., **29**, 55.
Webbink, R.F., 1976. Nature, **262**, 271.
Williams, R.E., 1982. preprint.
Williams, R.E., & Gallagher, J.S., 1979. Ap. J., **228**, 482.
Williams, R.E., Sparks, W.M., Gallagher, J.S., Ney, E.P., Starrfield, S.G., & Truran, J.W., 1981. Ap. J., **251**, 221.
Williams, R.E., Woolf, N.J., Hege, E.K., Moore, R.L., & Kopriva, D.A., 1978. Ap. J., **224**, 171.

DISCUSSION FOLLOWING J. MACDONALD'S TALK

FRIEDJUNG: I think there is a problem for old novae as far as the accretion rate is concerned, there is even a worse case of Nova Delphini where UV data indicate an accretion rate of $10^{-6}\ M_\odot/yr$, which probably will have to decrease. It is possible that quite a number of old novae which we see now, have accretion rates which are much more than the accretion rates required to give the explosion.

SCHATZMAN: I have three technical questions. The initial temperature in the white dwarf is, I suppose, the one which you deduce simply from the luminosity?

MACDONALD: Yes.

SCHATZMAN: The second question concerns the screening factors which you used.

MACDONALD: It is the weak screening which is important here.

SCHATZMAN: Can you describe the way in which the matter of the white dwarf interacts with the accreted matter?

MACDONALD: This probably happens via shear instabilities between the accretion disk and the white dwarf, this is probably where the CNO enhancement comes from, as has been suggested by Kippenhahn and Thomas. We also did some work on this and have shown that this mixing does occur, but probably not on such a quick timescale as Kippenhahn and Thomas suggested. It depends very much on the ratio of the thermal timescale and the accretion timescale. You expect that for high accretion rates you get less mixing.

EVANS: Are you in a position where you can say something about the relative abundances of CNO in the ejecta?

MACDONALD: We know that the CNO process will enhance nitrogen, so you would expect to have mostly nitrogen ejected. I think you are asking about the relative abundances of carbon and oxygen, I think that depends very much on what you put in, what the initial abundance of the white dwarf is, so there is nothing definite I can say.

SION: How did you obtain the recurrence times for the hydrogen shell flashes when you addressed the question of recurrent novae?

MACDONALD: It is just the accreted mass divided by the accretion rate.

SION: Did you actually follow the detailed evolution of the shell flashes?

MACDONALD: Well, the flash occurs over a time short compared to the accretion time.

SHAVIV: First, I am happy to see that flashes are obtained with solar abundances by other investigators and also that the plane (Menvelope, Z) has been adopted by the speaker. What I want to stress is that since the calculations are approximate, and I agree that this is the best way to explore the entire parameter space, you must be careful about the limits, because those are the places where the approximate treatment is bound to give you the largest errors.

SUGIMOTO: Concerning the accuracy, you can compare your solution with others. The accretion phase was computed by Nariai and Nomoto, the expansion phase was computed semi analytically by Fujimoto.

MACDONALD: I should say that my expansion phase results agree with those of Fujimoto. He also found that there is a critical pressure needed at the base of the interface to get a nova outburst and I get exactly the same numbers as he did.

WILLIAMS: I think it is interesting to note that of all the nova ejecta that we have analyzed, the two recurrent novae don't show any evidence for CNO enhancements, in contrast to all the classical novae ejecta which do. Secondly, I am not sure how much you should worry about a comparison between your ejection velocities for the envelope and the observed ones, because I think radiation pressure associated with the outburst will certainly accelerate a substantial fraction of the envelope to higher velocities and so it may well be that you could get from your calculations just a relatively small ejection and then the radiation will accelerate it to much higher velocities. There may be some observational evidence in support of this in the sense that when one looks at the ejection velocities deduced from the different absorption spectra, as time progresses the velocities all tend to increase, which is what you would expect if radiation pressure was important.

LIVIO: May I ask which opacities did you use?

MACDONALD: This is a good question, because I had this enhanced CNO. There are now opacities calculated by Cox for this sort of abundances, but I did not have those available at the time so I took the opacities of Cox and Stewart and Cox and Taylor and used a fitting formula suggested by Christi and produced my own opacity tables.

LIVIO: The reason for my question was that our experience, especially in the expansion phase, has been that the velocities that one gets depend very critically just on this, rather than on details that happen earlier.

SUGIMOTO: It is very difficult to calculate the expansion phase very accurately when the contribution of the radiation pressure becomes very large.

ROBINSON: I am delighted that the masses you derived are similar to the ones that I get. But after my talk earlier I would have to be the first one to point out that no matter what mass you would have gotten it would have agreed with somebody's mass.

MACDONALD: I think DQ Her is the only nova system in which you can make an attempt to get the mass of the white dwarf.

THE FORMATION OF OPTICAL CNO EMISSION LINES IN CATACLYSMIC VARIABLES

Robert E. Williams and Donald H. Ferguson
Steward Observatory, University of Arizona

ABSTRACT

Spectroscopic observations of CNO emission lines are presented for old nova systems, and possible excitation processes for the lines are considered. The Bowen fluorescence mechanism cannot generally be responsible for the strength of N III λ4640 because of the weakness of O III λ3429. Other CNO lines are observed which indicate that all of the lines are excited by resonance fluorescence of UV continuum radiation. Several nonfluorescent excited lines of carbon are also present in old novae, probably formed by recombination processes. The available data for the optical CNO lines suggest that non-solar CNO enhancements exist in quiescent novae, indicating that some of the binary systems may be evolved.

INTRODUCTION

The strongest emission lines observed in the optical spectra of cataclysmic variables are the hydrogen Balmer lines, He II λ4686, and He I triplet and singlet transitions. The strongest non-H or -He line is usually the N III - C III λ4640 - 50 complex. Its strength is variable from one object to another, and is frequently correlated with the intensity of the He II λ4686 line. The large intrinsic widths of the lines in most CVs cause the N III λ4640 and C III λ4650 multiplets to be blended, therefore the relative contributions of the N III and C III lines to the λ4645 feature are uncertain for most objects. In a few systems which have narrower lines or which have been studied at higher dispersions, the feature has been partially resolved to varying extents, and the λ4645 complex has consisted of comparable contributions from the N III and the C III. In HR Del (Hutchings 1979), DQ Her (Hutchings *et al*. 1979), and QU Car (Gilliland and Phillips 1982), for example, the dominant contribution appears to be from C III λ4650. On the other hand, in many CVs the strength of the N III λλ4097, 4103 multiplet, which must be emitted following emission of N III λ4640 since the 3d $^3P^o$ level has no other radiative avenue of decay (cf. Figure 1), and which is usually blended in the wings of Hδ, indicates that N III must make a dominant contribution to the λ4645 feature in these objects.

M. Livio and G. Shaviv (eds.), Cataclysmic Variables and Related Objects, 97–115.

The N III - C III complex is the most prominent emission from any of the CNO ions which occurs in the optical spectra of cataclysmics. No theoretical study of this or any other CNO emission has yet been made for nova systems. However, an analysis of the λ4645 feature in the spectra of X-ray binaries (XRBs) has been made by McClintock *et al.* (1975), motivated by the fact that λ4645 is consistently one of the strongest lines in the visible spectra of low-mass XRBs. Various mechanisms were considered for the formation of the λ4645 emission, and McClintock *et al.* concluded from models of the emitting region that it was due to N III λ4640, excited by the Bowen (1935) fluorescence mechanism. This conclusion was strengthened by spectroscopic observations of HZ Her by Margon and Cohen (1978), who found one of the emission lines involved in the Bowen fluorescence process, O III λ3444, to be present in the HZ Her spectrum with approximately the correct strength expected from fluorescence. Other than N III λ4640, the emission lines formed by the Bowen mechanism tend to be weaker than λ4640, and occur at wavelengths less than 3800 Å. Their detection in faint objects is not easy, and consequently little systematic follow-up study of their presence in interactive binaries has taken place, although Canizares *et al.* (1979) did tentatively report the presence of weak Bowen lines in several X-ray bursters. On the basis of these observational results and the theoretical work of McClintock *et al.*, the prevailing opinion has been that the λ4645 emission in XRBs is predominantly N III, and its unusual strength is caused by the selective excitation of the Bowen process.

N III BOWEN FLUORESCENCE IN CATACLYSMICS?

Since CVs and low-mass XRBs are believed to be very similar types of systems, and have similar optical spectra, it is suggestive that Bowen fluorescence may occur in cataclysmics. The energy level diagrams for He II, O III, and N III shown in Figure 1 illustrate how the Bowen process works. If He II λ304 photons are trapped by high optical depth, they scatter many times. A wavelength coincidence exists between the He II λ304 line and lines of the $2p^2\ ^3P - 3d\ ^3P^o$ resonance multiplet of O III, such that repeated He II scatterings eventually produce absorption by O III. For each absorption of O III λ304, there is ∿ 1% probability that a 3d - 3p transition takes place, one of which is the O III λ3429 multiplet, rather than re-emission of λ304. A λ374 A photon is emitted following the 3d - 3p emission, and sufficient scattering of the λ374 resonance line can then lead to absorption by N III $2p\ ^2P^o - 3d\ ^2D$ λ374, which coincidentally has the same wavelength as the O III line. With each N III λ374 absorption, there is a 0.5% likelihood that N III λ4640 will be emitted instead of λ374. The net effect of the entire scattering process is that He II λ304 (Ly-α) photons are converted via O III transitions into N III λ4640 emission. The fluorescence mechanism requires (1) high resonance line optical depths to produce trapping of the line radiation so conversion to optical transitions can take place, and (2) densities which are sufficiently low such that collisional de-excitation of the levels does not occur.

Conservation of photons in the scattering process imposes a limit on the relative intensities of the O III and N III lines which must be satisfied by Bowen fluorescence. Since escape of some resonance line photons must occur,

$$N(\mathrm{OIII}\ \lambda304) \geq N(\mathrm{OIII}\ \lambda\lambda3127,3429,2830) \geq N(\mathrm{NIII}\ \lambda374) \geq N(\mathrm{NIII}\ \lambda4640), \quad (1)$$

where $N = 4\pi J_\nu/(hc\nu)$ is the energy density of line photons. Using published transition probabilities (Wiese et al. 1966) which dictate the branching ratios for the O III λλ3127, 3429, and 2830 multiplets leads to the limit that $J(\mathrm{O\ III}\ \lambda3429) \geq 0.5\ J(\mathrm{N\ III}\ \lambda4640)$ for excitation by the Bowen mechanism. Realistically, the O III multiplet should be stronger than the N III multiplet since the lower limit to the O III/N III flux ratio requires 100% efficiency of the entire fluorescence process. However, in planetary nebulae the maximum efficiency found for the conversion of He II λ304 to O III λ374 is only 30% (Seaton 1960; Kaler 1967). And there is probably an even smaller efficiency for conversion of O III λ374 to N III λ4640 because of the smaller optical depths of the O III resonance lines relative to that of He II Ly-α. Because of the large velocity gradients which gas in CVs possesses, which drive down line optical depths, it would be surprising if Bowen fluorescence were more efficient in interacting binary systems than it is in the nebulae.

As an observational check on the Bowen fluorescence mechanism, we have obtained spectral scans of cataclysmic variables extending down to 3400 Å in an attempt to detect the O III λ3429 multiplet in those systems in which the N III - C III λ4645 feature is present. A representative sample of objects observed with moderately good signal-to-noise in the blue is shown in Figures 2 - 5. All of the objects are CVs, old novae and nova-like variables, having easily detectable λ4645 lines. For most of the systems, the N III λλ4097, 4103 multiplet seems to be present, appearing as unresolved bumps in the base of the Hδ line, substantiating the contribution of N III λ4640 to the λ4645 feature in these objects. Based on the comparative strengths of N III λ4099 and N III - C III λ4645, an appreciable fraction of λ4645 is due to N III.

Several points may be made from the spectral scans in Figures 2 - 5. First, with the exception of the magnetic AM Her-type variable PG 1550+19, none of these CVs shows O III λ3429 clearly. There is a strong suggestion of it in DQ Her, but the line there is probably too narrow to be real, and it does not appear on other scans of DQ Her. From the relative strengths of the λ4645 emission and (the upper limits to) O III λ3429 in these objects, the Bowen fluorescence mechanism can therefore essentially be ruled out as the cause of N III λ4640. Second, other CNO lines are present in the spectra of these CVs, and they have been identified in the scans. These lines are O III λ3712, C II λ4267, and C IV λ5805. The C II line occurs frequently in the old novae, and the C IV line is somewhat less common. Although the spectrum of BT Mon in Figure 2 does not extend out to λ5800 Å, the

C IV λ5805 line is present in other scans we have of this system (cf. Williams and Ferguson 1982, Figure 4).

An important question to be addressed concerning the spectra is: what conditions produce these lines, and what can be learned about CVs from the line strengths? The fact that certain CNO lines have intensities comparable to the H and He lines suggests that a selective excitation mechanism may be operative; however, we are convinced from the general weakness of O III λ3429 in objects which have prominent λ4645 that Bowen fluorescence is usually not responsible for the strength of N III λ4640. We suspect that the same conclusion may also apply to many of the low mass X-ray binaries.

CONTINUUM FLUORESCENCE OF CNO LINES

There is a similarity shared by the O III λ3712, N III λ4640, and C IV λ5805 lines which we believe to be significant: all of these multiplets have upper levels which are connected to the ground states of the respective ions by strong resonance transitions. That is, each of these lines is capable of being excited by fluorescent excitation. However, selective _line_ excitation such as the Bowen mechanism is unlikely to work for all of these transitions since they are all excited, or pumped, by different wavelengths. In fact, the Bowen mechanism does not even appear to be responsible for the N III line. Instead, we propose here that the above multiplets are formed in cataclysmics by _continuum_ fluorescence, i.e., photoexcitation of the upper levels by an extension of the strong UV continuum that CVs have been observed to have with the IUE satellite. It is unlikely that any other excitation process would produce these three particular multiplets so strongly in comparison with other CNO transitions. For example, these are not lines which would be excited by the recombination process since electron recapture tends to preferentially populate levels with high angular momentum. Similarly, collisional excitation of these transitions would also cause greater population of other transitions having lower excitation potentials, but which are not observed. Thus we believe the evidence favors continuum fluorescence for the excitation of λ3712, λ4640, and λ5805.

In order to check on our assumption of continuum fluorescent excitation, we have determined from the energy level diagrams of CNO ions (Bashkin and Stoner 1975) all those optical transitions which could occur from this process and which might be expected to be present in the spectra of CVs. The situation in which fluorescent excitation occurs involves three levels, one of which is the ground state of the ion, and is schematically depicted in Figure 6. Absorption of continuous radiation at wavelength λ_{12} takes place by a resonance line, always in the UV for CNO ions. Following each 1→2 UV absorption, there is a finite probability, typically $\sim$ 1% depending on the relative transition probabilities, that emission of a subordinate optical transition 2→3 may occur rather than re-emission of the UV resonance line. The emissivity of the optical line can be written as

$$j_{23} = N_2 A_{23} \frac{h\nu_{23}}{4\pi} \quad , \tag{2}$$

where N_2 is the number density of ions in level 2. In a steady-state situation, the equilibrium population of level 2 is determined by the rate of resonance absorptions into the level and the rate of spontaneous transitions out, i.e.,

$$N_1 \, B_{12} \, J_{\nu_{12}} = N_2 \, (A_{21} + A_{23}). \tag{3}$$

The stimulated Einstein coefficient B_{12} is related to the spontaneous transition probability A_{21} by the relation $B_{12} \propto A_{21}\lambda^3_{12}\, g_2/g_1$, where the g_i are the statistical weights of the levels. Substituting eqn. (3) into (2) therefore gives for the emissivity of a line excited by fluorescence,

$$j_{23} \propto N_1 \, J_{\nu_{12}} \frac{\lambda^3_{12}}{\lambda_{23}} \, A_{23} \, \frac{g_2}{g_1} \quad , \tag{4}$$

where we have assumed that $A_{21} >> A_{23}$, as is usually the case when $2 \rightarrow 1$ is a UV resonance line and $2 \rightarrow 3$ is an optical transition. In most situations N_1 may be taken to be the ion density, and therefore the intensity of a line excited by fluorescence is directly proportional to the ion abundance, the mean intensity of exciting radiation at the UV pumping frequency, and atomic constants of the transitions. We have compiled a list of the strongest CNO lines capable of being photo-excited by UV continuum radiation, and give them in Table 1 together with relevant atomic data. The final column gives the relative emission strengths of the lines, assuming all ions to have the same abundance and the UV continuum radiation to have a flat distribution.

The observed lines excited by continuum fluorescence in CVs originate in a region where the CNO are doubly and triply ionized. This region must have a lower density than the accretion disk in order for radiative and scattering processes to dominate over collisions, because LTE conditions prevail in the disk (Williams and Ferguson 1982), and therefore the CNO lines are probably not formed in the main part of the disk. On the other hand, it is unlikely that they originate very far from the central UV continuum-emitting region of the disk since a larger flux of radiation enhances the formation of the lines. Thus, the optimal environment for the line formation is in a lower density "chromospheric" region immediately above and below the disk, or in an outflowing wind in the vicinity of the disk.

It should be noted that the multiplets O III λ3429, O III λ3712, N III λ4640, and C IV λ5805 which have been observed in CVs, are among the strongest ones expected from fluorescent excitation of these ions. Given the presence of these lines, the following lines from these same elements might also be expected to be present on the basis of the information in Table 1: C III λ5250, N IV λ6381, and O IV λ3409. None of these lines has been detected in any of our scans, although the N IV multiplet is not expected to be very strong. Their absence may be due to several factors, including (a) the ionization may be sharply peaked such that C^{+3}, N^{+2}, and O^{+2} are much more abundant than other ionization stages where the fluorescence occurs, or (b) the computed transition probabilities for the unseen lines may have been overestimated. An additional possibility, that the continuum flux responsible for exciting the unobserved λλ3409, 5250, 6381 lines is substantially less than that pumping the observed λλ3429, 3712, 4640, 5805 lines is very unlikely since most of these transitions are excited by resonance lines having similar wavelengths, around λ300 Å.

An example of the very useful type of information about cataclysmics that is potentially available from analysis of the CNO emission lines comes from applying eqn. (4) to the observed fluorescence excited lines under the simplistic assumptions that the lines are formed together in the same region and that the UV continuum flux is constant over the ∿ 75 Å interval from 300 - 375 Å that serves to excite the lines. The abundance ratios of the ions, by number, are related to the line fluxes by

$$\frac{C^{+3}}{N^{+2}} = 2.1 \ \frac{F(\lambda 5805)}{F(\lambda 4640)} \ , \tag{5}$$

$$\frac{N^{+2}}{O^{+2}} = 1.3 \ \frac{F(\lambda 4640)}{F(\lambda 3712)} \ . \tag{6}$$

For CP Pup, where the intensities are roughly in the ratio F(λ5805)/F(λ4640)/F(λ3712) $\cong$ 1/2/2, allowing for some contribution of C III to λ4645, the deduced ion abundances are $C^{+3}/N^{+2} \cong 1$ and $N^{+2}/O^{+2} \cong 1$. If these ions are the predominant ones of CNO in the emitting regions, then C $\cong$ N $\cong$ O, which is non-solar and may be evidence for evolved gas. More extreme examples of non-solar, evolved CNO abundances for the emitting gas may be indicated for objects such as RR Pic, which have moderately strong N III λ4640 with no detectable C IV or O III lines. Presumably, the emitting gas is situated sufficiently far out in the potential well of the white dwarf that it represents mass lost from the secondary star rather than gas somehow ejected from the degenerate companion.

NON-FLUORESCENT EXCITATION OF CNO LINES

With the exception of C II λ4267 and C III λ4650, the optical CNO emission lines observed in cataclysmics can be explained in terms of fluorescent excitation. However, neither λ4267 nor λ4650 couple directly to the ground states of C II or C III via electric dipole resonance lines, consequently these lines must be excited by a different process. Both lines have moderately high excitation potentials (> 20 eV), and thus are unlikely to be strongly collisionally excited. C II λ4267 is expected to be the most intense optical line in the recombination spectrum of C II (Williams 1982) because electrons recaptured to most of the high (n, l) levels cascade through λ4267 on the way to the ground state. On the other hand, the upper level of C III λ4650 is not likely to be heavily populated by electron recapture in comparison to a number of other optical transitions in the C III spectrum. Thus, we encounter a dilemma in trying to explain the strength of λ4650 in CVs: both collisional excitation and recombination should produce other lines with greater strength than λ4650, but which are not observed in the spectra of catalysmics. Apparently, some selective excitation mechanism is operative for this line. This dilemma is not unique to the spectra of CVs, but appears to be a general situation when C III λ4650 is observed in emission. For example, in their investigation of the emission spectra of Wolf-Rayet stars, Castor and Nussbaumer (1972) considered the excitation of a number of levels of C III from a variety of processes, and yet still failed to produce the observed λ4650 intensity in their models by factors of 1 to 2 orders of magnitude. The large discrepancy they found between their calculations and the observations led them to suggest that some process involving the continuum of C III, such as dielectronic recombination, populates the 3p $^3P^o$ level and causes λ4650 to be enhanced. We suspect this same situation occurs in cataclysmics and causes C III λ4650 to contribute to the N III - C III λ4645 feature.

The excitation of C II λ4267 is more straightforward, almost certainly involving only electron recapture processes. There is a possibility of dielectronic recombination populating the $2s^2 4f\ ^2F^o$ upper level of λ4267 by radiative decay from the 2s2p4f ^2D, ^2F, ^2G autoionizing levels that lie ∿ 3 eV above the C II ionization limit. Storey (1981) has calculated the reaction rates and shown that emission of λ4267 will be enhanced over the 2-body recombination rate for temperatures $\geq 2 \times 10^4$ K. Assuming the chromospheric temperatures in CVs to be less than this value, the presence of λ4267 in the spectra of old novae with the characteristic flux seen in these systems, of the order of ≥ 10% of the Hβ flux, may require a rather high C/H abundance in the emitting gas. With Hβ also formed by recombination in the same region as λ4267, the relation between line strengths and ion abundances is

$$\frac{F(\lambda 4267)}{F(H\beta)} = \frac{N(C^{+2})}{N(H^+)} \; \frac{\alpha(\lambda 4267)}{\alpha(H\beta)} \; \frac{4861}{4267} , \tag{7}$$

where the ratio of the effective line recombination coefficients has been computed to be $\alpha(\lambda 4267)/\alpha(H\beta) = 8.2$ (Seaton 1978). The resulting abundance is $C^{+2}/H^+ \cong 10^{-2}$ in the old nova systems in which the C II line has been observed.

SUMMARY

The N III - C III λ4645 line is the strongest optical emission feature from CNO ions in cataclysmic variables, and there is evidence that both N III and C III contribute to the line. The mean wavelength of the feature usually falls between the wavelengths of the two multiplets. In some objects the feature appears resolved into individual N III and C III components, and in many CVs where the feature is relatively strong, the blending of N III $\lambda\lambda$4097, 4103, which is emitted following λ4640 emission, with Hδ is clearly evident in the base of the Balmer line.

The weakness or absence of O III λ3429 emission in old novae in which N III is present rules out the Bowen process as the excitation mechanism. Instead, the presence of other emission lines such as O III λ3712 and C IV λ5805, which can be excited by fluorescence at wavelengths other than those involved in the Bowen mechanism, is strong circumstantial evidence that all of these lines are emitted following resonance scattering of continuum radiation. The relative intensities of the lines in several systems indicate comparable abundances for the C, N, and O. Even greater departures from solar abundances are indicated for objects like RR Pic which show N III λ4640 but not O III, suggesting $N/O > 1$. Such systems may have nuclear-processed gas being transferred from an evolved secondary onto the degenerate dwarf companion.

Further evidence for the presence of evolved gas in cataclysmics may come from the existence of C II λ4267 and C III λ4650 emission. These lines are not directly excited by fluorescence, and their observed strengths suggest a high abundance of carbon because the effective reaction rates for their formation, involving electron recombination, are smaller than those for fluorescent excitation.

Several problems remain concerning the interpretation of the optical CNO emission in CVs. First, the absence of C III λ5250 in objects in which C II λ4267, C IV λ5805, and N III λ4640 emission are present requires explanation. The λ5250 line is pumped by UV wavelengths very near those exciting the C IV line, and its transition probability is such that whenever the C II and C IV lines are observed, the abundance and ionization of carbon should be optimum for λ5250 to be emitted with an intensity comparable to that of C IV λ5805. The computed transition probability for λ5250 may simply be in error. Or perhaps the C II recombination line is formed in a larger, separate region from the fluorescent-excited C IV line, in which case the region where carbon is doubly ionized would be more distance from the source of exciting continuum radiation, causing λ5250 to be much fainter than λ5805. Second, any analysis of CNO emission in close binaries should also include the UV lines which can be observed with the IUE satellite. There is a good possibility that the optical and UV lines are formed together, since the lines observed in both spectral regions require a scattering medium. Analyses of UV emission-line

TABLE 1

OPTICAL CNO LINES EXCITABLE BY RESONANCE FLUORESCENCE

Ion	Multiplet	λ_{12}(A)	A_{23} (10^8 sec^{-1})	$10^4 A_{23} g_2 \lambda_{12}^3/(g_1 \lambda_{23})$
C II	λ3920 3p $^2P^o$ -4s ^2S	636	1.87	4.1
	λ4638 4p $^2P^o$ -6d ^2D	530	0.030[a]	0.16
	λ6259 4p $^2P^o$ -5d ^2D	560	0.021[a]	0.10
	λ7234 3p $^2P^o$ -3d ^2D	687	0.45	3.4
C III	λ3170 4s ^1S -5p $^1P^o$	291	0.32	0.75
	λ5250 4d ^1D -5p $^1P^o$	291	0.52	0.73
	λ8500 3s ^1S -3p $^1P^o$	386	----	----
C IV	λ4786 5d ^2D -6p $^2P^o$	212	0.34	0.20
	λ5805 3s ^2S -3p $^2P^o$	312	0.32	0.50
C V	λ3479 2s ^1S -2p $^1P^o$	40	0.16	0.001
N II	λ4794 3p ^3D -3d $^3D^o$	534	0.36	1.9
	λ5001 3p ^3S -3d $^3P^o$	530	0.75	2.2
	λ5938 3p ^3P -3d $^3D^o$	534	0.56	2.4
N III	λ4640 3p $^2P^o$ -3d ^2D	374	0.57[a]	1.07
N IV	λ6381 3s ^1S -3p $^1P^o$	247	0.19	0.13
N V	λ4609 3s ^2S -3p $^2P^o$	209	0.41	0.24
O II	λ3292 3p $^4P^o$ -4s ^4P	419	0.85	5.7
	λ3754 3p $^4S^o$ -4s ^4P	419	0.26	1.5
	λ4152 3p $^4P^o$ -3d ^4P	430	1.01	5.8
O III	λ3429 3p ^3P -3d $^3P^o$	304	0.79	0.65
	λ3712 3p ^3P -3d $^3D^o$	306	1.10	1.4
O IV	λ3409 3p $^2P^o$ -3d ^2D	238	1.15	0.76
O V	λ5114 3s ^1S -3p $^1P^o$	172	0.25	0.075

[a]Transition probabilities from Kurucz and Peytremann (1975). All others from Wiese *et al*. (1966).

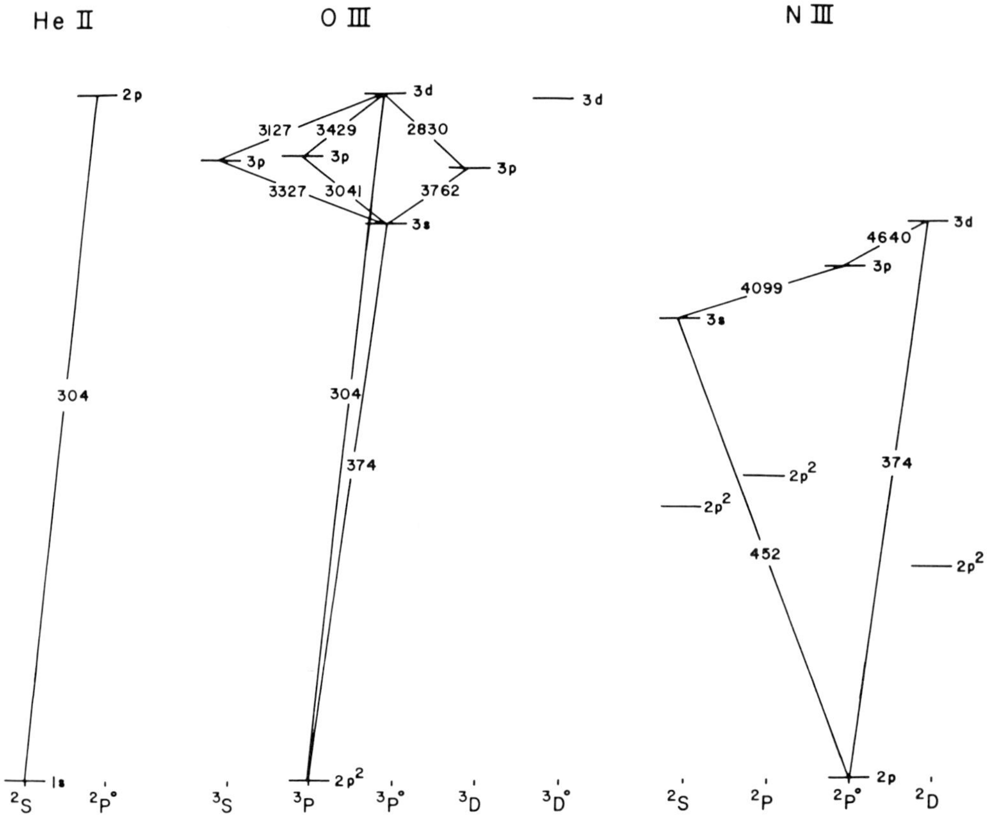

Figure 1 -- Partial energy level diagrams for He II, O III, and N III showing the transitions involved in the Bowen fluorescence mechanism.

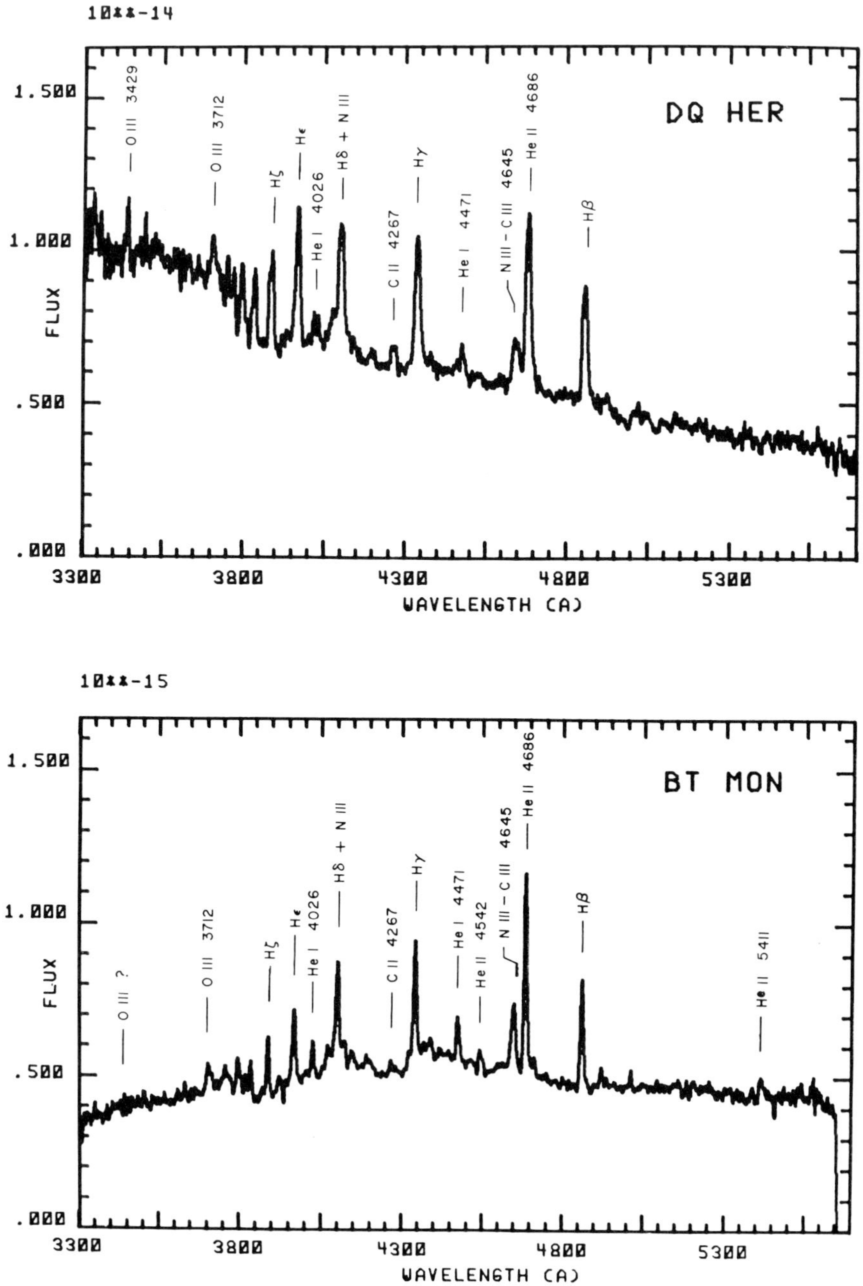

Figure 2 -- Spectral scans of the old novae DQ Her and BT Mon obtained with the Steward Observatory 2.3 m telescope and intensified Reticon scanner. The monochromatic flux F_λ is in units ergs/(cm^2 sec A).

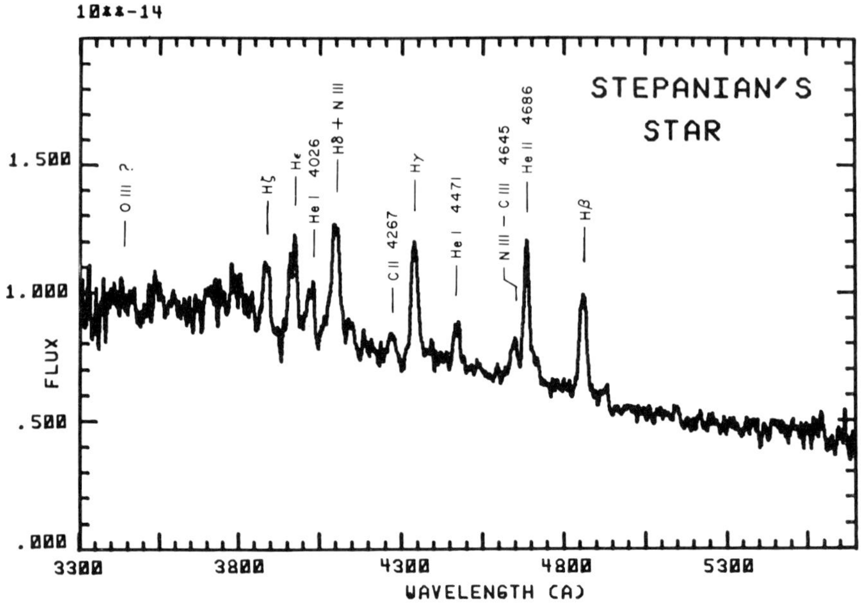

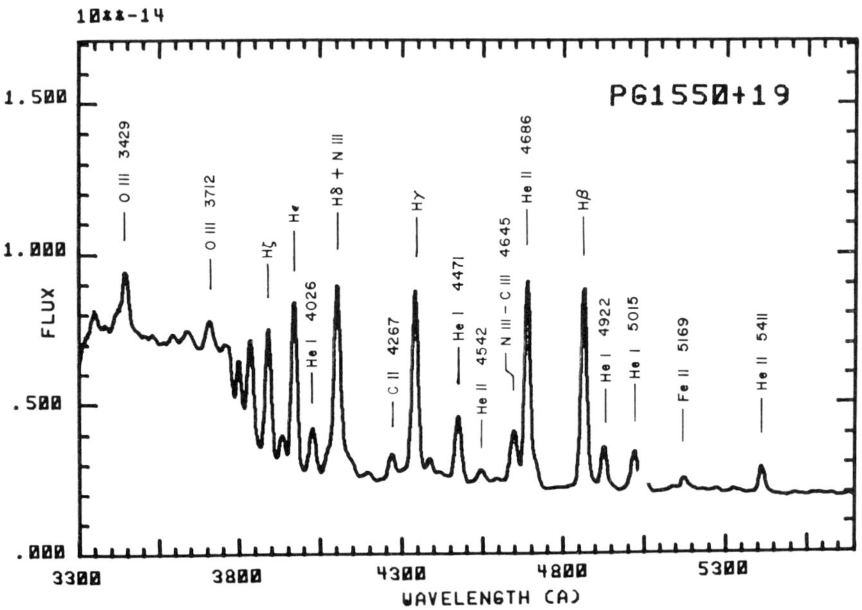

Figure 3 -- Spectral scans of the nova-like variables Stepanian's Star and PG 1550+19, obtained with the Steward Observatory 2.3 m telescope.

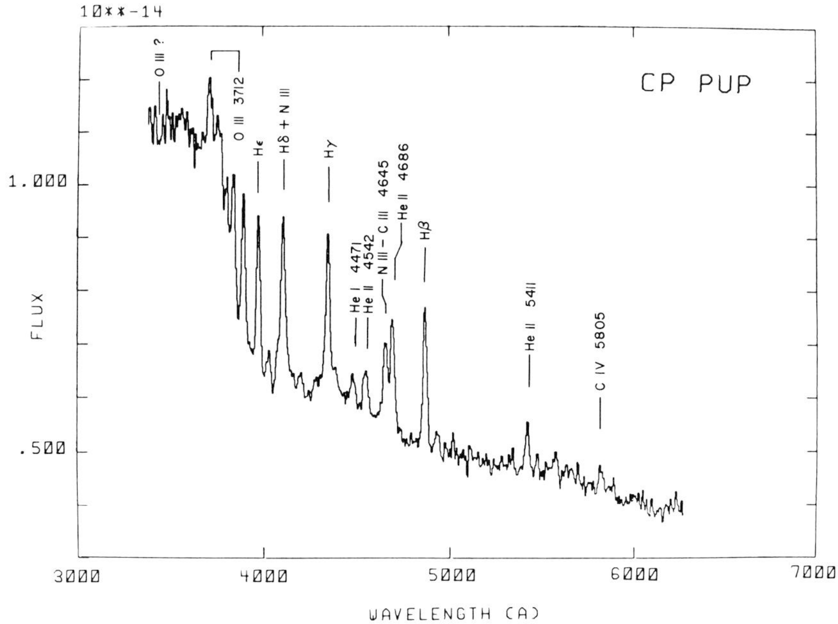

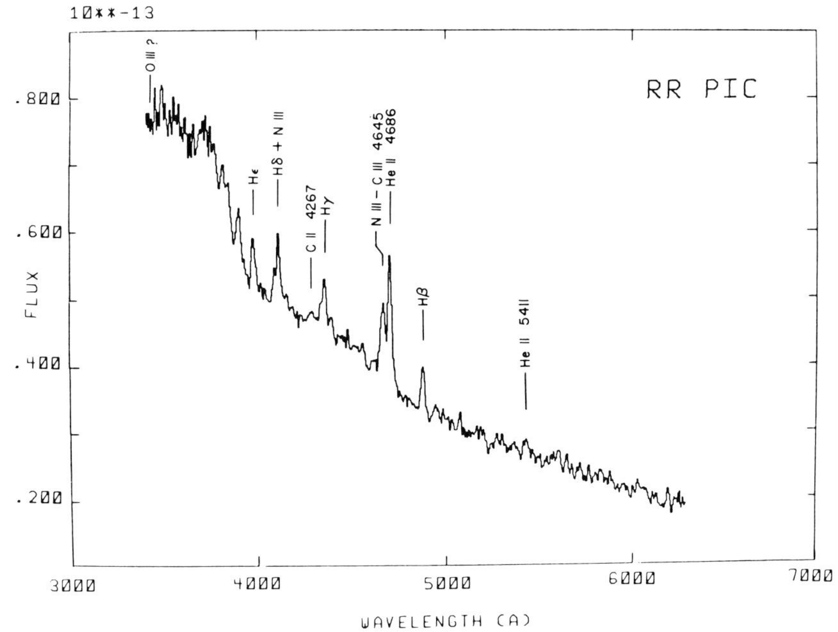

Figure 4 -- Spectra of the old classical novae CP Pup and RR Pic, obtained with the Cerro Tololo 4 m telescope and SIT Vidicon in April 1982. Units of flux are the same as Figures 2 and 3.

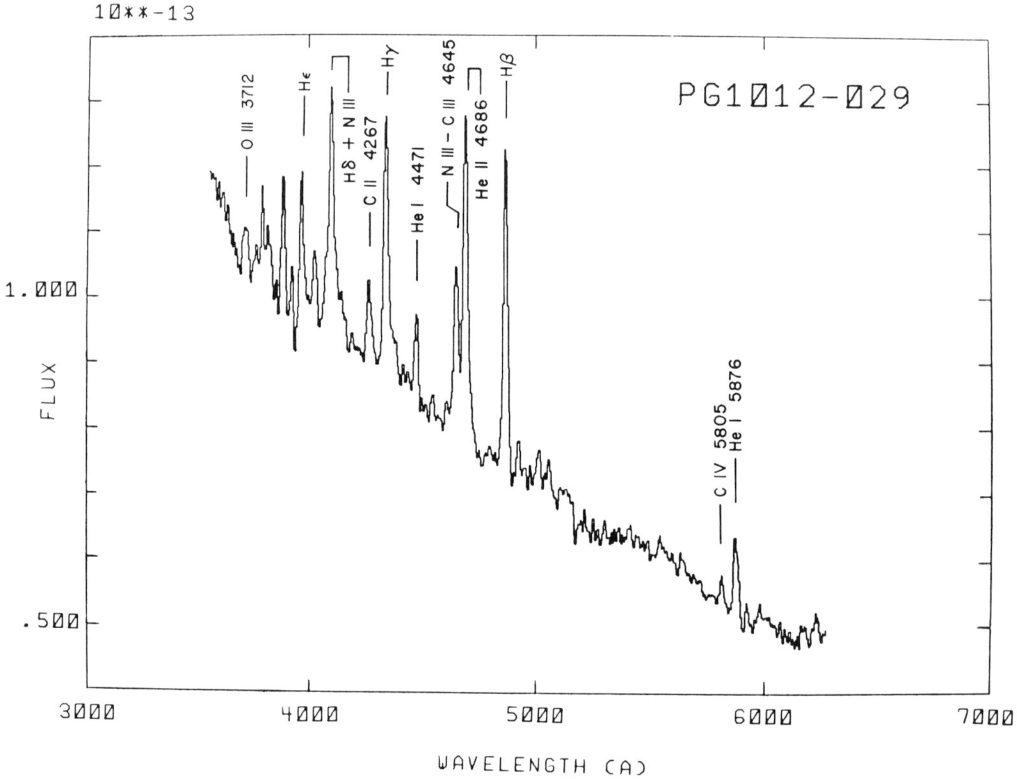

Figure 5 -- Spectral scan of the nova-like variable PG 1012-03, obtained with the Cerro Tololo 4 m telescope and SIT Vidicon.

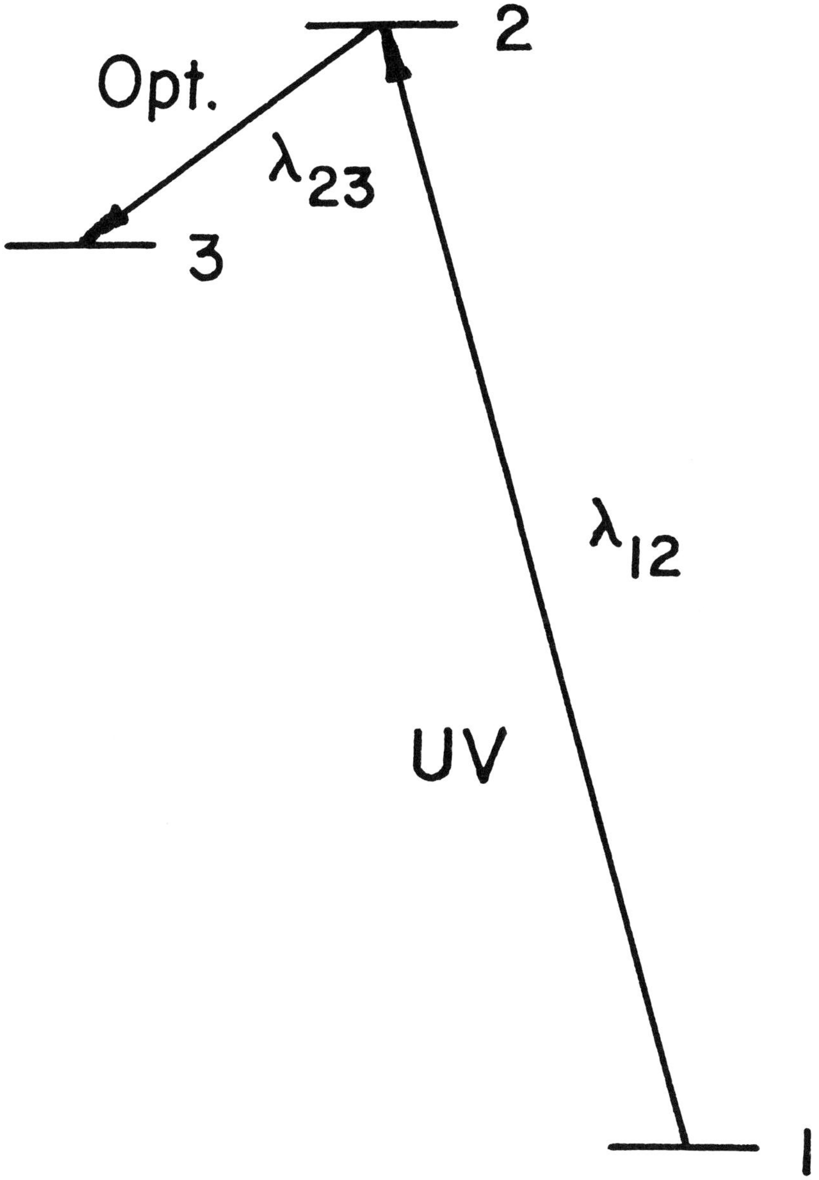

Figure 6 -- Schematic energy level diagram for transitions excited by resonance fluorescence of continuum radiation.

spectra have been performed for AE Aqr (Jameson et al. 1980) and V603 Aql (Ferland et al. 1982), but they have not shown clear evidence in these two systems for the magnitude of non-solar CNO abundances that our optical line interpretation has indicated for some CVs in the present investigation. The ultimate resolution of the question of the evolutionary status of interacting binaries should eventually result from the combined analysis of both optical and UV emission lines in representative systems.

REFERENCES

Bashkin, S., and Stoner, J. O. 1975, Atomic Energy Levels and Grotrian Diagrams (Amsterdam: North-Holland).

Bowen, I. S. 1935, Ap. J., 81, 1.

Canizares, C. R., McClintock, J. E., and Grindlay, J. E. 1979, Ap. J., 234, 556.

Castor, J. I., and Nussbaumer, H. 1972, MNRAS, 155, 293.

Ferland, G. J., Lambert, D. L., McCall, M. J., Shields, G. A., and Slovak, M. H. 1982, Ap. J., 260, in press.

Gilliland, R. L., and Phillips, M. M. 1982, Ap. J., in press.

Hutchings, J. B. 1979, Ap. J., 232, 176.

Hutchings, J. B., Cowley, A. P., and Crampton, D. 1979, Ap. J., 232, 500.

Jameson, R. F., King, A. R., and Sherrington, M. R. 1980, MNRAS, 191, 559.

Kaler, J. B. 1967, Ap. J., 149, 383.

Kurucz, R. L., and Peytremann, E. 1975, A Table of Semiempirical gf Values, Smithsonian Astrophys. Observ. Special Report No. 362.

Margon, B., and Cohen, J. G. 1978, Ap. J. (Letters), 222, L33.

McClintock, J. E., Canizares, C. R., and Tarter, C. B. 1975, Ap. J., 198, 641.

Seaton, M. J. 1960, Repts. Prog. Phys., 23, 313.

Seaton, M. J. 1978, in IAU Symp. No. 76: Planetary Nebulae, ed. Y. Terzian (Dordrecht: Reidel), p. 131.

Storey, P. J. 1981, MNRAS, 195, 27P.

Wiese, W. L., Smith, M. W., and Glennon, B. M. 1966, Atomic Transition Probabilities (Washington: USGPO), Vol. 1, NSRDS-NBS 4.

Williams, R. E. 1982, Ap. J., 261, in press.

Williams, R. E., and Ferguson, D. H. 1982, Ap. J., 257, 672.

DISCUSSION FOLLOWING R. WILLIAMS' TALK

EVANS: The ejecta abundances, how are these likely to be effected by the fact that you may have condensations of graphite grains and secondly is there any prospect of determining whether there is any change in carbon abundance before and after the nebular stage?

WILLIAMS: First of all, grain formation may be relevant, we know certainly in FH Ser that there was grain formation, just from the infrared within a month after the outburst. What that would tend to do is to cause the actual CNO to hydrogen abundances to be greater than those we talk about here, in which case we should perhaps qualify our statement about the CNO to hydrogen abundances after the outburst being similar to those before. It is an order of magnitude statement in any event. I cannot say anything definitely because I have not observed in the infrared but certainly it is something that is potentially important. About the second part of your question, Ferland and I are working on that problem right now, perhaps you are aware of the paper he wrote with Jim Truran that suggested that the conventional analysis of the spectra of DQ Her since its outburst indicates that the carbon to hydrogen abundance has changed, we now think that perhaps that conclusion can be explained in terms of conventional nebular analysis rather than dust formation but that doesn't rule out dust formation.

KING: Doesn't your fluorescence mechanism rather tightly constrain the continuum in that it's got to be able to lift the thing from the ground state to the third level, but not ionize it?

WILLIAMS: In fact, it has to ionize it also, because I think that is probably what is producing the OIII and the NIII. This whole fluorescence process is predicated on the fact that you have a scattering situation and that doesn't occur in the conventional disk model because the densities are too high, therefore I suspect that these lines are not coming from the inner part of the disk where the hydrogen and helium lines are coming from, but from some photoionized chromosphere or a wind or something like that, in which case I would say that the continuum is what is producing the ionization and also giving the fluorescence.

KING: Do you have to arrange it very cunningly so that you don't kill the element you are actually using?

WILLIAMS: No, it is not different from any other photoionization calculation.

WARNER: This morning we saw a spectrum of BV Cen which had a very well developed solar type spectrum, in other words, the secondary could be seen, is there any evidence of CH or CN enhancement in that spectrum, do you know?

WILLIAMS: I don't know if one sees it.

WARNER: There is the possibility for an independent witness for your story.

BIANCHINI: Is it possible that collisional excitation, with the interstellar medium, plays a role in favour of certain lines?

WILLIAMS: Only for GK Per. Because that is the only nova shell spectrum that indicates an extremely hot gas, the electron temperatures are 40000°K. It has a spectrum that looks just like a supernova remnant and it is the only really asymmetric shell, so the case can be made for GK Per.

NETZER: Surely the equation you used is only for the optically thin case, is that procedure correct? You are talking about a resonant line.

WILLIAMS: It is valid for optically thin optical transitions and optically thin or thick UV.

NETZER: You can take the number of continuum photons, you can easily extrapolate from the IUE data and see if the number you get there is about the same number you expect to get in the optical by conversion.

WILLIAMS: Yes, that is a good point.

NETZER: If you do explain the CII line by recombination, then have you done the same exercise for the other lines assuming that you know roughly the N and O abundance? because I would expect them to have about an equal strength.

WILLIAMS: Not quite. I have done something else, and that is, assuming that you have recombination, I have computed what intensities you would expect for those lines relative to other lines in the same ion, CIV 5805 would be weaker than some other CIV lines if it were due to recombination because they are not high angular momentum states, so it was on that basis that I have ruled it out. I should say that there is a possibility that 4267 may have a di-electronic recombination contribution.

DID NOVA AQUILAE 1982 POSSESS A SUPERCRITICAL WIND ?[1]

M. Friedjung
Institut d'Astrophysique, 98 bis Bd Arago, F-75014 Paris

Abstract

The extremely high velocity absorption components observed in the ultraviolet with IUE, could have been produced by a supercritical wind. In this case most radiation would have been emitted shortwards of Lyman α, even at the time of the first IUE observations.

Nova Aquilae 1982 was discovered on January 27, but was difficult to observe at first because of closeness to the sun. A number of IUE spectra were obtained by Blades and Snijders from February 24, and the results were communicated to members of the ESA + SERC nova target of opportunity team. A preliminary report concerning these spectra has been made by Snijders, Seaton and Blades (1982).

The most surprising result obtained up to now is perhaps that concerning the high velocity absorption components seen on February 24 and March 2. The low resolution of all the well exposed spectra shows the presence of absorption components of two absorption systems at about 3000 and 7000 km s^{-1}, denoted systems a and b respectively. The violet edges of systems a and b lie at velocities at the order of 4000 and 9000 km s^{-1} respectively. The latter velocity if interpreted as a terminal velocity is enormous; velocities of this order are only known for supernovae. The largest classical nova Orion velocity given by Payne-Gaposchkin (1957) is 3820 km s^{-1} in her table 10.7. It is because of the enormous velocity, that doubts have been cast on whether Nova Aquilae 1982 is a classical nova.

The model which appears best to fit the observations of classical novae involves a rapid ejection of about 10^{-4} $M_\odot$, followed by a diminishing continued ejection (see Friedjung, 1977). The latter is most

1: Based on observations made with the International Ultraviolet Explorer (IUE) at the Villafranca satellite tracking station operated by the European Space Agency.

M. Livio and G. Shaviv (eds.), Cataclysmic Variables and Related Objects, 117–120.

easily considered to be a wind pushed by the radiation pressure of an object above the Eddington limit (Friedjung, 1966 ; Bath and Shaviv, 1976). In a recent study of mine (Friedjung, 1981), conditions are given for this type of wind. When the object is significantly above the Eddington limit the trapping of photons in the acceleration region of the wind gives an order of magnitude relation between the photospheric radiation flux and the kinetic energy flux, as well as one for the wind terminal velocity :

$$\chi_{Ts} \; L \simeq \frac{64}{3}\pi \; V_s^3 \; r_s \; \frac{V_{ac}}{V_s} \tag{1}$$

Here χ_{Ts} is the total opacity at the photosphere, L the luminosity, V_s the terminal velocity, V_{ac} the velocity mean where the wind is accelerated, and r_s is the photosphere radius. Taking $V_{ac} = 0.5V_s$ one obtains numerically when electron scattering dominates the opacity.

$$\frac{L}{r_s} = 100 \; V_s^3 \tag{2}$$

Comparison with observed velocities for FH Serpentis suggests that this should be replaced by

$$\frac{L}{r_s} \simeq 10 \; V_s^3 \tag{3}$$

in the later stages.

The IUE observations of Nova Aquilae 1982 on February 24 were made when the nova had faded at least 4 magnitudes, so it is reasonable to suppose that it was then in the later stages of continued ejection. Observations of novae as well as equation (3) indicate an increase of V_s as a nova fades, because r_s decreases while L only decreases a small amount. It will therefore be supposed that system b had a velocity characteristic of the wind around February 24, and system a consisted of material ejected earlier, and being swept up by system b. Such an interpretation is supported by the fact that both systems a and b are seen for the CIV and SiIV resonance lines, while system a is only seen for the lower ionization AlIII resonance doublet, and system b only for an excited NIV line (letters of T. Snijders to the target of opportunity team). This is consistent with the envelope being photoionized and being more excited near the photosphere, while system a is formed further than system b from the latter.

Equation (3) can then be applied to the observations of February 24, so as to see whether the large velocity of system b could have been produced by the type of wind considered here. Taking the mean velocity of 7000 km s^{-1} (the justification for taking this rather than the velocity of the violet edge will be given later),

$$\frac{L}{r_s} \; 3.43 \times 10^{27} \text{ erg s}^{-1} \text{ cm}^{-1} \tag{4}$$

If $L = 1.55 \times 10^{38}$ erg/s, twice the Eddington limit of a half solar mass object, $r_s = 4.5 \times 10^{10}$ cm. The corresponding photospheric temperature is $1.0 \times 10^{5\circ}$ and if the nova is at a distance of D kpc, the predicted flux at 1250 Å is $1.2 \times 10^{-10}/D^2$ erg cm^{-2} s^{-1} Å^{-1}. The observed flux on February 24 near 1250 Å corrected for reddening corresponding to E(B-V) = 0.55 (Snijders, Seaton and Blades, 1982) is near 1.5×10^{-11} erg cm^{-2} s^{-1} Å^{-1}. Hence for the assumed luminosity too large a flux would be observed for D less than 3 kpc.

The energy distribution given by Snijders, Seaton and Blades as well as in letters of T. Snijders to the target of opportunity team is not far from a classical Lynden-Bell steady state accretion disk distribution, with a total luminosity near 1×10^{36} D^2 erg s^{-1} on February 24. This is considerably less than the super Eddington luminosity required here for the wind, most of which is in wavelength regions unobservable with IUE. It seems that within the framework of the model proposed here, it is easiest to suppose that the flux in the wavelength regions observed came from material between the components of the binary and perhaps also the secondary, all being heated by radiation from the photosphere. The fairly rapid flux variations on February 24 involving a flux decrease of up to 25% in some wavelength regions in 5 1/2 hours, lead one however to cast some doubt on the interpretation in terms of a steady state disk.

Lines formed near the photosphere would be broadened both by electron scattering and by radiation damping. It is easy to show that the latter could explain the CIV system b width, if all carbon near the photosphere was in the form of CIV, and there was an overabundance of only 5, with respect to the cosmic abundances of Allen (1973). For larger overabundances the same effect would, occur if most carbon was in other ionization stages. In any case there seems little justification in measuring the violet edges of absorption components and calling the corresponding velocity a "terminal" velocity.

Another test of the self consistency of the explanation of the system b velocity proposed here, can be made from equation (12) of my recent study (Friedjung, 1981). In the electron scattering dominant case $\alpha = 1$, and the radius of the region of acceleration r_{ac} is given by

$$\frac{r_{ac}}{r_s} = \frac{16}{3} \frac{V_s}{c} \frac{V_{ac}}{V_s} \tag{5}$$

This gives $r_{ac} = 3 \times 10^9$ cm, which is larger than the radius of a white dwarf. Therefore acceleration can be considered as occurring above the surface of a perhaps somewhat bloated white dwarf; the result would have been embarassing had r_{ac} been smaller than a typical white dwarf radius.

The explanation given here appears consistent though other possible explanations could exist. It should be noted that the system b velocity is of the order of the escape velocity from a white dwarf. Hence alternative explanations should in any case involve winds from a white dwarf.

Acknowledgements

I wish to thank T. Snijders, M.J. Seaton, and M.J. Blades for their extremely speedy communication of observations to the other members of the target of opportunity team. In addition I had a discussion with Professor Seaton in March 1982.

References

Allen,C.W.: in Astrophysical Quantities, ed. University of London, The Athlone Press, 1973.

Bath,G.T. and Shaviv,G.: 1976, Mon. Not. R. Astr. Soc. 171, p.311.

Friedjung,M.: 1966, Mon. Not. R. Astr. Soc. 132, p.317.

Friedjung,M.: 1977, in "Novae and Related Stars", ed. M. Friedjung, Reidel, Dordrecht, p.61.

Friedjung,M.: 1981, Acta Astron. 31, p.373.

Payne-Gaposchkin.: 1957, "The Galactic Novae, North Holland Publishing Compananany".

Snijders,M.A.J., Seaton, M.J., and Blades,J.C.: 1982, in Proceedings of third European IUE Conference, p.177.

INFRARED OBSERVATIONS OF NOVA AQUILAE 1982

M.F. Bode[1], J.K. Davies,[2] A. Evans,[1] and D.C.B. Whittet[2]
(1) Department of Physics, University of Keele, Staffordshire, U.K.
(2) Division of Physics and Astronomy, Preston Polytechnic, U.K.

Abstract

We describe infrared photometry and spectroscopy of Nova Aquilae 1982. The broadband observations suggest that the dust shell of Nova Aquilae was anomalous, while the spectroscopic observations point to the presence of molecules in the nova environment.

Nova Aquila was discovered (Honda 1982) on 1982 January 27th. The visual light curve, based on visual estimates (IAU circulars), IUE FES counts (derived assuming (B-V) = 0) and photometry at the South African Astronomical Observatory (SAAO), suggest a fast nova with visual decay rate $\dot{m}_V = 0.3$ mag d^{-1}. Spectral development and outburst amplitude suggest that the nova was discovered at maximum.

An infrared excess was first detected, 37d after outburst, by Williams and Longmore (1982) and subsequent infrared photometry has been carried out at SAAO (reported here), UKIRT and by Gehrz et al (1982). The light curve at 2.2 μm, based on these data, together with the above visual light curve, is shown in Fig. 1. Assuming that Williams and Longmore's photometry caught the nova dust shell near maximum infrared luminosity, the optical depth in the dust shell

$$\tau \simeq L\ (\text{IR max})/L\ (\text{vis max}) = 0.15$$

consistent with the apparent absence of a break in the visual light curve.

The infrared excess of Nova Aquilae seems to be anomalous for a fast nova on two counts: (i) the amount of dust present is surprisingly large; (ii) dust formation occurred anomalously early. If the time of maximum infrared flux t_{max} = 37d, the grain growth model (Clayton and Wickramasinghe 1976) implies that grain condensation occurred at t_0 = 16d. Even if t_{max} = 60d, condensation would have had to occur at t_0 = 25d. For "normal" novae, obeying the usual relationships

M. Livio and G. Shaviv (eds.), Cataclysmic Variables and Related Objects, 121–125.

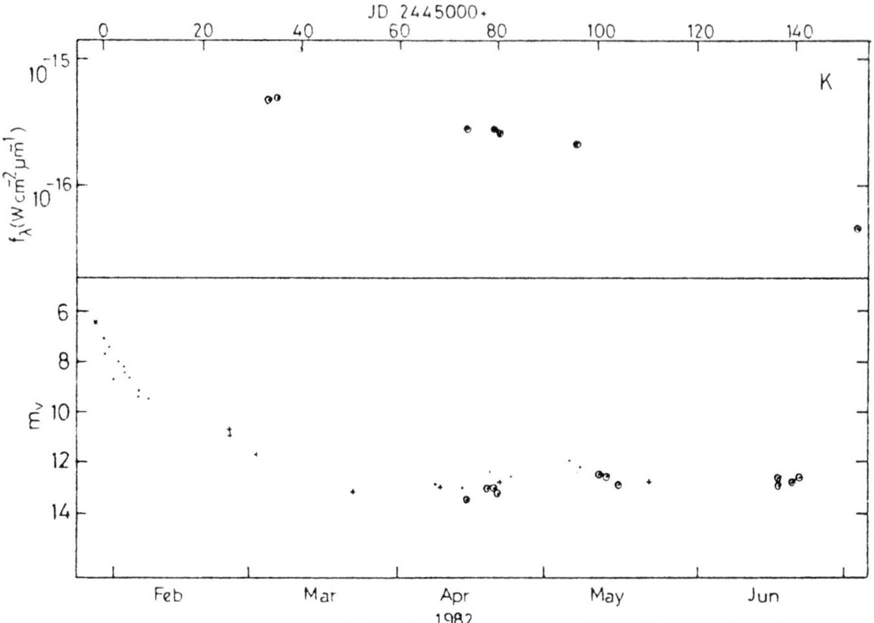

Fig. 1 K(2.2 μm) and visual light curves; visual magnitudes compiled from visual estimates (points), IUE FES (crosses) and photoelectric photometry (circles)

between speed class, bolometric luminosity and principal ejection velocity, we would expect a nova with $\dot{m}_V$ = 0.3 mag d^{-1} to condense grains only after $t_0 \gtrsim 80$ d (Bode and Evans 1982). This discrepancy can be circumvented if the material from which grains formed was moving at ∿ 10^4 km s^{-1}, or the nova is a factor ∿ 25 underluminous for its speed class. Although evidence for high velocity material is provided by IUE observations (Snijders et al 1982; Friedjung 1982) it is unlikely that this gave rise to grains. There is no discrepancy if the grains predate the outburst.

In the near infrared ($\lesssim$ 5 μm), the dust shell spectra can be well described by blackbodies cooling from ∿ 1100 K (day 37) to ∿ 800 K (day 156). At long wavelengths ($\gtrsim$ 10 μm) however, the photometry of Gehrz et al (1982) reveals an excess relative to an extrapolation of the near infrared spectra and in particular there is a feature around ∿ 10 μm. Such a feature is unique amongst dusty novae. On the basis of 8-13 μm spectroscopy, Gehrz et al attribute this feature to SiC at 11.5 μm; Aitken et al (1982), however, interpret the feature as the 10 μm 'silicate' feature. The resolution of this issue has obvious implications for the grain condensation process; also it could tell us something about the C/O ratio before condensation, and hence about the outcome of the thermonuclear runaway that gave rise to the eruption. Also, the spectrum of the nova dust shell is much broader than blackbody, or there may be another feature at ∿ 20 μm.

The 2-4 μm CVF spectrum obtained at SAAO on 1982 April 18th is shown in Fig. 2. Several weak absorption features are apparent in the 3-4 μm band (the Brackett lines α and γ are of no great interest in the present context).

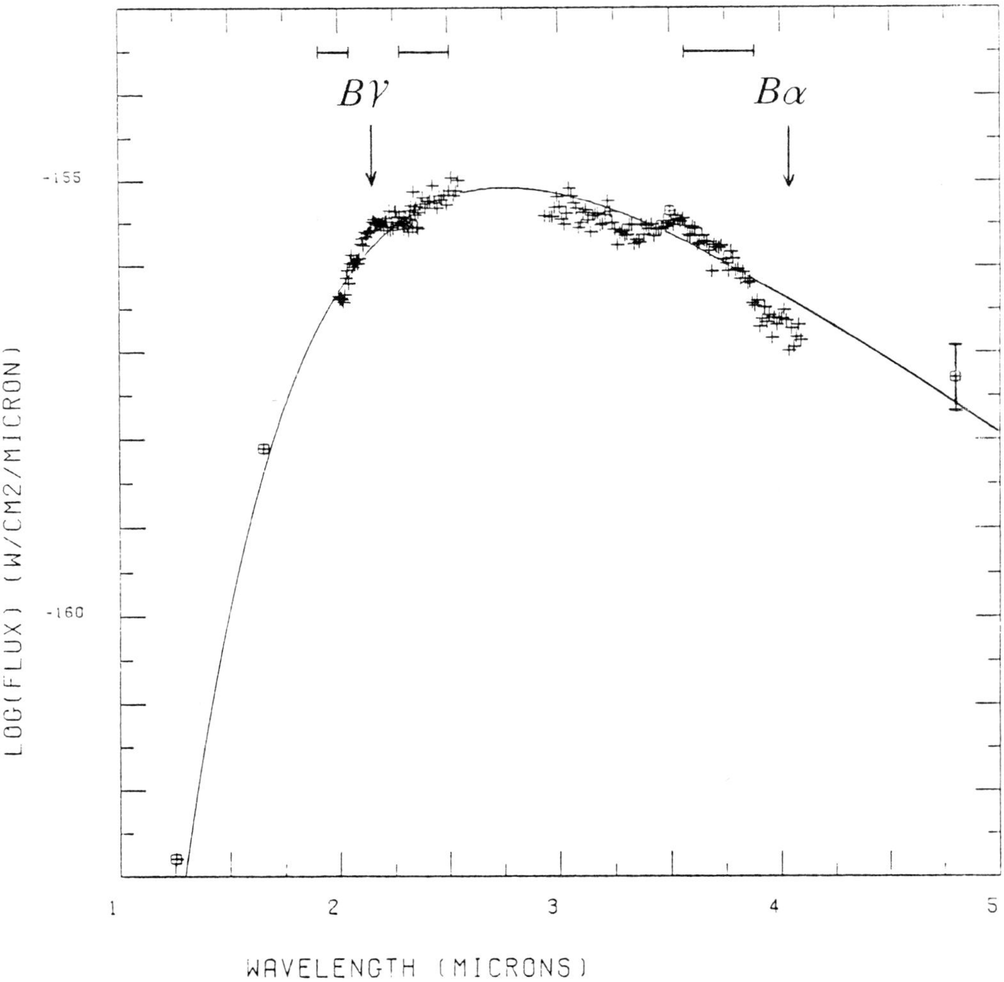

Fig. 2 2-4 μm spectrum of Nova Aquilae. JHKLM photometry included as squares. Curve is a blackbody with temperature 1050 K; horizontal lines are assumed continuum. Location of Brackett α and γ lines are indicated.

A least squares fit of a blackbody to continuum data in the CVF spectrum gives a dust shell at 1050 K, consistent with the 5 μm photometry. This procedure leads to two absorption features, centred on ∿ 3.2 μm and 4.0 μm, a combination which, together with emission at ∿ 10 μm, is a common signature of carbon-rich stars with extensive dust shells (in which case of course the '10 μm' feature is identified with SiC). In such objects the 3.1 μm feature is identified with C_2H_2/HCN, while the 3.9 μm band is identified with CS/C_2H_2. The origin of the spectral features is uncertain, but they do point to the existence of molecules in the circumnova environment. Molecules (CN) have previously been detected - briefly - in the very slow nova DQ Her (Payne-Gaposhkin 1957). Whether these molecules are in the gas phase, or attached to grain surfaces, only more detailed observations can tell. The fact that they exist at all in the nova environment is extremely remarkable.

Acknowledgements

We thank Dr. R. Argyle for providing a finding chart, and Drs. D. Kilkenny (optical) and P.M. Williams and I.S. Glass (infrared) for communicating data in advance of publication. MFB is supported by the SERC.

References

Aitken, D., Roche, P. and Whitmore, B., IAU Circ. 3717, 3724
Bode, M.F. and Evans, A., 1982, MNRAS, 200, p.175
Clayton, D.D. and Wickramasinghe, N.C., 1976, Ast. Space Sci., 42, p.463
Friedjung, M., 1982, this volume
Gehrz, R.D., Hackwell, J.A. and Grasdalen, G.L., 1982, IAU Circ. 3711
Honda, M., 1982, IAU Circ. 3661
Payne-Gaposhkin, C., 1957, "The Galactic Novae", Dover.
Snijders, M.A.J., Seaton, M.J. and Blades, J.C., 1982, in Proceedings of Third European IUE Conference, p. 177
Williams, P.M. and Longmore, A.J., 1982, IAU Circ. 3676

DISCUSSION FOLLOWING A. EVANS' TALK

FRIEDJUNG: Was the first dust formed later than the time when the high velocity systems were formed?

EVANS: The high velocity material was first seen on February the 24 which is day 31, the infrared excess was first picked up on day 37, but if that was at maximum then condensation should have occurred at day 16. Even if we are generous and suppose that maximum infrared luminosity was around about day 60, which is consistent with the patchy infra red light curve, we still need condensation at day 25 which again is before the high velocity material.

SHAVIV: Do you have any explanation as to how come that the grains survive the hot surroundings in the ejecta?

EVANS: The kind of model that Bode and I have been plugging for a couple of years now, involves not grain condensation in the ejecta but grains existing before the outburst.

WILLIAMS: Just an observational fact, grains exist in the hot environment of planetary nebulae.

MITCHELL: You tend to favour the idea of absorption around 3.5μ on the basis that the hotter black body curve doesn't fit the 5μ point well. Supposing that you have a shell with a significant temperature gradient then you are going to get more cool dust, therefore the long wavelength end is going to be enhanced so that it further confuses the question of whether these things are absorptions or emissions.

EVANS: That's right. The only thing we've got to go on is that so far all spectra of nova dust shells have been as near black body as makes no difference. There is another aspect to this, perhaps we shouldn't hang too much on the 5μ point because in previous dust novae there has been a 5μ emission which people have attributed to CIII molecules and these have been perhaps suggested as nucleation centers, so you are right, perhaps we shouldn't attach too much emphasis on the 5μ-photometry.

PIIROLA: I have just one comment which supports the formation of dust particles around novae, in Nova Cygni 1978 the polarization changed quite remarkably 30 days after maximum light and I don't think there is any other explanation than formation of dust particles.

SUGIMOTO: Is is possible to estimate the total amount of dust?

EVANS: Yes. It depends somewhat on the distribution of grains in the shell. Typically, for a very dusty nova like say, NQ Vul it is around about several $10^{-5} M_{\odot}$, it certainly indicates that the carbon in the ejecta has to be condensed into grains pretty efficiently.

SUGIMOTO: Is it possible to estimate the amount of gas ejected?

WILLIAMS: The shell mass for Nova Cygni 1978 was $8 \times 10^{-5} M_{\odot}$ and the carbon to hydrogen was about 10^{-2} by number, so it was less than $10^{-5} M_{\odot}$.

THE CONTINUUM ENERGY DISTRIBUTION OF THE OLD-NOVA GK PER (1901)

A. Bianchini and F. Sabbadin
Asiago Astrophysical Observatory
36012 Asiago, Italy

ABSTRACT The observed X-ray to IR continuum energy distribution of the old-nova GK Per, corrected for the contribution of the late-type secondary and for i.s. extinction,is found to be consistent with the model of an accreting magnetic white dwarf. Furthermore, we discuss plausible modifications of the standard disc structure caused by the presence of magnetic field lines threading the disc and evaluate the contribution to the observed radiation field from several physical mechanisms.

1. INTRODUCTION

Amongst classical old-novae, GK Per(1901) can be considered an exceptional object (Bianchini and Sabbadin 1982, Bianchini et al. 1981, Bianchini et al. 1982, Rosino et al. 1982). In a recent paper Bianchini and Sabbadin (1982), hereafter called BS, showed that the continuum energy distribution of the nova is peaked at λ 3600 and demonstrated that this can be explained by the presenceof a magnetic field sufficiently strong to disrupt the hotter inner parts of the accretion disc thus leading to radial inflow along the field lines on to the magnetic poles of the white dwarf. The observed UV continuum of the nova shortward of λ 2000 was interpreted as the low energy tail of black-body radiation from the surface of the white dwarf heated by thermal conduction inside the accretion column. Here we analyze the continuum energy distribution of GK Per from X-ray to infrared wavelengths and discuss the relative importance of the different physical mechanisms responsible for the observed radiation field. In what follows we will adopt the GK Per binary system parameters given by BS. The UV and optical fluxes given in this paper are twice as much as those derived by BS as they considered only one hemisphere around the star.

2. ACCRETION DISC MODELS

The emergent spectrum from an optically thick accretion disc having inner radius $R_i > R_{WD}$ can be derived by applying a modification of the standard disc model of Shakura and Sunyaev (1973), Bath et al. (1974), Lynden-Bell and Pringle (1974) and Bath (1976) in which the disc does not extend to

M. Livio and G. Shaviv (eds.), Cataclysmic Variables and Related Objects, 127–131.

the star but maintains at any radius the original radial temperature distribution. The best fit to the observed UV to visual flux distribution, corrected for the contribution of the K2 IV secondary (Gallagher and Oinas 1974), is obtained at $\dot{M}=1\times10^{18}$g s^{-1} and $R_i=1.3\times10^{10}$cm (hereafter called solution A); the resulting magnetic field at the surface of the star is $B=2.5\times10^{8}$G. A second solution (solution B) is obtained following Ghosh and Lamb (1979) who demonstrated that, due to threading of magnetic field lines across the disc, the temperature decreases in the inner parts of the disc whilst it increases in the outer regions compared to the standard disc model. For a white dwarf of moderate "rotation fastness" ($\omega_s=0.3$) the best fit to the corrected UV to visual continuum is obtained for $\dot{M}=2\times10^{16}$g s^{-1} and $R_i=1.2\times10^{9}$cm; this yields $B=5\times10^{5}$G.

3. DISCUSSION

To discriminate between solutions A and B it is necessary to analyze the continuum energy distribution of the old-nova from X-ray to infrared wavelengths. The spectrum produced by an accreting magnetic degenerate dwarf generally has six components (King and Lasota 1979, Kylafis and Lamb 1979):

1) the spectrum produced by the accretion disc
2) a black-body limited UV, visual or infrared cyclotron component produced by the hot, post-shock emission region
3) a hard X-ray bremsstrahlung component also produced by the hot emission region
4) a hard UV or soft X-ray black-body component produced by cyclotron and bremsstrahlung photons which are absorbed by the stellar surface and re-emitted
5) a hard UV or soft X-ray black-body component emitted by the white dwarf polar caps which are heated by thermal conduction of the electrons inside the accretion columns
6) a hard UV or soft X-ray black-body component emitted by polar caps heated by collision of the infalling matter

In general we have:

$$L_{acc} = L_D + L_{ff} + L_{Cyc} + L_{BB} \qquad (1)$$

L_{acc} being the total luminosity of the system due to accretion; L_D the luminosity of the disc; L_{ff} the luminosity of the bremsstrahlung emission; L_{Cyc} the luminosity of the cyclotron component ; L_{BB} the sum of components 4,5 and 6. The observed UV, optical and infrared (work in preparation) fluxes indicate that L_D is of the order of 1×10^{34}erg s^{-1}. The X-ray luminosity, L_{ff}, obtained by folding a 100 KeV (for a 1.3 $M_\odot$ white dwarf) bremsstrahlung spectrum having a column density $NH=7\times10^{20}$cm^{-2} through the Einstein instrument response (Cordova et al. 1981), is 6.6×10^{33}erg s^{-1}. The cyclotron luminosity will strongly depend on the values of B and $\dot{M}$ (King and Lasota 1979). The black-body emission from polar caps must satisfy the following conditions:

a) Flux at λ 1300 $\cong 6\times10^{28}$erg s^{-1} A^{-1}
b) Flux at λ 10 $\leq 5.6\times10^{30}$erg s^{-1} A^{-1}

Condition a) is derived from the difference between the observed and the calculated (disc models) fluxes at λ 1300; condition b) means that the black-body flux at λ 10 (where extinction is negligible) must not exceed that expected from a 100 KeV bremsstrahlung thermal spectrum having $L_{ff}=6.6\times10^{33}$ erg s^{-1}. Fig. 1 compares the observed continuum of GK Per with that predicted by solutions A and B.

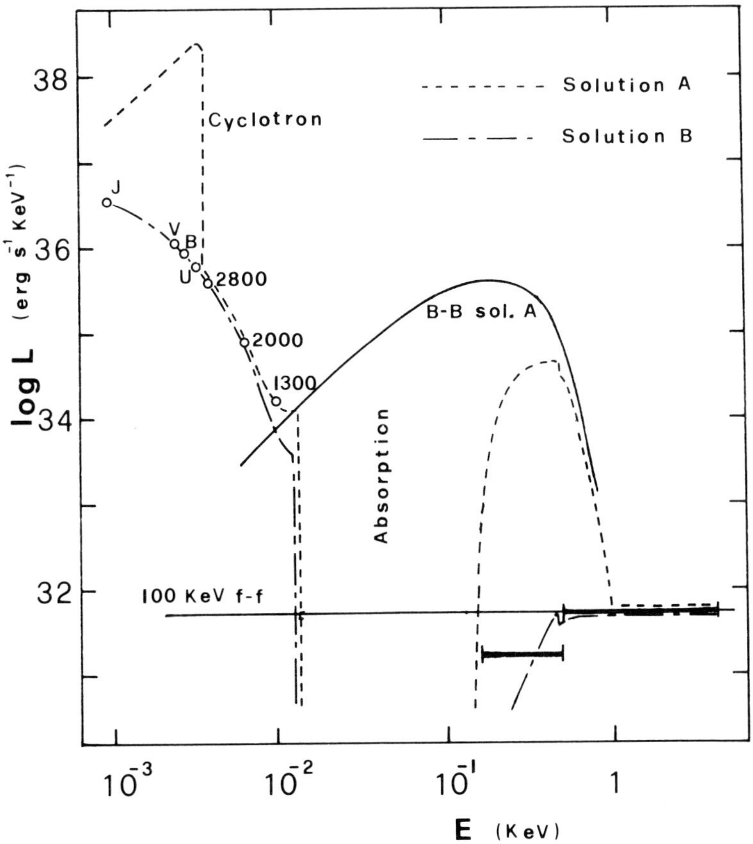

Figure 1. Comparison of the observed X-ray to infrared continuum energy distribution of GK Per with that predicted by solutions A and B. Open circles indicate fluxes at UBVJ and selected UV wavelengths; bars refer to the mean fluxes observed in the two spectral ranges of the Einstein instrument. Both maximum BB and minimum Cyc components are indicated. The extinction in the far UV and X-ray region is for a column density $NH=10^{20} cm^{-2}$. UV and UBVJ data are corrected as indicated by BS.

In solution A $L_{acc}=5.8\times10^{35}$ erg s^{-1} and conditions a) and b) are satisfied only for $L_{BB}<0.4L_{acc}$. From equation (1) we obtain $L_{Cyc}>0.6L_{acc}$, but this is impossible since the Rayleigh-Jeans tail of the black-body component produced by self-absorbed cyclotron emission would be in this case (i.e. with $B=10^6$ G) limited at λ 3500, which corresponds to three times the fundamental cyclotron frequency (King and Lasota 1979). This would then produce much larger fluxes than observed in the optical re-

gion. In the case of solution B we have $L_{acc}=1.2\times10^{34}$ erg s^{-1} and, for the given $\dot{M}$ and B values, $L_{Cyc} << L_{ff}$. From equation (1) we obtain $L_{BB}\cong 0$. Thus, solution B is not completely satisfactory since a hot black-body component is very likely present in our spectra.
Our results suggest that disc models involving mass transfer rates and magnetic fields which are intermediate between those of solutions A and B should be discussed. While steady, optically thick disc models adequately represent the spectrum of a number of cataclysmic variables, there are many cases in which the observed UV continua are better interpreted as the superposition of two black-body sources. In conclusion, the observed deviations from a "steady-disc distribution", which often take the form of optical and/or IR excess, might be due to the contribution from the bright spot as well as to the presence of magnetic fields sufficiently strong to modify the geometry and the radial temperature distribution of the "standard disc".

References

Bath,G.T.,Evans,W.D.,Papaloizou,J.,Pringle,J.E.:1974,Monthly Notices Roy. Astron. Soc. 169,447

Bath,G.T.:1976, in Novae and Related Stars, Ed. Friedjung,M.,D. Reidel, Dordrecht,Holland

Bianchini,A.,Hamzaoglu,E.,Sabbadin,F.:1981, Astron. Astrophys. 99,392

Bianchini,A.,Sabbadin,F.,Hamzaoglu,E.:1982, Astron. Astrophys. 106,176

Bianchini,A. and Sabbadin,F.:1982, Proceedings of the Third European IUE Conference, Madrid 1982, ESA SP-176, 187

Cordova,F.A.,Mason,K.O.,Nelson,J.E.:1981,Astrophys. J. 245, 609

Gallagher,J.S.,Oinas,V.:1974,Publ. Astron. Soc. Pacific 86, 952

Ghosh,P.,Lamb,F.K.:1979,Astrphys. J. 232, 259

King,A.R.,Lasota,J.P.:1979,Monthly Notices Roy. Astron. Soc. 188, 653

Kylafis,N.D.,Lamb,D.Q.:1979,Astrophys. J. Letters 228, L 105

Lynden-Bell,D.,Pringle,J.E.:1974, Monthly Notices Roy. Astron. Soc. 168, 603

Rosino,L.,Bianchini,A.,Rafanelli,P.:1982,Astron. Astrophys. 108,243

Shakura,N.I.,Sunyaev,R.A.:Astron. Astrophys. 24, 337

Wargau,W.,Drechsel,H.,Rahe,J.:1982, Proceedings of the Third European IUE Conference, Madrid 1982, ESA SP-176, 215

DISCUSSION FOLLOWING A. BIANCHINI'S TALK

MEYER: I wonder whether somebody could perhaps comment on the following point in Ghosh and Lamb's model. If you pick out angular momentum, of course you do work against the torque by the differential motion in the disk and since this is differential work I guess it probably has to be dissipated and it might well be that a part of that dissipation would come back into the disk. That could be a source of heating the disk, so it might matter in the model whether you include such terms or not.

BIANCHINI: No, what we really did is just an attempt, more qualitative than quantitative. Just considering Ghosh and Lamb's work you can find the temperature distribution is changing in that way.

LAMB: I would also like to try to respond to that, noting, of course, that this is Fred Lamb's work and not my own, so I am not an expert on it. Your point is well taken, but is already reflected in what has been done here. Namely, the dissipation has been moved from the innermost part of the disk to further out by the magnetic field lines that thread the disk. For that reason, the temperature is lower than usual in the innermost part and then somewhat higher farther out. But the dissipation is still there.

SHAVIV: If I understood you correctly the main effect is that you have an R_i, you cut off the disk from touching the star. So if someone would have come to you with a different type of boundary layer between the disk and the star, your result would not really reject it immediately. Then, the model leading to an on pole accretion by the star, rather than along the equator is not really absolutely compulsory out of your results, is that correct?

BIANCHINI: Yes. I can say one more thing, the solution B which seems the best one, needs a fractional area of 0.02 which is quite large it is larger than the 10^{-3} which you get from the theory for a magnetically driven accretion column.

LAMB: What you are just discussing concerns more the black body tail, and what the spectrum is like in the hard X-rays and the soft X-rays. However, if we consider just the optical and near UV energy distribution, there may still be a significant difference. If the disk is disrupted fairly far out from the white dwarf, then it will strongly affect the energy distribution in the blue and near UV. The effect is more than just changing the boundary layer, which presumably mostly affects much shorter wavelengths. Instead of comparing models A and B, what does the flux distribution look like if you take just a normal disk flux distribution?

BIANCHINI: If you take a normal disk you don't get the peak of the energy distriubtion, which is at 3600 Å, this is the problem. I want to mention one more point, IUE spectra of the star during the outburst in 1981 have been taken, I saw the spectra and actually the peak is moving to the UV, to near 1000 Å and I don't know if these models will fit such data, qualitatively speaking everything is going in the right direction, you just increase the accretion rate.

THE 'ISOTHERMAL' PHASE OF NOVA DUST SHELLS

R.M. Mitchell[1], A. Evans[2] and M.F. Bode[2]

(1) Dept. of Applied Mathematics, Queen Mary College, London, U.K.
(2) Dept. of Physics, University of Keele, Staffordshire, U.K.

Abstract

We describe a model for the evolution of the infrared spectrum of the dust shell of nova NQ Vul. The effects of nucleation and grain growth, together with extended, but diminishing, mass loss from the nova, are included. The variations in the effective temperature of the dust shell that occur near infrared maximum may be understood in terms of varying optical depth in a dust shell having significant temperature gradient. However, a more consistent picture is shown to combine interrelated optical depth and grain size variations.

The development of an infrared excess in classical novae some weeks after visual maximum is well known. The excess appears at the time of nova transition and for novae having dust shells that are optically thick in the infrared, its appearance also coincides with the break in the visual light curve. This behaviour led Clayton & Wickramasinghe (1976; CW) to consider the nucleation and growth of graphite grains in nova ejecta.

For NQ Vul and LW Ser, the dust shell temperatures deduced from the photometry of Ney & Hatfield (1978;NH) and Gehrz et al (1980), reveal an initial decrease prior to infrared maximum, followed by a rise, a 'levelling out' and finally a decline after $\sim$ 200 d. Gehrz et al (1980) interpreted this behaviour in terms of an initial phase of grain growth, followed by grain destruction and finally a period of cooling as the dust shell expands. However this scenario merits closer inspection because (a) in a dust shell that is optically thick to the illuminating radiation, temperature gradients arise even if the shell is geometrically thin and (b) grain destruction may be expected to occur sooner rather than later as the nova environment presumably becomes less hostile with time.

Here we present a preliminary report of a detailed treatment of the dust shell of NQ Vul, accurately solving the radiative transfer

M. Livio and G. Shaviv (eds.), Cataclysmic Variables and Related Objects, 133–138.

problem for a series of quasi-static spherically symmetric models. These are based on the quasi-diffusion method of Leung (1975, 1976), as described by Mitchell & Robinson (1978). Our work was prompted by the hope that changes in the infrared opacity of the expanding dust shell would suffice to account for the temperature behaviour of the dust shell, for the following reason. In an optically thick shell, the dust shell temperature will essentially be that at the surface of unit optical depth; at first this surface will be located well away from the nova but as the shell expands and thins out, the surface will contract and the dust shell temperature will rise.

The expansion velocity v of the principal spectrum (from which the grains would be expected to condense) was $\sim$ 1000 km s^{-1} for NQ Vul (Klare & Wolf 1978), while dust condensation began $\sim$ 60 d after visual maximum (NH); the condensation distance $R_{min} \approx 5.2 \times 10^{14}$ cm. We assume (Shara 1981), that the mass loss rate declines exponentially with decay constant k. Hence the gas density ρ in the circumnova environment varies as

$$\rho(r,t) \propto r^{-2} \exp\{-k\,(t-r/v)\}\ . \tag{1}$$

We assume that grain nucleation and growth occur instantaneously at R_{min} (cf. CW). The ρ^2 dependence of nucleation rate at R_{min} leads to a grain number density distribution

$$n(r,t) \propto r^{-2} \exp\{-2k\,(t-r/v)\} \tag{2}$$

while the grain size distribution is

$$a(r,t) \propto \exp\{-k(t-r/v)\}\ . \tag{3}$$

We assume graphite grains with the following dependence of absorbing efficiency Q_{abs} on wavelength λ:

$$\begin{aligned} Q_{abs} &= \text{const.} && (\lambda \lesssim 2\pi a) \\ &\propto \lambda^{-1.8} && (\lambda > 2\pi a) \end{aligned} \tag{4}$$

The bolometric luminosity of the nova is assumed constant throughout the period of interest (Bath & Shaviv 1976), with the value $L = 9 \times 10^{37}$ erg s^{-1}. At day 60, the nova temperature is expected to be $\sim 4 \times 10^4$ K and to increase thereafter. For such temperature behaviour, the radiative power of the nova is emitted mainly at wavelengths at which the dust opacity is flat; thus the temperature behaviour is not critical and we assume a constant value of 5×10^4K.

We use the visual light curve of NH to help deduce the time-dependence of the optical depth of the dust distribution. This procedure demands prior knowledge of the time dependence of the unreddened visual flux. We have considered several forms for the latter,

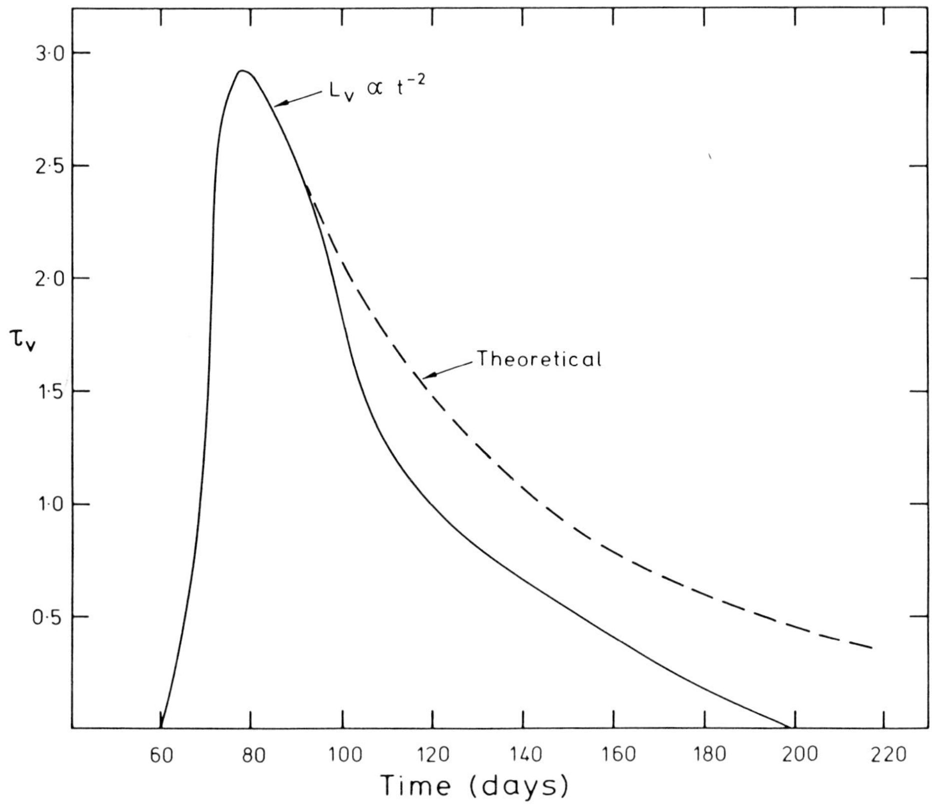

Fig. 1 Time variation of visual optical depth, based on (i) observed light curve and assumed t^{-2} dependence of unreddened visual flux (solid curve) and (ii) theoretical variation of grain number density and radius in dust shell (see text)

but the most satisfactory behaviour comes from a dependence of the form $L_V \propto t^{-2}$, characteristic of the free-free emission from an optically thin, expanding gas shell of uniform thickness. Indeed, this behaviour has been observed in a number of other novae. Fig. 1 shows the time dependence of optical depth derived from this assumption, together with the theoretical variation which follows from the grain size and number density distribution given in Eqs. (2) and (3) above.

From the work of NH, six sets of photometry were selected to cover all phases of dust-shell temperature variation. Grain radius at the outer edge of the dust shell, a_{max}, was adjusted to provide the best fit to the data at infrared maximum (80.8 d after visual maximum). Model computations performed assuming the theoretical $\tau_V(t)$ dependence shown in Fig. 1 together with a_{max} = 0.6 μm, succeed in reproducing the initial temperature fall and subsequent rise after infrared maximum.

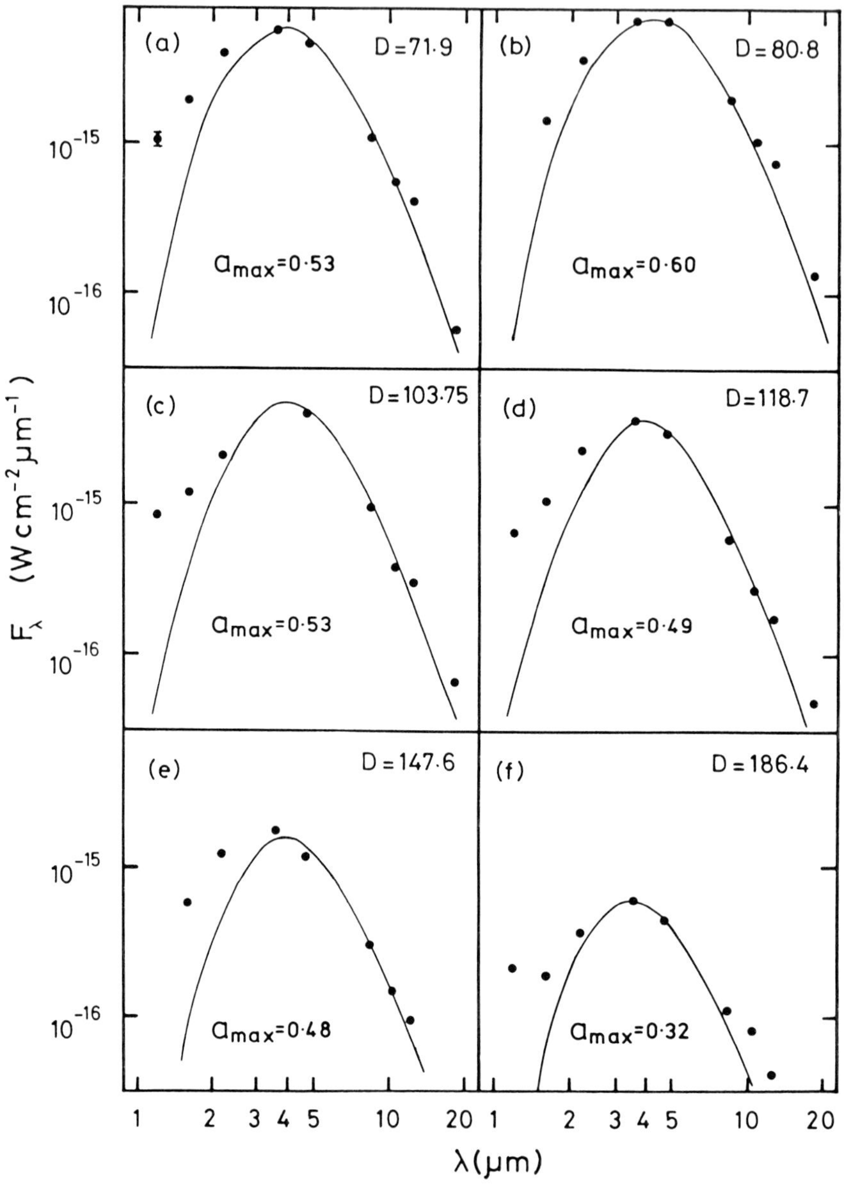

Fig. 2 Model fits to the evolving infrared energy distribution for NQ Vul; time in days (D) and a_{max} (μm) values as indicated

However, the model dust shell was found to cool more rapidly than the observational data suggest is the case for $t \gtrsim 120$ d.

The required temperature behaviour at later times may be achieved if reductions in grain size occur after $t \gtrsim 120$ d. Gehrz et al (1980) invoked grain destruction to explain the dust temperature rise in LW Ser; here the effect is required to arrest the rate of cooling of the shell at later times. A reduction in grain size from $a_{max} = 0.6$ μm to $a_{max} = 0.4$ μm would produce the effect required. The impact of subsequently ejected, higher velocity material (e.g. 'Orion' ejecta) on grains condensed from the principal ejecta would result in grain sputtering, but probably not of the magnitude required. The interaction of outflowing grains with gas in the nova environment is currently being investigated.

If we interpret the data in terms of a change in grain size, an indication of the reduction required may be derived from the optical depth: if grain radii are reduced throughout the shell by the factor α the optical depth is reduced by α^2; α may be derived from the ratio of empirical ($L_V \propto t^{-2}$) to theoretical optical depth (cf. Fig. 1). A sequence of models based on changing grain size and the empirically derived optical depth variation of Fig. 1 is shown in Fig. 2; the good agreement between theory and observation is evident (the excess flux at short wavelengths is due to contamination by free-free emission). This suggests that the following phases occurred in the ejecta of NQ Vul : (a) a short ($\sim$ 20 d) period of grain growth, leading to (b) an optically thick shell of grains, ranging in grain size up to 0.6 μm, followed by (c) an extended period of grain destruction, resulting in $\sim$ 50% reduction in grain size. In this scenario, the temperature rise following infrared maximum is due to the combined result of (a) the higher temperatures attained by the smaller grains and (b) the shrinking of the infrared pseudophotosphere, primarily caused by the interrelation of optical depth and grain size discussed above.

It seems that the establishment of an infrared 'pseudophotosphere' as discussed here, can indeed provide a satisfactory explanation of the optically thick dust shells of novae. On the other hand, a complete explanation seems to require some interaction of the outflowing grains with circumnova gas. This work emphasises again (Bode & Evans 1981) the potential usefulness of a detailed treatment of the infrared development of novae in understanding the eruption as a whole.

RMM and MFB are supported by the SERC.

References

Bath, G.T. and Shaviv, G., 1976, Mon.Not.Roy.Astron.Soc., 175, p.305

Bode, M.F. and Evans, A., 1981, Mon.Not.Roy.Astron.Soc., 197, p.1055

Clayton, D.D. & Wickramasinghe, N.C., 1976, Astrophys. Space Sci., 42, p.463
Gehrz, R.D., Grasdalen, G.L. & Ney, E.P., 1980, Astrophys.J., 237, p.855
Klare, G. & Wolf, B., 1978, Astron.Astrophys.Supp., 33, p.327
Leung, C.M., 1975, Astrophys.J., 199, p.340
Leung, C.M., 1976, J.Quant.Spect.Rad.Transfer, 16, p.559
Mitchell, R.M. and Robinson, G., 1978, Astrophys.J., 220, p.841
Ney, E.P. & Hatfield, B.F., 1978, Astrophys.J., 219, p.L111
Shara, M.M., 1981, Astrophys.J., 243, p.126

DISCUSSION FOLLOWING R. MITCHELL'S TALK

KING: You had a persistent excess of flux at short wavelength which you said was due to free-free, I think you have a possible way of checking that, because at least if the Balmer lines are optically thin you can relate any Balmer emission to the free-free you would expect from the system.

MITCHELL: Yes.

INFRARED SPECTRA OF NOVA DUST SHELLS

M.F. Bode and A. Evans
Department of Physics, University of Keele, Staffordshire, U.K.

Abstract

Results of fitting blackbody and pseudoblackbody functions to infrared photometry for three classical novae are briefly summarised.

Several classical novae have developed an infrared excess due to emission by dust some weeks after the visual outburst. Two models have been put forward to account for this behaviour, namely (i) the nucleation and growth of grains in the nova ejecta (Clayton and Wickramasinghe 1976) and (ii) the existence of a dust shell that predates the nova eruption (Bode and Evans 1980). In either case, the expected flux spectrum of the dust shell is expected to be of the form

$$f_{\nu} = \frac{A\, x^{\varepsilon+3}}{\exp(\eta x) - 1} \qquad (1)$$

where x is a frequency parameter, ε is determined by grain emissivity and/or grain temperature gradients in the dust shell and η is characteristic of the dust shell temperature.

We have fitted Equation (1) to infrared photometry for the three novae NQ Vul, LW Ser and V 1668 Cyg, with the following conclusions:

(a) the data are consistent with blackbody fits for all times for all three novae;

(b) the dust shell temperatures of LW Ser and NQ Vul rose after an initial decline, subsequently levelled out and finally fell (cf. Fig. 1 for NQ Vul);

(c) neither pre-existing grain, nor grain growth models, in their original forms, adequately account for the time-dependence of A and η (Eq. (1))

The grain growth model is rather more versatile than the pre-existing grain model in that several options are available for

M. Livio and G. Shaviv (eds.), Cataclysmic Variables and Related Objects, 139–142.

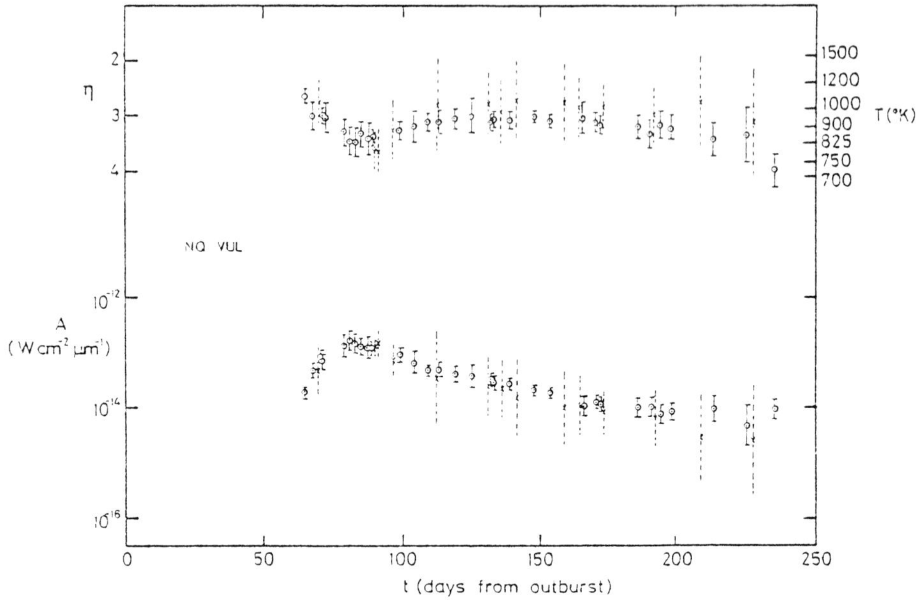

Figure 1 The variation of A and η for NQ Vul, based on a fit of Equation (1) to the photometry of Ney & Hatfield (1978) (circles) and Sato et al (1978) (crosses). Error bars are 3σ

developing the original treatment. For example,

(a) the ejecta may not be spherically symmetric and ejecta geometry may affect rate at which grains nucleate and grow;

(b) ejection in classical novae occurs on an extended timescale, ∿ days to weeks depending on speed class, whereas on grain growth models essentially instantaneous ejection is assumed;

(c) grain growth may be accompanied by concurrent destruction processes;

(d) for dust shells that are optically thick in the infrared, as is the case for NQ Vul and LW Ser, we get the establishment of what is essentially an infrared "pseudophotosphere". Details of this interpretation are given in this volume by Mitchell et al (1982).

The full text of this paper is being published in the Monthly Notices of the Royal Astronomical Society.

References

Bode, M.F. & Evans, A., 1980, Astron.Astrophys., 89, p.158

Clayton, D.D. & Wickramasinghe, N.C., 1976, Astrophys. Space Sci., 42, p.463

Mitchell, R.M., Evans, A. and Bode, M.F., 1982, this volume.

Sato, S., Kawara, K., Kabayashi, Y., Maihara, T., Oda, N. and Okuda, H., 1978, Pub.Astron.Soc., Japan, 30, p.419.

DISCUSSION FOLLOWING A. EVANS' TALK

FRIEDJUNG: I think it is quite reasonable to suppose that dust is formed in the principal shell which would contain most of the mass and will be relatively cool, the continued ejection probably would not have very much effect on the dust formation. So I don't think you need to take account of the continued ejection process.

EVANS: Yes, but the instantaneous ejection is a little oversimplified.

FRIEDJUNG: Of course, but most of the mass will be in the fairly dense part which comes off near maximum and the dust will be mainly formed in it.

MITCHELL: What we are saying is that the early phase, the principal spectrum is in itself extended.

WILLIAMS: As I understand it, one of the conventional dogmas is, that you have instantaneous grain formation after the outburst, because there is usually a simultaneity between the decrease in the visual light curve and the inception of infra red radiation and I am just wondering according to your picture how would you account for the simultaneity of the two? Is it just a coincidence?

EVANS: I guess that after the recovery, to the secondary maximum if you like, the nature of the visual flux is totally different from the nature of the visual flux before the transition minimum and I guess you can ascribe the break in the light curve just to a shut-off in the mass loss rate and a comparatively sudden decrease of the pseudo-photosphere.

WILLIAMS: Are there any novae which undergo this decrease in the visual for which the infra red radiation does not occur at the same time? You don't have in your picture a natural explanation for the decrease in the visual light, associated with properties of the dust.

EVANS: They are connected in that the dust only feels the outburst when the temperature of this pseudo-photosphere goes up.

BATH: As I understand your work, there is a correlation between the optical light curve and the infra red development of the infra red light curve because one feeds the other, the optical light curve plus the UV shift is being transformed into the infra red by the absorption properties of the grains.

WILLIAMS: The absorption properties are not related to the temperature though, whereas the emission properties are, that is my point, you can have the dust there at all times and it is absorbing, but it does not emit. It seems to me that the *relative fraction of the visible radiation* that the grains are absorbing is independent of temperature, it is the same at all times. I am saying the following: suppose you have pre-existent grains which have the same size and all their properties are constant, except the temperature, (which is irrelevant to their absorptive properties) and you have this expanding shell, it seems to me that the fraction of light that they absorb at one wavelength is constant.

MITCHELL: There is a coincidence in this model, because it is the break in the light curve, which in this model is due to the underlying source, has got to be coincident with the energy distribution peaking in the region of 800 Å.

<u>WILLIAMS</u>: So according to your model that <u>is</u> a coincidence?
<u>MITCHELL</u>: Yes.

ON EVOLUTIONARY MODELS FOR SNI PROGENITOR STARS

Jeffrey D. Colvin
EG&G/Los Alamos Operations

ABSTRACT

It has been proposed that the Type I supernova (SNI) explosion starts in a white dwarf (WD) star that is accreting mass from a companion star in a close binary orbit. Others' computations have described this explosion, but how an accreting WD can ever become an SNI if it first ejects the accreted envelope in one or more nova outbursts is still unresolved. My calculations of WD evolution include the effects of mass accretion, convective mixing, nuclear burning, and gravitational settling. These calculations test the Starrfield, Truran, and Sparks (1981) proposal that at low accretion rates, settling of the CNO nuclei will lead to steady-state hydrogen burning, not nova outburst. The preliminary results of this study, consisting of two evolutionary sequences of an initially cool pure ^{12}C 1.0 $M_\odot$ WD show that nearly half of the carbon near the base of the accreted layer, initially of solar composition, gravitationally diffuses out in 2 x 10^5 years.

INTRODUCTION

Many researchers have proposed that the Type I supernova (SNI) explosion originates in a white dwarf (WD) star that is undergoing mass accretion from a companion star in a close binary orbit. Others' hydrodynamic computations describe the final result -- deflagration, off-center or central detonation leading to total disruption, growth toward the Chandrasekhar limit and gravitational collapse -- that depends primarily on the mass and composition (usually assumed uniform in shells) of the progenitor WD, the mass accretion rate, and the composition of the accreting matter. Nomoto (1980) has summarized the possible explosion scenarios (see his Table 1). All these computations assume that the SNI progenitor is an accreting WD, and thus address neither the question of how good the evidence is that SNIs come from an old population, nor whether the uniformity of observed characteristics of SNIs demand a single progenitor type.

M. Livio and G. Shaviv (eds.), Cataclysmic Variables and Related Objects, 143–148.

These computations also assume the structure of the pre-explosion WD. We still must resolve the following questions. 1) How does a given WD evolve to the pre-explosion configuration? 2) What is the most likely pre-explosion configuration, hence explosion mechanism? 3) How can an accreting WD ever evolve to a SNI if it first ejects the accreted envelope in one or more nova outbursts? Starrfield, Truran, and Sparks (1981) have proposed that because at low mass-accretion rates, gravitational settling of the CNO nuclei out of the accreted H envelope can occur on the accretion time scale or faster (Fontaine and Michaud 1979, Muchmore 1981), only steady-state hydrogen burning will occur, not nova outburst.

I have performed WD evolutionary calculations in an attempt to test the Starrfield *et al*. suggestion and to clarify these questions. The results of these calculations do not tell us anything fundamentally new about the SNI explosion mechanism, but do provide some much-needed detail about the composition morphology of the progenitor star, and apply some recent results on gravitational diffusion in WD stars to test directly the Starrfield *et al*. suggestion.

CALCULATIONS AND RESULTS

I used a modified version of Lamb's (1974) white dwarf evolution code based on a variant of the Henyey method described by Kippenhahn, Weigert, and Hofmeister (1967). The structure equations are integrated outward from the center, with mass fraction as the independent variable. The boundary conditions at logarithmic mass fraction -10^{-6} (approximately $10^{-6}\,M_\odot$ into the star) are computed by the method of triangles (Kippenhahn, Weigert, and Hofmeister 1967), starting with a grey atmosphere calculation, for which a radiative model atmosphere is integrated inward until it reaches a mass fraction chosen so that it coincides with an optical depth of about 1 to 10, or until it becomes convectively unstable according to the Schwarzchild criterion. The pressure and temperature at this depth constitute the surface boundary conditions for the envelope integrations used in the method of triangles.

I modified the code extensively to calculate the changing composition profile throughout the WD that results from mass accretion, convective mixing, nuclear burning, and gravitational settling. The initial model is a pure ^{12}C WD of mass 1.00 $M_\odot$, a luminosity of $L = 1.8 \times 10^{-4} L_\odot$, and an effective temperature of 7620K. The equation of state used for the fully ionized pure carbon shells in the interior of the star is taken from Lamb's (1974) work. It includes the effects of crystallization and the phase transition, electron exchanges, and Coulomb interactions in the liquid and solid phases, in addition to the usual expressions for semirelativistic, partially degenerate electrons. For the mixed composition shells in the partially ionized outer layers of the star (on either side of the boundary at logarithmic mass fraction -10^{-6}), I modified the code to compute thermodynamic properties for the mixed compositions by linearly interpolating in log T and log P in the equation of state tables of Fontaine, Graboske, and Van Horn (1977) and combining quantities as described by Colvin *et al.* (1977). In this study, the Iben I (X_H = 0.999), Iben V (X_{He} = 0.999), and Weigert V (X_C = 0.999) chemical compositions are treated as if they were pure H, He, and C, respectively. The table for the nearly pure carbon mixture is used to represent all elements collectively except H and He.

Mass accretion is accounted for by computing the amount of added mass given the time step and mass accretion rate, adding a new surface mass shell with this amount of mass, and rescaling all the other mass fractions accordingly. The composition of the accreted matter is specified independently.

Each mass shell is checked for convective instability by the Schwarzchild criterion, and, if convective, the temperature gradient computed from a standard mixing length theory, with the mixing length chosen as one half the pressure scale height. Composition in the convectively mixed regions of the star is computed as the mass fraction average of the individual compositions of the mass shells in the convection zone, including the accreted mass for the surface convection zone. This is a proper procedure as long as the convective time scale is short compared to the evolutionary time step, as it is for our models.

Energy generation is computed for all mass shells by the nuclear reaction network described by Starrfield et al. (1972). It includes energy losses resulting from neutrino emissivity (Beaudet, Petrosian, and Salpeter 1967). The energy generation and loss terms are included in the energy conservation equation, and composition changes resulting from nuclear reactions are also computed. Composition changes resulting from gravitational diffusion are computed using the scheme given by Fontaine and Michaud (1979). Diffusion velocities computed more recently by Muchmore (1981), by numerically integrating collision cross-section integrals including Debye potentials are in reasonable agreement with those of Fontaine and Michaud.

This is the first WD evolutionary calculation that includes all of the physical processes thought to be responsible for the composition morphology, while making use of the best available equation of state for WD matter.

The preliminary results of this study consist of two evolutionary sequences computed from the same starting model, one with gravitational diffusion and one without. Matter of solar composition (X=0.700, Y = 0.285) was assumed to accrete with spherical symmetry onto the WD at a rate of 10^{-11} $M_\odot$ yr^{-1}.

Note in Fig. 1 that the approximately isothermal core interior to the boundary at logarithmic mass fraction -10^{-6} remains unaffected by 2 x 10^5 years of accretion. In the outer layers of the star the effect of the accretion is to increase the pressure near the surface and to increase the temperature near the base of the accreted layer. The structure shown in Fig. 1 is for the sequence without gravitational diffusion. With diffusion the star is 10 K warmer and 10^{-6} $L_\odot$ brighter at the surface after 2 x 10^5 years because of the different opacities of the progressively hydrogen-rich material. This is a small effect on the star structure, but it has an interesting effect on the gravitational diffusion itself, as is seen in the next two figures.

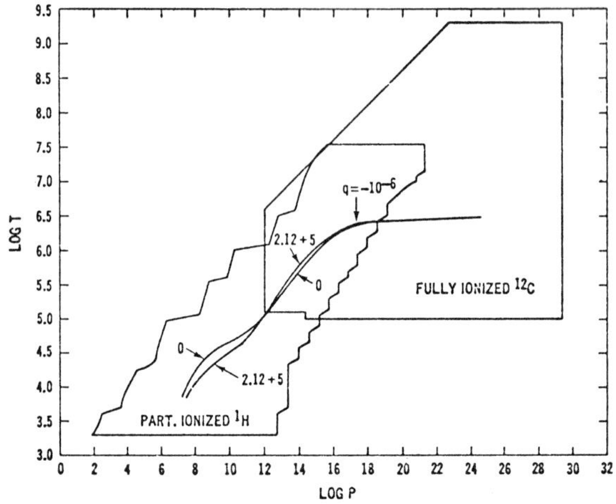

Figure 1. Star structure and EOS grids in log T - log P plane. $q = \log(M_r/M_\star)$ is logarithmic mass fraction, with M_r the mass internal to radius r in a star of mass $M_\star$.

In the first of these, Fig. 2, we see that in the initial stages of accretion, the effect of the added hydrogen-rich material is to suppress in part the convection. Thus, the convection zone retreats toward the surface as the accretion boundary moves inward, leaving behind regions of mixed composition, as first pointed out by Colvin et al. (1977). As the accretion boundary moves farther into the star we are left (for the case of no diffusion) with a growing region below the convection zone of matter containing 1.5 percent (by mass) carbon. This region is shown in much greater detail for the sequence including gravitational diffusion in Fig. 3. Here we see that in only 2 x 10^5 years nearly half the original carbon near the base of the accreted layer has diffused out. Thus, as diffusion proceeds, the opacity changes with the changing composition below the convection zone, the temperature goes up in the region, and because the diffusion coefficient increases with the 5/2 power of the temperature, the diffusion is hastened. What this means is that, at these low accretion rates, the CNO nuclei will settle out of the base of the accreted layer long before the approximately 10^{-4} $M_\odot$ of hydrogen-rich matter necessary for a nova outburst can accumulate, so the star can be expected to evolve to a SNI without going through a nova stage. This supposition, of course, needs to be tested with detailed calculations going to much later times.

As the accreted matter is pushed to higher densities in a WD as cool as the one considered here, however, the screening approximation on which the calculation of the diffusion velocities depends begins to break down, so the results (even for this case) may not be very accurate. It would be better to start with a hotter WD model, or, better still, to do a numerical plasma simulation to determine the transport properties of the cooler WD matter.

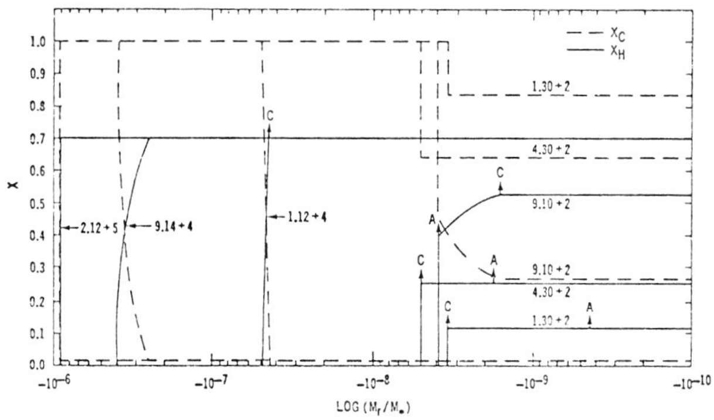

Figure 2. Hydrogen and carbon composition profiles vs logarithmic mass fraction at several times (no gravitational diffusion). "C" marks the position of the base of the convection zone, and "A" marks the position of the accretion boundary.

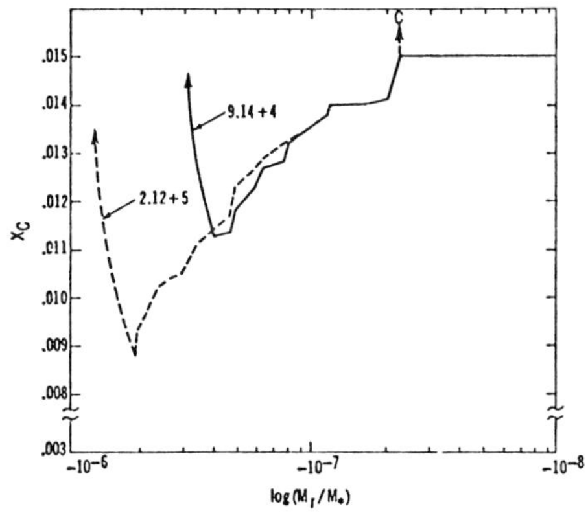

Figure 3. Details of carbon composition vs logarithmic mass fraction at several times, including gravitational diffusion.

Beaudet, G., Petrosian, V., and Salpeter, E.E. 1967, Ap. J., 150, 979.

Colvin, J.D., Van Horn, H.M., Starrfield, S.G., and Truran, J.W. 1977, Ap. J., 212, 791.

Fontaine, G., and Michaud, G. 1979, Ap. J., 231, 826.

Fontaine, G., Graboske, H.C., Jr., and Van Horn, H.M. 1977, Ap. J. Suppl., 35, 293.

Kippenhahn, R., Weigert, A., and Hofmeister, E. 1967, Meth. Comput. Phys., 7, 129.

Lamb, D.Q. 1974, Ph.D. Thesis, Univ. of Rochester, N.Y.

Muchmore, D. 1981, preprint.

Nomoto, K. 1980, Proceedings of the Texas Workshop on Type I Supernovae (Austin: University of Texas Press).

Starrfield, S.G., Truran, J.W., and Sparks, W.M. 1981, Ap. J., 243, L27.

Starrfield, S.G., Truran, J.W., Sparks, W.M., and Kutter, G.S. 1972, Ap. J., 176, 169.

DISCUSSION FOLLOWING J. COLVIN'S TALK

WILLIAMS: At the beginning of your talk you mentioned that if you have a carbon-oxygen white dwarf and a high accretion rate and I think it was $5x10^{-8}M_{\odot}/yr$, you would expect helium shell flashes, can you elaborate?

COLVIN: This is taken from a paper that Nomoto presented at the type I supernova workshop in Austin two years ago. So this doesn't represent calculations that I have performed.

WILLIAMS: Does that mean that triple α occured in accretion onto a degenerate dwarf?

COLVIN: Yes, I believe that's what this means.

WILLIAMS: I have been asking other people if you can get that in that situation and they said, no.

SUGIMOTO: I think the answer should be as follows: When hydrogen accretes, a hydrogen shell flash takes place, hydrogen continues to accrete so that many hydrogen shell flashes are obtained. Then a part of the hydrogen is converted into helium and the helium zone grows until a helium shell flash takes place.

WILLIAMS: What if it was mainly helium that was accreted?

SUGIMOTO: The growth of the helium zone in mass by conversion of hydrogen by nuclear burning is the same as the accretion of helium.

SHAVIV: I would like to comment that the story with with CNO settling down is not clear. Because, if it is mixed by shear instabilities with the core, to give rise to a high CNO abundance, then the whole settling wouldn't occur. If the settling does occur and you get a layer of hydrogen which is CNO enriched at the bottom and CNO poor at the surface you might enhance the rate at which the thermonuclear runaway goes. Now will come the question what is seen in the ejecta. If the CNO is fully mixed we will observe solar abundance in most of the ejecta detonated by a small layer enriched with CNO. Hence the CNO abundance will change in time. The question of gravitational settling is not solved, because it will start to build up gradients that will try to annihilate it.

COLVIN: I agree. Comparing to observations depends upon the depth of the mixing. The settling is going to take place only below the mixed layer, so it is crucial how far down that mixed layer goes.

LAMB: If I understand how the gravitational diffusion occurs, to the extent that there is complete mixing of the outer layer of the star, the settling is enhanced. The settling rate increases as you go to higher densities. Therefore, if the outer layer of the star is mixed and brings unsettled material rapidly down to high densities and temperatures, then the heavy elements will actually settle out more quickly than if they had to diffuse all the way in from the outermost layers.

COLVIN: That's correct.

PHYSICAL CONDITIONS INSIDE WHITE DWARFS AND TYPE I SUPERNOVAE

SCHATZMAN, E.
Observatoire de NICE, France

1. After the first suggestion by Schatzman (1963) that accreting white dwarfs, close to the limiting mass, might be the progenitors of type I Supernovae, the problem was not studied until Schatzman (1974) described the competition between collapse dominated by beta decay, which would lead to neutron star formation and collapse dominated by nuclear reactions leading to stellar disruption.

Obviously, the chemical composition determines the behaviour of the accreting white dwarf and, for example a pure carbon white dwarf, is likely to explode as an effect of accretion. This has been considered by Wheeler (1974), Taam (1980), Starrfield et al. (1981), Nomoto (1980, 1981), Sugimoto and Nomoto (1980), Chevalier (1981), Woosley et al. (1981). However the consequences of the physical conditions (solid state versus liquid state for example), considered by Canal and Schatzman (1978) have been discussed in a parallel series of papers by Canal and Isern (1980), Canal, Isern and Labay (1980a, b), Mochkovitch (1980), whereas others, like Nomoto and Miyaji (1980) studied the collapse of white dwarfs with a special chemical composition. The purpose of this paper is to discuss the role of the physical conditions inside a collapsing white dwarf, close to the instability limit, and to discuss briefly the present uncertainties concerning the exact physical conditions in a white dwarf.

It should be reminded that, at the beginning of the collapse, a large fraction of the interior of the white dwarf is likely to be in a solid state. The temperature of solidification at $\rho = 2 \,.\, 10^{10}$ g cm^{-3} is $6.5 \,.\, 10^7$ °K for pure Carbon and $1 \,.\, 10^8$ for pure Oxygen. At these densities, for a random mixture of Carbon and Oxygen, the thermal runaway takes over the β capture, except for very low Carbon concentration. (Canal and Schatzman, 1976). It is quite clear that if we are not dealing with a random mixture the situation can be quite different.

M. Livio and G. Shaviv (eds.), Cataclysmic Variables and Related Objects, 149–153.

2. We can consider the following possible situations for a mixture of Carbon and Oxygen.

(A) Liquid mixture. Carbon and Oxygen are fully mixed. The number of pairs of Carbon nuclei is proportional to the square of the concentration. There is no thermonuclear runaway when the carbon concentration is sufficiently low (Canal and Schatzman, 1976).

(B) Solid mixture. The possibility of an eutectic has been considered by Stevenson (1980). Then we have the following possibilities:

(B)(a). No Eutectique.

(B)(a)(i). Oxygen solidifies first and sinks in the center of the star. This bas been considered by Mochkovitch (1980).

(B)(a)(ii). The solidification produces a mixture of randomly distributed Carbon and Oxygen nuclei. We can either have a mixture of small Oxygen and Carbon crystals, as has been considered by Kovetz and Shaviv (1970) as a complete random distribution. In the first case, the energy production goes like the concentration X_c; in the second case, as checked recently by Savedoff (1981), the energy production goes like the square X_c^2 of the concentration.

(B)(a)(iii). There is an ordered mixture of Carbon and Oxygen nuclei in the solid state. This case will be considered in the next section.

(B)(b). There is an Eutectic. This bas been considered by Mochkovitch (1980), and by Canal et al. (1980).

(B)(b)(i). If the concentration in number of Oxygen is larger than 33.2 %, solid oxygen falls to the center and a mantle of eutectic solidifies. This would favour collapse by β-capture.

(B)(b)(ii). If the concentration in number of Oxygen is less than 33.2 %, solid Carbon falls to the center and a mantle of eutectic solidifies. This would favour thermonuclear runaway. This case is not really different from the pure Carbon white dwarf case.

3. THE ORDERED MIXTURE.

Following a suggestion of Jancovici (1982), it can be shown that the free energy of two Carbon nuclei, embedded in solid Oyxgen at T = 0 is lower than the free energy of a chain of two Carbon atoms imbedded in the same solid. Therefore it is quite clear that there will be a tendency for the Carbon nuclei to keep away from each others. This seems to be true, but to a lesser extent, for a mixture of an equal number of Carbon and Oxygen atoms (Dyson, 1971). We are then facing a situation already considered by Kirschnits (1960) where the rate of pycno-nuclear reactions can be strongly decreased by the crystalline structure of the mixture. As a consequence, we can expect that solid white dwarfs with $N_c \lesssim N_o$ could collapse by electron capture on oxygen, whereas for $N_c \gtrsim N_o$ the energy liberation per C - C reaction would favour explosive processes.

Let us now consider more closely the situation. Assume that we have a lattice of solid Oxygen and introduce 2 Carbon atoms. We compare the free energy of the system for the case where the two Carbon atoms are far apart, and for the case where the arc close together and form a chain of two nuclei.

The energy of the space charge background is the same in the two configurations. We then have to consider the difference in the electrostatic energy between the nuclei.

We first assume with Jancovici (1982) that the lattice is not disturbed by the replacement of one or two Oxygen nuclei by one or two Carbon nuclei.

. One Carbon of charge Z_2 is surrounded by 6 Oxygen of charge Z_1. Then

$$F_{CO} - F_{OO} = \frac{6\ Z_1\ Z_2\ e^2}{a} - \frac{6\ Z_1^{\ 2}\ e^2}{a}$$

. Two Carbons are surrounded by 8 Oxygens. Then

$$F_{C^2} - F_{OO} = \frac{Z_2^{\ 2} + 10\ Z_1 Z_2 - 11\ Z_1^{\ 2}}{a}\ e^2$$

and we have

$$\Delta F = -\ (Z_1 - Z_2)^2\ (e^2/a)$$

$$= -\ \frac{(Z_1 - Z_2)^2}{Z_1^{\ 2}}\ \Gamma_1\ kT = -\ 0.0625\ \ \Gamma_1\ kT$$

with $\Gamma_1 = (Z_1^{\ 2}\ e^2/a\ kT)$

where a is the edge of the sc lattice.

The result can be slightly improved by moving the Oxygen atoms, and by moving the two carbon atoms in the chain. This gives

$$\Delta F = - 0.0538 \; \Gamma_1$$

The case of an equal mixture of two species has been considered by Dyson (1971). It should be noticed that the most bound C - O lattice ((bec lattice with Z_1 and Z_2 sites forming a (sc) sublattice) is more bound that the mixture of two (bc) lattices respectively of Carbon and Oxygen, the difference in the free energies being only, for each pair, only

$$\Delta F \cong - 3.95 \; 10^{-4} \; \Gamma_1 \; kT$$

We see that for a low concentration, the difference in free energy introduces for the reaction CC + OO → CO + CO a Boltzman factor which is $\exp(- f\Gamma_1)$ where f is of the order of 0.05. With f = 0.054 and Γ_1 = 160 (crystallization of Oxygen), the Boltzman factor is of the order of 2 . 10^{-4}. We shall conclude that the suggestion of Kirshnits (1960) is verified and that for a Carbon concentration $N_c \lesssim (1/2)(N_c+N_o)$ the rate of nuclear reactions is appreciably reduced by the potential barrier.

4. ASTROPHYSICS.

This result opens the possibility that, according to the chemical composition, and even for fast cooling and solidification of a Carbon-Oxygen mixture, a solid white dwarf could collapse, following electron capture on Oxygen, or experience some sort of explosive process due to the Carbon-Carbon nuclear reaction. The consequence of this sort of bifurcation, as far as the statistics of neutron stars and type I supernovae are concerned has been already discussed by Canal et al. (1980).

BIBLIOGRAPHY.

Canal, R., Isern, J.: 1980, in "White Dwarfs and Degenerate Stars", Fourth Annual Workshop, H. van Horn, V. Weidemann, Ed. University of Rochester, p. 52-55

Canal, R., Isern, J., Labay, J.: 1980a, Astrophys. J. Lett. L 33, 241

Canal, R., Isern, J., Labay, J.: 1980b, Space Sci. Rev. 27, 595
Canal, R., Schatzman, E.: 1976, Astron. Astrophys. 46, 229
Chevalier, R.A.: 1981, Astrophys. J. 246, 267
Dyson, F.J.: 1971, Ann. Physics 63, 1
Jancovici, B.: 1982, Report to an Informal Meeting on Dense Matter, Les Houches (unpublished)
Kirchnits, D.A.: 1960, Soviet Phys. JETP 11, 362
Kovetz, A., Shaviv, G.: 1970, Astron. Astrophys. 8, 398
Mochkovitch, R.: 1980, Thèse de 3ème Cycle, Université de Paris VII (to be published)
Nomoto, K.: 1980, Space Sci. Rev. 27, 563
Nomoto, K.: 1981, in "Fundamental Problems in the Theory of Stellar Evolution", IAU Symposium n° 93, D. Sugimoto, D.Q. Lamb, D.N. Schramm ed., Reidel, p. 295
Nomoto, K., Miyaji, S.: 1981, in "White Dwarfs and Degenerate Stars", Fourth Annual Workshop, H. Van Horn, V. Weidemann ed., University of Rochester, p. 56
Savedoff, M.: 1981, private communication
Schatzman, E.: 1963, in L. Gratton ed. Star Evolution, Proc. Intern. School of Physics Enrico Fermi, Academic Press, N.Y., p. 389
Schatzman, E.: 1974, in "The Nuclei of Galaxies, Black Holes and Collapsed Matter", Course III, Inter-School of Cosmology and Gravitation, "Ettore Majorana", Center for Scientific Culture, Erice, Italy
Starrfield, S., Trurau, J.W., Sparks, W.N.: 1981, Astrophys. J. Lett. L 27, 243
Stevenson, D.J: 1980, J. Phys. Suppl. n° 3, vol. 41, C2-53
Sugimoto, D., Nomoto, K.: 1980, Space Sci. Rev. 25, 155
Taam, R.E.: 1980, Bull. American Astron. Soc. 12, 465
Wheeler, J.C.: 1974, Astrophys. J. 187, 337
Woosley, S.E., Weaver, T.A., Taam, R.E.: 1981, in "Type I Supernovae", Proceedings of the Texas Workshop on Type I Supernovae, J.C. Wheeler, ed., University of Texas, p. 96

DISCUSSION FOLLOWING E. SCHATZMAN'S TALK

SHAVIV: One has also to consider the possibility that in the liquid phase, before it solidifies it might very well be that for certain ratios of carbon to oxygen the one is not immiscible in the other and then things may start earlier than the phase you are talking about.

SCHATZMAN: This has been considered by J.P. Hansen and co-workers and for that kind of Z ratio which is 8:6 the non immiscibility in the liquid phase takes place at densities which are much lower than those we have to consider here. So it is only in the solid phase that it can possibly take place.

THE INTERMEDIATE POLARS

Brian Warner
Department of Astronomy, University of Cape Town

INTRODUCTION

Until 1976, cataclysmic variable star research proceeded with few requirements for the inclusion of magnetic fields in theoretical models. Although models for low-mass X-ray binaries stressed the importance of magnetic fields (Lamb *et al.* 1973) and there was an increasing number of known magnetic single white dwarfs (Angel 1977), and a magnetised white dwarf had been one of the models proposed to explain the rapid oscillations in DQ Her (Herbst *et al.* 1974, Katz 1975), there was no anticipation of the more general role that magnetic fields now seem destined to play. The two major reviews of the time (Robinson 1976, Warner 1976) scarcely considered the presence of magnetic fields.

The discovery that AM Her combines the characteristics of cataclysmic variable, low-mass X-ray binary and magnetic white dwarf (Cowley and Crampton 1977, Priedhorsky 1977, Szkody and Brownlee 1977, Tapia 1977), and the subsequent discovery of several similar objects (reviewed by Tapia in this volume), revolutionised the subject. It rapidly became evident that, as large magnetic fields demonstrably dominate the flow pattern of accreted material onto synchronously rotating primaries in the polars (AM Her-type objects: Kruszewski 1978), lesser magnetic fields may play important roles in other cataclysmic variables, with detectable effects in many (perhaps all) of them. In other words, we may suspect that in the polars we see only the extremum of a range of magnetic field strengths and that, as with an iceberg, the visible tip implies a large body of a similar nature but less immediate detectability.

There are many cataclysmic variables for which there is as yet no clear need to invoke magnetic fields, but there is also an increasing number of objects in which magnetic fields do appear to play a role which, although less spectacular than that seen in the polars, is sufficient to introduce the classification "intermediate polar".

M. Livio and G. Shaviv (eds.), Cataclysmic Variables and Related Objects, 155–172.

IMPOTENT POLARS

In a polar the magnetic field strength of the degenerate star is sufficiently large that synchronous rotation results through magnetic interaction with its companion (Joss *et al.* 1979). Furthermore, the accretion flow is completely controlled by the geometry of the field, so that material funnels from the companion into an accretion column at one or both magnetic poles. No accretion disc would be expected in such a system. Most of the radiation in the system arises in the accretion columns which, with synchronous rotation, leads to the observed orbitally phase-locked large circular and linear polarization. Field strengths of $1\text{-}3 \times 10^7$ gauss have been detected in the polars VV Pup (Wickramasinghe and Visvanathan 1980) and AM Her (Schmidt *et al.* 1981, Latham *et al.*, 1981).

Senescent polars, in which the field is decaying, will break synchronism as the degenerate star is spun up by the accretion torque. The degree of asynchronism will depend on the rate of mass transfer, the strength of the field and the time lapsed since rotation ceased to be synchronous. Adolescent polars in which the field is strong but the system is relatively young, will not yet have reached synchronism. Systems born with smaller fields will be more or less braked by magnetic interaction and may show any degree of asynchronism.

All of these systems must appear under the classification of intermediate polar; it remains to discover whether we can distinguish between these different types and whether representatives of all classes can be found. For systems in which the field is large enough to establish an extensive magnetosphere, with its attendant accretion columns, some similarities with the polars themselves may be anticipated: e.g. weak polarization or X-ray emission. As is evident from this qualitative discussion, demonstrable asynchronism is the primary classification criterion for intermediate polars.

INTERMEDIATE POLARS DISCOVERED

The precise position of the X-ray source 2A0526-328, determined by Schwartz *et al.* (1979), enabled Charles *et al.* (1979) to identify it with a previously unknown star of the cataclysmic variable type - the first to have been discovered by means of its X-ray emission. It was noted that the spectrum of 2A0526-328 (now designated TV Col) is almost identical to that of the polars AM Her and EF Eri (2A0311-227). High speed photometry (Warner 1980a) showed that the system has the colours and rapid flickering of a cataclysmic variable.

Extensive photometry (Motch 1981) and spectroscopy (Hutchings *et al.* 1981) established the singular property that whereas the orbital period (from radial velocity modulation) is $5^h29^m\!.2$, the principal photometric period is $5^h11^m\!.5$. The beat period between these ($4^d\!.024$) shows as a modulation of the light curve, i.e. both periods are present as luminosity variations.

The model proposed for TV Col by Hutchings *et al.*, in analogy with an older model developed for DQ Her (discussed below), is that the binary contains a magnetic degenerate star, rotating progradely once in $4^{d}.024$ relative to the binary revolution. A beam of light from the accretion column rotates with the degenerate star, producing more or less illumination according to the distribution of material in the system.

Simultaneous with these studies of TV Col, the identification and elucidation of another source was in progress. H2252-035 was identified with a previously unrecognised 13th magnitude variable by Griffiths *et al.* (1980). Its spectrum resembles that of a dwarf nova at minimum light.

High speed photometry showed the system to have prominent luminosity variations with a 14.3 min period and a mean brightness orbitally modulated with period 3.6 hrs (Warner 1980b, Patterson and Price 1980). White and Marshall (1980, 1981) found that the X-ray flux is modulated at a period of 13.4 min; this period was also found at low amplitude in optical photometry (Warner and O'Donoghue 1980, Warner *et al.* 1981).

The combination of photometric and spectroscopic observations led to a model (Patterson and Price 1981, Warner *et al.* 1981, Hassall *et al.*, 1981) in which an X-ray beam emitted from the rotating degenerate star (period P_D = 13.4 min) illuminates either the companion (Patterson and Price 1981) or the gas in the extended hot spot region (Hassall *et al.* 1981) and is converted to optical emission. The period P_{opt} (=14.3 min) of the latter would then be associated with P_D and the orbital period P_{orb} by $P_{opt}^{-1} = P_D^{-1} - P_{orb}^{-1}$:- a relationship borne out by observations (Warner *et al.* 1981).

The similarity of the models for TV Col (for which, however, X-ray modulation with the $5^{h}11^{m}.5$ period still has to be observed) and H2252-035, in both of which slowly rotating (but asynchronous) degenerate stars are postulated, established that intermediate polars had been observed. Neither of the stars shows optical polarization, indicating magnetic fields at least an order of magnitude lower than in the polars.

With the steady production of precise X-ray positions, the past year has seen the identification of several more polars and intermediate polars:-

TABLE 1

The Intermediate Polars

Variable Star	X-Ray	Orbital Period	Other Period(s)
TV Col	2A0526-328	$5^{h}29^{m}.2$	$5^{h}11^{m}.5$
	H2215-086	4 01.5	20.9 (22.9)
AO Psc	H2252-035	3 35.5	14.3 (13.4)
V1223 Sgr	4U1849-31	3 22.8	13.2
	3A0729+103	3 14.2	15.2

SPECTRA AND ENERGY DISTRIBUTIONS

The emission line spectra of the five established intermediate polars (Table 1), typical examples of which are shown in Figures 1 and 2, closely resemble those of polars (Raymond *et al.* 1979) or nova remnants (Warner 1976). The great strength of HeII 4686Å and the presence of the 4640Å CIII/NIII blend distinguishes them from dwarf novae. Observations in the range 1000-2000Å with IUE (Hassall *et al.* 1981, Bonnet-Bidaud *et al.* 1982) show strong CIV emission and the presence of CII, SiIV and NV.

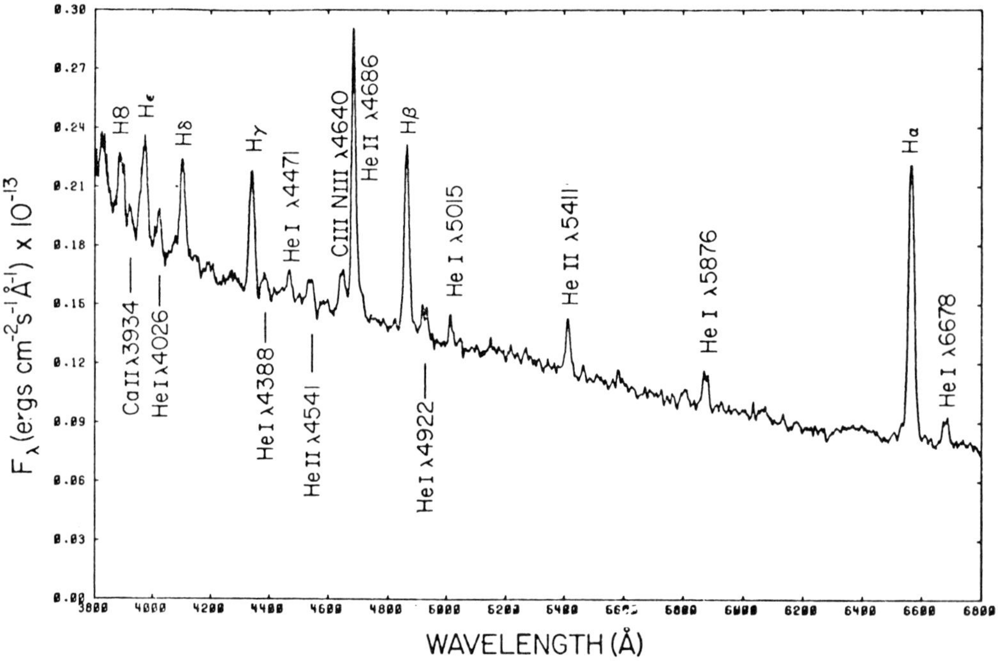

Figure 1. Spectrum of H2215-086. (Reproduced by permission from Shafter and Targan, *Astron. J. 87*, 655, 1982)

All five objects have broad lines, with FWZI ∿1000-1200 km s^{-1} (Shafter and Targan 1982, Hassall *et al.* 1981, Watts *et al.* 1980, Warner *et al.* 1982, Steiner *et al.* 1981). Variability of line width and line splitting is seen in 2A0526-328 (Watts *et al.* 1980); Hutchings *et al.* (1981) find line widths up to 3000 km s^{-1} on occasion in this star and large variations in intensity uncorrelated with either of the periodicities.

In contrast to the situation in the polars, where the flux distribution shows the presence of two components with temperatures $\gtrsim$10 keV

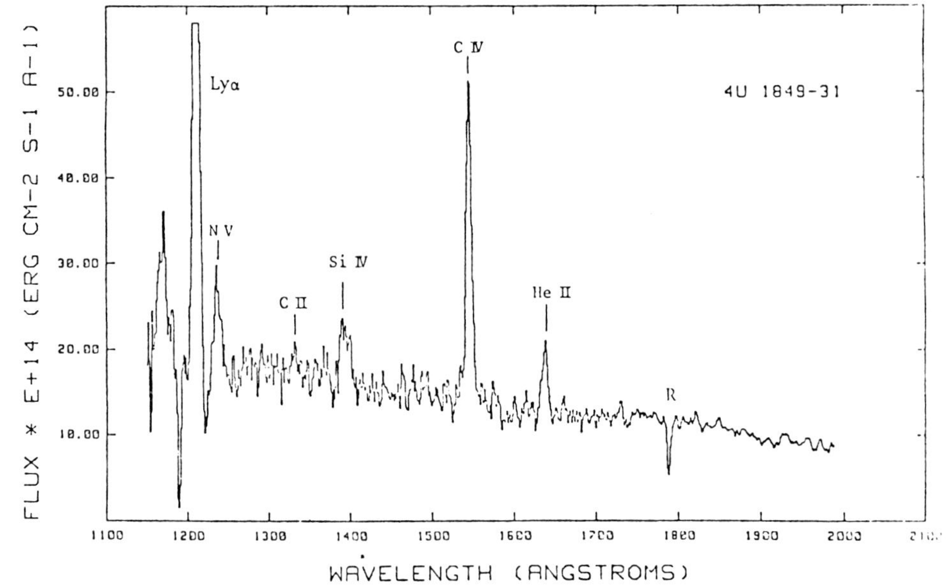

Figure 2. Spectrum of V1223 Sgr. (Reproduced by permission from preprint of Bonnet-Bidaud, Mouchet and Motch)

and $\lesssim$0.1 keV respectively (e.g. Kylafis and Lamb 1982) the fluxes in H2215-086 (Shafter and Targon 1982), H2252-035 (Griffiths *et al.* 1980) and V1223 Sgr (Bonnet-Bidaud *et al.* 1982) are in general quite well described by $F_\lambda \propto \lambda^{2.2 \pm 0.2}$ which is characteristic of an accretion disc. However, in H2252-035, Hassall *et al.* (1981) and Cordova (1981) obtain a better fit with a composite of disc and 12 000 K blackbody. The latter is attributable to a hot spot. Wickramasinghe *et al.* (1982) find an S-wave distortion in the spectra of this star, characteristic of that seen from the hot spots in dwarf novae. This, together with the broad emission lines, is a strong indicator that at least a partial disc with a hot spot is present in the intermediate polars.

X-RAY EMISSION

The polars typically possess both hard (kT$\sim$40 keV) and soft (kT$\lesssim$0.1 keV) X-ray emission components (Kylafis and Lamb, 1982, and references therein). The recognition of the five identified intermediate polars has resulted from their appearance in surveys for hard X-rays. The emissions from V1223 Sgr and H2252-035 are characterised by either thermal emission with kT > 10 keV or a power-law model of photon index $\sim$1.4 (Steiner *et al.* 1981, White and Marshall 1981). None of the five stars has been detected in the soft X-ray region.

In H2252-035 the X-ray flux is highly modulated with a period of 805 secs (Patterson and Garcia 1980, White and Marshall 1981). This period is also seen in the optical (Warner *et al.* 1981) and has been shown (Motch and Pakull 1981) to have a steeper spectrum in the ultraviolet than the 859 sec reprocessed component. These authors show that the source of the 805s optical pulsation is beamed in the same direction as the X-ray emission and can be attributed to the Rayleigh-Jeans tail of a hot blackbody. This could be a result of X-ray heating of the degenerate star below the accretion column, or from X-ray reprocessing as the beam sweeps round the uniform parts of the accretion disc. The existence of an X-ray and optical beam argues strongly for an accretion column close to the surface of the degenerate star.

3A0729+103 is the only other of these stars for which an X-ray modulation has been detected: McHardy and Pye (1982) find evidence for a period in the range 900-1000 secs, similar to the 913 sec optical period.

POLARISATION

No positive detection of circular polarisation has yet been made in the intermediate polars. Quoted limits are 0.07% in 2A0526-328 and H2252-035 (Stockman *et al.* 1982). Williams and Johns (1980) give an upper limit of 0.2 percent for modulated linear polarisation in H2252-035.

In the polars, circular polarisation is attributed to cyclotron emission from the accretion column (e.g. Wickramasinghe and Visvanathan 1980). The wavelength dependence of the polarisation shows the correct behaviour - increasing in the red region of the spectrum. In the polars we are fortunate that the field strengths ($\sim 10^7$ gauss) are just sufficient to produce significant polarisation in the optical region. In order to confirm the lower fields postulated for the intermediate polars, polarimetric observations in the infrared will be required.

LONG-TERM OPTICAL BEHAVIOUR

The polars are well-known for possessing both high and low states of luminosity, differing by 2-3 magnitudes and with gradual transitions between them. Historical light curves, derived from the Harvard plate collection, have been determined for two intermediate polars by Belsepene (1981) which show that V1223 Sgr spends most of its time in a high state but suffered 5 transitions to low states (2-3 magnitudes fainter) and a twelve year interval of irregular brightness at intermediate magnitudes. H2252-035 has shown only one drop of a magnitude below its normal high state.

RAPID OPTICAL BEHAVIOUR

All five intermediate polars show the rapid flickering activity characteristic of cataclysmic variables (Warner 1976). Short lived quasi-periodic oscillations have been seen in H2252-035 with periods ∿100 secs (Warner *et al.* 1981). The striking feature of this group, however, is the large amplitude of modulations associated with rotation of the degenerate component, as illustrated in Figures 3 and 4.

In H2252-035, as discussed earlier, the principal optical period (859 secs) differs from the X-ray period and demonstrates that the modulated optical flux originates from X-ray heating of a feature that is fixed in the rotating frame of the binary system. Patterson and Price (1981) suggested that this object is the secondary star, but Hassall *et al.* (1981) rejected this in favour of reprocessing in the region of the hot spot. More recent work (Motch and Pakull, 1981) shows that the flux distribution of both the 859 sec and the orbitally modulated components are similar and consistent with calculated X-ray illumination of the atmosphere of the secondary (Milgrom 1976). The relative phasing of the 805 and 859 sec modulations supports this conclusion (Motch and Pakull 1981, Patterson 1981a). The amplitude of the optical 805 sec modulation varies and is at times too small to be detected (Wickramasinghe *et al.* 1982). These same authors show that the quantitative arguments against the heated secondary hypothesis in H2252-035, made by Hassall *et al.*, are negated when allowance is made for the contribution made by disc luminosity and shadowing effects.

In H2215-086 the situation is the reverse of that in H2252-035. The large amplitude 20.9 min oscillations in brightness (Shafter and Targan 1982, Patterson and Steiner 1981) are accompanied by a low amplitude oscillation at 22.9 mins which is the side band produced by the orbital period of $4^{h}_{\cdot}03$ (Patterson 1982). In this case, therefore, either we see mostly the direct effects of heating of the axially symmetric parts of the accretion disc and the secondary intercepts only a small fraction of the beam, or, most unlikely, the degenerate star is rotating retrogradely. An X-ray period is required to distinguish between the possibilities.

3A0729 + 103 appears to have one or more periodicities longer than the 15.2 min principal oscillation, but these are not clearly connected with an orbital sideband (Warner *et al.*: in preparation). In TV Col both "synodic" and "sidereal" periods show themselves in the photometry (Motch 1981); their amplitudes are about equal.

WHITE DWARF OR NEUTRON STAR?

The X-ray luminosity of H2252-035 is estimated to be 4 x 10^{32} x $(d/220\ pc)^2$ ergs sec^{-1} (White and Marshall 1981). This, together with the fact that L_x/L_{opt}∿1, typical for cataclysmic variables known to contain a white dwarf, led Patterson and Price (1981) to argue that

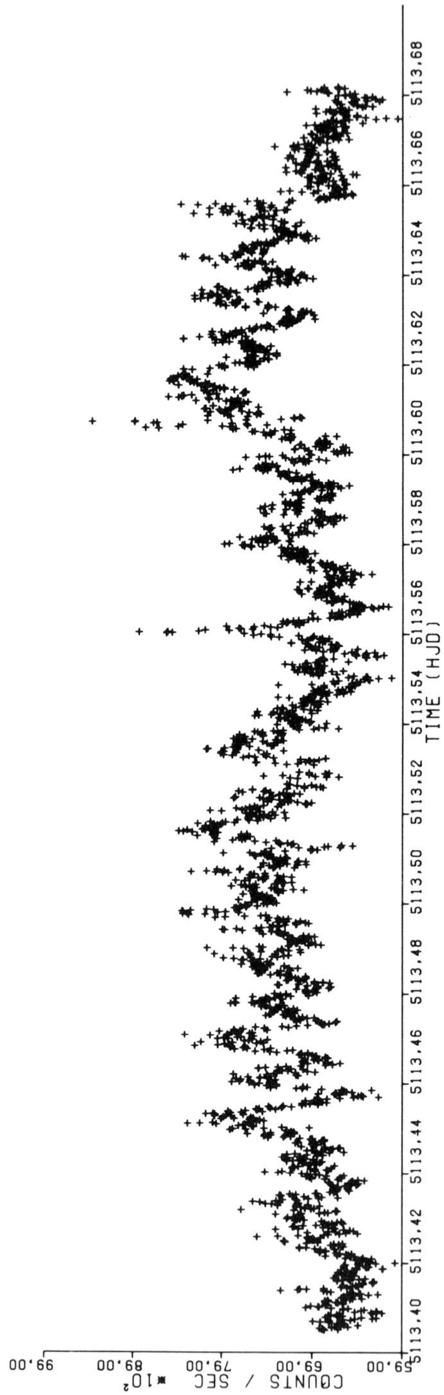

Figure 3. Light curve of V1223 Sgr. 23 May 1982.

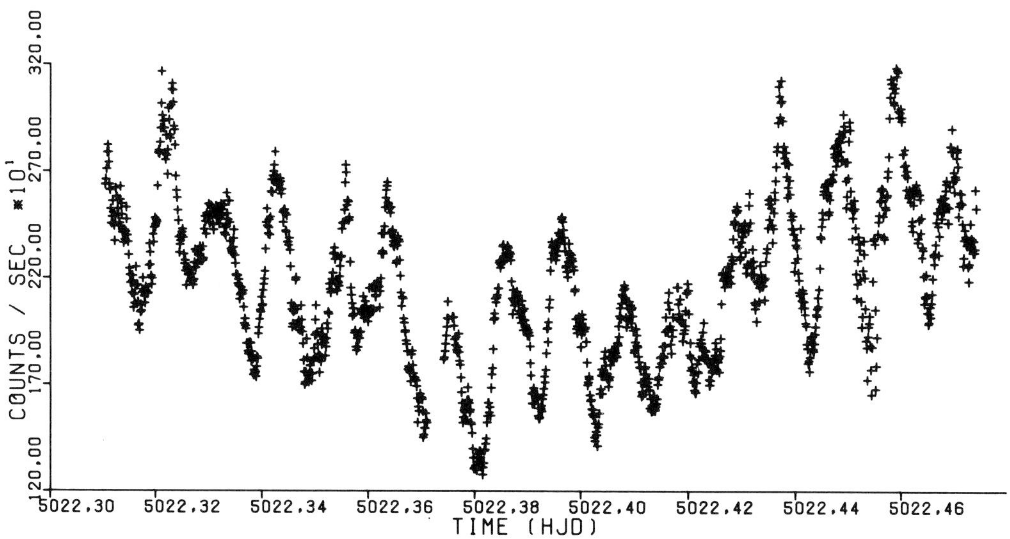

Figure 4. Light curve of 3A0729+103 21 February 1982

H2252-035, and by analogy the other intermediate polars, contain a white dwarf. The more comprehensive discussion by White and Marshall, however, leaves the situation open: they show that most of the light in the system probably comes from reprocessing off the disc; the luminosity required implies a large (unobservable) soft X-ray or far-UV flux, which is independent of the nature of the compact object and can readily bring L_x/L_{opt} down to ~ 1 instead of the ~ 300 normally expected for a neutron star. In addition, although $L_x \sim 10^{32}$ ergs sec^{-1} is five orders of magnitude lower than typical values for neutron stars, the X-ray neutron pulsar X Per has $L_x \sim 10^{32}$ ergs sec^{-1} and similar rotation period to H2252-035.

A possible method of distinguishing between white dwarf and neutron star primaries is to look at the rate of spin-up produced by the accretion torque. White and Marshall (1981) show that the expected Alfven radius in either case (for magnetic fields $\lesssim 10^{5 \text{ or } 6}$ gauss for the white dwarf) implies slow rotation in the definition of Ghosh and Lamb (1979); the latters' equations (15)-(18) and their Figure 8 may therefore be used. Then

$$\frac{-\dot{P}}{P} \simeq 8 \times 10^{-5} \; \mu_{30}^{2/7} \; R_6^{6/7} \left(\frac{M}{M_\odot}\right)^{-3/7} I_{45}^{-1} \; P \; L_{37}^{6/7} \quad \text{yr}^{-1} \tag{1}$$

where μ,R,I and L are respectively the magnetic moment, radius, moment of inertia and luminosity (in cgs) of the degenerate star. For a neutron star $R_6 \sim 1$, $I_{45} \sim 2$ and for a $1M_\odot$ white dwarf $R_6 \sim 10^3$ and $I_{45} \sim 10^5$. The weak dependence on μ is convenient, as is the fact that a neutron star with a field $\sim 10^{12}$ gauss and a white dwarf with field $\sim 10^5$ gauss have similar magnetic moments.

For X Per, P = 836s, L = 1 x 10^{34} ergs sec^{-1} (Mason 1977) and equation (1) predicts $\dot{P}/P \sim -9 \times 10^{-5}$ yr^{-1} for a neutron primary, which compares moderately well with the observed value of -1.7×10^{-4} yr^{-1}.

In the case of the intermediate polars, where P $\sim 10^3$s, equation (1) gives ($M = 1M_\odot$)

$$\left.\begin{aligned} \frac{-\dot{P}}{P} &\sim 1.5 \times 10^{-5} \left(\frac{P}{10^3}\right) L_{34}^{6/7} \quad yr^{-1} \text{ for neutron star} \\ &\sim 1.1 \times 10^{-7} \frac{P}{10^3} L_{34}^{6/7} \quad yr^{-1} \text{ for white dwarf} \end{aligned}\right\} \qquad (2)$$

For a given luminosity and period, therefore, we have a factor of approximately 100 difference in timescale for the two processes. Furthermore, the neutron star case is large enough to be detectable in observations made over a baseline $\sim$1 year, whereas the white dwarf case requires $\sim$10 years.

None of the intermediate polars in Table 1 has been observed long enough yet to distinguish definitely between the two possibilities. Observations of V1223 Sgr (Warner *et al.*: in preparation) show the same period in 1981 and 1982 to within $\sim$0.007 sec, but a full analysis of the data is still in progress.

POSSIBLE INTERMEDIATE POLARS

(i) TT Arietis

TT Arietis possesses many of the properties of an intermediate polar. Its spectroscopic orbital period is well-established by Cowley *et al.* (1975) to be 3^h18^m, but the photometric period appears to be 3^h11^m (Smak and Stepien 1975). The effect of the slippage between the two periods is clearly seen in Figures 2 and 3 of the paper by Mardirossian *et al.* (1980).

The long-term optical behaviour (Fuhrmann 1981) shows that the star is normally at $m_V \sim 10^m.8$ but there have been occasional drops to $\sim 12^m.3$ (in 1956, 1958 and 1967) and most recently to $14^m.5$ in November 1980.

Stockman *et al.* (1982) gives an upper limit of 0.02 percent circular polarisation in TT Ari.

Cordova *et al.* (1981) found hard X-ray emission from TT Ari with L_X(0.1-4.5 keV) $\sim 1.4 \times 10^{32}$ erg sec^{-1}. Later observations (Mason and Cordova 1982) found good correlation between optical and X-ray brightness variations, both for the orbital modulation and for faster flickering and quasi-periodic variations.

Although TT Ari shows the broad emission lines of HeII and NIII/CIII in the optical (Cowley *et al.* 1975) and CIV, NIV, NV in the far-UV

(Krautter *et al.* 1981) the optical region also shows broad absorption lines characteristic of an optically thick accretion disc (Cowley *et al.* 1975). In this manner, TT Ari differs substantially from the established intermediate polars and may represent a system undergoing a higher rate of mass transfer than the latter objects.

(ii) EX Hya

EX Hya, classified from its light curve as a dwarf nova, although an unusual one (Warner 1976), is uncharacteristic in showing weak NIII emissions in its optical spectrum (Cowley *et al.* 1981). The total line width of $\sim$7000 km sec^{-1} (Cowley *et al.*) indicates an accretion disc reaching close to the surface of the degenerate star (if it is a white dwarf).

The most remarkable aspect of EX Hya, however, is the presence of a brightness modulation with period 67.0 min (Vogt *et al.* 1980), quite distinct from the orbital period of 98.3 min which is established both spectroscopically and from the presence of eclipses.

Warner and McGraw (1981) found from the morphology of the eclipses that the 67-min optical modulation may arise either from periodically variable rate of mass transfer or from the rotation of an illuminating beam in the manner of an intermediate polar. They pointed out that if the optical modulation arises in the same manner as in H2252-035 (i.e. as X-ray heating of the hot spot) then a rotation period of 40 mins would be required for the degenerate component.

EX Hya is one of the brightest of the cataclysmic variables in hard X-rays (Watson *et al.* 1978). Kruszewski *et al.* (1982) find that the X-ray spectrum contains two components, the lower energy (kT $\lesssim$ 1.8 keV) is modulated with the 67 min period but the higher one is not. It is clear, therefore, that the optical modulation is derived from X-ray heating of axially symmetric parts of the accretion disc, and not from the hot spot or secondary.

The interpretation given by Warner and McGraw assumed that the EX Hya eclipses are of a hot spot on the outer edge of the accretion disc. The observed variations in eclipse depth would then be explicable as grazing eclipses of a spot whose radius vector varies and whose luminosity is dominated and modulated by X-ray heating from the rotating beam. In view of the 67 min, rather than 40 min, X-ray period this model is no longer tenable. However, the discovery (Warner and Cropper 1982) that in the dwarf nova V2051 Oph, whose orbital period of 90 mins puts it in the same class as EX Hya, the flickering and most of the optical flux from the system arises in the *inner disc* and not from the hot spot, provides the solution to the otherwise incompatible EX Hya observations.

In EX Hya, as in V2051 Oph, the disc region close to the degenerate component is now proposed as the source of most of the optical flux. Much of this luminosity derives from heating from an X-ray beam emitted by the 67-min rotation of the primary. The broad emission lines, arising from this same region, are excited by the beam thus causing the 67 min modulation in line intensities found by Gilliland (1982). The latter also found a line intensity modulation with period 33.5 min, attributable to beams emitted from accretion columns at both magnetic poles of the primary.

The optical eclipses must now be attributed to grazing eclipses of the bright central region of the accretion disc. In the absence of significant luminosity from the outer disc regions or a bright spot this is as compatible with the photometry as was the former model. As is the case in V2051 Oph (Warner and Cropper 1982), the central disc region varies both in size and brightness, causing variations in eclipse profile. The absence of X-ray eclipses (Kruszewski *et al.*) implies that the primary itself is not eclipsed - only the nearside of the inner parts of the accretion disc.

Gilliland (1982) finds that the 67 min period is changing at a rate $P/\dot{P} \sim -2.8 \times 10^6$ yrs. The bolometric luminosity of EX Hya is $\sim 6 \times 10^{33}$ ergs sec^{-1} (Bath *et al.* 1980) which in equations (2) give $P/\dot{P} \sim -2.6 \times 10^4$ yrs for a neutron star and $P/\dot{P} \sim -3.5 \times 10^6$ yrs for a white dwarf. The agreement with prediction for the white dwarf case is very satisfactory.

THE DQ HERCULIS STARS

The nova remnants DQ Her (Herbst *et al.* 1974 and references therein) and V533 Her (Patterson 1979a) and the nova-like variable AE Aqr (Patterson 1979b) have optical brightness oscillations of 71.1, 63.6 and 33.1 secs respectively. DQ Her is not detected in the X-ray region (Cordova *et al.* 1981), probably because of its high inclination. V533 Her similarly has yet to be detected in X-rays (but the upper limit, $L_X \lesssim 3 \times 10^{31}$ ergs sec^{-1} is a factor of ten higher than in DQ Her: Mason and Cordova 1982). However, AE Aqr not only has measurable flux in the 0.1 - 4 keV range but shows large modulation of this flux in phase with the optical oscillations (Patterson *et al.* 1980). All three objects show high stability: DQ Her $P/\dot{P} = -2.7 \times 10^6$ yr (Patterson *et al.* 1978), V553 Her $|P/\dot{P}| > 8 \times 10^6$ yr (Patterson, quoted by Cordova and Mason 1982) and AE Aqr $|P/\dot{P}| > 4 \times 10^7$ yr (*Ibid.*). The high stability suggests a phenomenon associated with the degenerate component, rather than with structures in the accretion disc. With luminosities $10^{34} - 10^{35}$ ergs sec^{-1} for nova remnants (Warner 1976) and AE Aqr (Patterson 1979b), equations (2) require that the primaries be white dwarfs.

In the case of DQ Her, eclipses of the white dwarf and disc provide a probe of the oscillations. The phase shift of the 71 sec oscillations during the eclipse (Warner *et al.* 1972; Patterson *et al.* 1978) have been successfully interpreted in terms of a beam reflected (or reprocessed off of the accretion disc (Petterson 1980 and references therein).

The similarity with the longer period intermediate polars is increased by the presence of circular polarisation at ∿0.6%, synchronised with the 71 secs oscillations (but with a fundamental period of 142 secs: Swedlund *et al.* (1974), Kemp *et al.* (1974).

AE Aqr is circularly polarised in the optical region by ∿0.6 percent (Szkody *et al.* 1982).

The model of DQ Her deduced by Chanon *et al.* (1978) (see their Figure 4) and the model of AE Aqr given by Patterson (1979b) (see his Figure 12) are closely analogous to the model we have described for the longer period systems.

The properties of these systems are more familiar than the newly discovered longer period intermediate polars, so we will not discuss them fully. It is clear, however, that they have the appearances of rapidly rotating magnetised white dwarf primaries. The decreasing period of DQ Her suggests a late stage of evolution of a once slowly rotating intermediate polar.

RELATED SYSTEMS

(i) The Short Period Systems

Passing from the high stability of the DQ Her stars we find a range of similar systems of lower stability. UX UMa has an oscillation period near 29 secs, which can change by up to 0.4 percent in a night but which shows a phase shift through eclipse similar (but in the opposite sense) to that in DQ Her (Nather and Robinson 1974). This has been interpreted in terms of a beam model by Petterson (1980).

The dwarf nova HT Cas has ∿20 sec oscillations during outburst which show similar phase variations through eclipse (Patterson 1981b) and which therefore establish the beamed nature of all the oscillations seen in dwarf novae during outburst. The lack of long term stability ($P/\dot{P}$ ∿ days in many cases), however, excludes the pure rotation model that we have discussed so far.

The basic model, however, may not be so very different. Suppose, for example, the primaries in dwarf novae are slow magnetic rotators and that outbursts result from an instability in the outer accretion disc (or ring) in accordance with recent conclusions (e.g. Cannizzo *et al.* 1982). The increased mass flow during an outburst reduces the size of the Alfvén radius but later, as the flow diminishes, allows it to expand again. Inhomogeneities around the Alfvén circumference would modulate the accretion flow down the magnetic poles and produce luminosity variations modulated at a frequency given by the difference in frequencies of the accretion disc at the Alfvén radius and the rotation of the primary. Such modulation would show the phase jitter seen in soft X-rays oscillations of outbursting dwarf novae (Cordova *et al.*

1980), although maintaining a basic underlying period. This period would vary in the manner seen in dwarf nova outbursts (i.e. $\dot{P} < 0$ on the rising branch and $\dot{P} > 0$ on the descending branch: Warner and Brickhill 1978).

In this manner, or in any other in which magnetic fields may be deduced to be present and determining the accretion geometry, all of the dwarf novae may eventually be considered as intermediate polars.

(ii) The Longer Period Objects

MV Lyr (Robinson *et al.* 1981, Schneider *et al.* 1981) is a nova-like variable with an orbital period of 3^h12^m derived from radial velocity observations. Its claim to be considered as a possible intermediate polar derives from the long-term light curve behaviour: it is normally at magnitude 12.0 - 14.0 but occasionally fades to 18.0 or fainter, where it can stay for as long as $1\frac{1}{2}$ yrs (Robinson *et al.* 1981 and references therein). No periodicity in the light curve, which could provide the rotation period of the primary, has yet been found. Robinson *et al.* (1981) dismiss the suggestion by Voikhanskaya *et al.* (1978) that MV Lyr is a polar, but their arguments still allow the possibility that it is an intermediate polar.

The hard X-ray emission from MV Lyr (Becker 1981) is not exceptional for a cataclysmic variable.

Stepanian's star is another possible candidate, having shown a low state (Liller 1980). Radial velocities provide the orbital period of 3^h48^m (Margon *et al.* 1980), confirmed by the eclipse period determined by Horne (1980) but the latter found no other orbitally modulated light variations. If the system is like TV Col or EX Hya then longer photometric runs would be required to detect an asynchronous periodicity. Szkody (1981) has drawn attention to the fact that Stepanian's star does not have the energy distribution of a typical cataclysmic variable.

VY Scl, which shows occasional low states, but for which no orbital or rotation period is yet known (Warner and Van Citters 1974), is another possible candidate.

A whole range of possibilities have been uncovered by Schoembs' (1982) discovery that the otherwise normal dwarf nova CN Ori possesses a doubly-periodic light curve with humps recurring at periods of $3^h54^m.9$ (the supposed orbital period) and $3^h49.7$. The connections between this behaviour, the similar phenomenon seen in TV Col (Table 1) and the appearance of superhumps (with periods a few percent *longer* than the orbital periods) during outbursts of the SU UMa class of dwarf novae (Warner 1976) are not yet clear.

ACKNOWLEDGEMENTS

I would like to take the opportunity to acknowledge that much of the stimulus for my interest in the polars and intermediate polars, as opposed to systems without magnetic fields, came from my attendance at the hugely successful workshop at Santa Cruz in August 1981. I also acknowledge the observational and reductional assistance provided by Mark Cropper and Sue Allen.

REFERENCES

Angel, J.R.P. 1977, *Ap. J. 216*, 1.
Bath, G.T., Pringle, J.E. and Whelan, J.A.J. 1980, *M.N.R.A.S. 190*, 185.
Becker, R.H. 1981, *Ap. J. 251*, 626.
Belserene, E.P. 1981, *B.A.A.S., 13*, 524.
Bonnet-Bidaud, J.M., Mouchet, M. and Motch, C. 1982, *Astr. Astrophys.*, in press.
Cannizzo, J.K., Ghosh, P. and Wheeler, J.C. 1982, preprint.
Chanan, G.A., Nelson, J.E. and Margon, B. 1978, *Ap. J. 226*, 963.
Charles, P.A., Thorstensen, J., Bowyer, S. and Middleditch, J. 1979, *Ap. J. Letts. 231*, L131.
Cordova, F.A. 1981. Reported at Santa Cruz Workshop.
Cordova, F.A., Chester, T.J., Tuohy, I.R. and Garmire, G.P. 1980, *Ap. J. 235*, 163.
Cordova, F.A., Mason, K.O. and Nelson, J.E. 1981, *Ap. J. 245*, 609.
Cowley, A.P. and Crampton, D. 1977, *Ap. J. 216*, L45.
Cowley, A.P., Crampton, D., Hutchings, J.B. and Marlborough, J.M. 1975, *Ap. J. 195*, 413.
Cowley, A.P., Hutchings, J.B. and Crampton, D. 1981. *Ap. J. 246*, 489.
Fuhrmann, B. 1981, *I.B.V.S.* No. 1905.
Ghosh, P. and Lamb, F.K. 1979, *Ap. J. 234*, 296.
Gilliland, R.L. 1982, *Astrophys. J.*, in press.
Griffiths, R.E., Lamb, D.Q., Ward, M.J., Wilson, A.J., Charles, P.A., Thorstensen, J., McHardy, I.M. and Lawrence, A. 1980, *M.N.R.A.S. 193*, 25P.
Hassall, B.J.M., Pringle, J.E., Ward, M.J., Whelan, J.A.J., Mayo, S.K., Echerarria, J., Jones, D.H.P., Wallis, R.E., Allen, D.A. and Hyland, A.R. 1981, *M.N.R.A.S. 197*, 275.
Herbst, W., Hesser, J.E. and Ostriker, J.P., 1974, *Ap. J. 193*, 679.
Horne, K., 1980, *Ap. J. Letts. 242*, L167.
Hutchings, J.B., Crampton, D., Cowley, A.P., Thorstensen, J.R. and Charles, P.A., 1981, *Ap. J. 249*, 680.
Joss, P.C., Katz, J.I. and Rappaport, S.A., 1979, *Ap. J. 230*, 176.
Katz, J.I., 1975, *Ap. J. 200*, 298.
Kemp, J.C., Swedlund, J.B. and Wolstencroft, R.D., 1974, *Ap. J. Letts. 193*, L15.
Krautter, J., Klare, G., Wolf, B., Wargau, W., Dreschel, H., Rahe, J. and Vogt, N. 1981, *Astr. Ap. 98*, 27.
Kruszewski, A., 1978 in Nonstationary Evolution of Close Binaries, p. 55, Warsaw.

Kruszewski, A., Mewe, R., Heise, J., Chlebowski, J., Van Dijk, W. and Bakker, R., 1982, *Space Sci. Rev.* in press.
Kylafis, N.D. and Lamb, D.Q., 1982, *Ap. J. Suppl. 48,* 239.
Latham, D.W., Liebert, J. and Steiner, J.E., 1981, *Ap. J. 246,* 919.
Lamb, F.K., Pethick, C.J. and Pines, D. 1973, *Ap. J. 184,* 271.
Liller, M.H. 1980, *I.B.V.S.* No. 1743.
Mardirossian, F., Mezzetti, M., Pucillo, M., Santin, P., Sedmak, G. and Giurcin, G. 1980, *Astr. Astrophys. 85,* 29.
Margon, B., Downes, R. and Szkody, P. 1980, *I.A.U. Circ.* No. 3465.
Mason, K.O., 1977, *M.N.R.A.S. 178,* 81P.
Mason, K.O. and Cordova, F.A. 1982, preprint.
McHardy, I.M. and Pye, J.P. 1982, *I.A.U. Circ.* No. 3687.
Milgrom, M. 1976, *Ap. J. 208,* 191.
Motch, C. 1981, *Astr. Astrophys. 100,* 277.
Motch, C. and Pakull, M.W. 1981, *Astr. Astrophys. 101,* L9.
Nather, R.E. and Robinson, E.L. 1974, *Ap. J. 190,* 637.
Patterson, J. 1979a, *Ap. J. 233,* L13.
Patterson, J. 1979b, *Ap. J. 234,* 978.
Patterson, J. 1981a, Reported at Santa Cruz Workshop.
Patterson, J. 1981b, *Ap. J. Suppl. 45,* 517.
Patterson, J. 1982, private communication.
Patterson, J. and Garcia, N. 1980, *I.A.U. Circ.* No. 3514.
Patterson, J. and Price, C. 1980, *I.A.U. Circ.* No. 3511.
Patterson, J. and Price, C. 1981, *Ap. J. 243,* L83.
Patterson, J. and Steiner, J. 1981. *B.A.A.S. 13,* 817.
Patterson, J., Robinson, E.L. and Nather, R.E. 1978, *Ap. J. 224,* 570.
Patterson, J., Branch, D., Chincarini, G. and Robinson, E.L. 1980, *Ap. J. 240,* L133.
Petterson, J.A., 1980, *Ap. J. 241,* 247.
Priedhorsky, W.C. 1977, *Ap. J. 212,* L117.
Raymond, J.C., Black, J.H., Davis, R.J., Dupree, A.K., Gursky, H., Hartmann, L. and Matilsky, T.A. 1979, *Ap. J. Letts. 230,* L95.
Robinson, E.L., 1976, *Ann. Rev. Astr. Astrophys. 14,* 119.
Robinson, E.L., Barker, E.S., Cochran, A.L., Cochran, W.D. and Nather, R.E. 1981, *Ap. J. 251,* 611.
Schmidt, G.D., Stockman, H.S. and Margon, B. 1981, *Ap. J. 243,* L157.
Schneider, D.P., Young, P. and Shectman, S.A. 1981, *Ap. J. 245,* 644.
Schoembs, R., 1982, *Astr. Astrophys.,* in press.
Schwartz, D.A., Bradt, H., Briel, U., Doxsey, R.E., Fabbiano, G., Griffiths, R.E., Johnston, M.D. and Margon, B. 1979, *A.J. 87,* 655.
Shafter, A.W. and Targan, D.M. 1982, *A.J. 87,* 655.
Smak, J. and Stepien, K. 1975, *Acta Astr. 25,* 379.
Steiner, J.E., Schwartz, D.A., Jablonski, F.J., Busko, I.C., Watson, M.G., Pye, J.P. and McHardy, I.M. 1981, *Ap. J. Letts. 249,* L21.
Stockman, H.S., Liebert, J. and Moore, R.L. 1982, *Ap. J.* in press.
Swedlund, J.B., Kemp, J.C. and Wolstencroft, R.D. 1974, *Ap. J. Letts. 193,* L11.
Szkody, P. 1981, *P.A.S.P. 93,* 456.
Szkody, P. and Brownlee, D.E. 1977, *Ap. J. 212,* L113.
Szkody, P., Michalsky, J.J. and Stokes, G.M. 1982, *P.A.S.P. 94,* 137.
Tapia, S. 1977, *Ap. J. 212,* L125.

Vogt, N., Krzeminski, W. and Sterken, C. 1980, *Astr. Astrophys. 85*, 106.
Voikhanskaya, N.F., Gnedin, Yu. N., Efimov, Yu. S., Mitronov, I.G. and Shakhovskoi, N.M. 1978, *Pis'ma Astron. Zh. 4*, 272.
Warner, B. 1976, *I.A.U. Symp. 73*, 85.
Warner, B. 1980a, *M.N.R.A.S. 190*, 69P.
Warner, B. 1980b, *I.A.U. Circ.* No. 3511.
Warner, B. and Van Citters, G.W., 1974, *Observatory 94*, 116.
Warner, B. and Brickhill, A.J. 1978, *M.N.R.A.S. 182*, 777.
Warner, B. and O'Donoghue, D. 1980, *I.A.U. Circ.* No. 3525.
Warner, B. and McGraw, J.T. 1981, *M.N.R.A.S. 196*, 59P.
Warner, B. and Cropper, M. 1982, *M.N.R.A.S.*, in press.
Warner, B., Peters, W.L., Hubbard, W.B. and Nather, R.E. 1972, *M.N.R.A.S. 159*, 321.
Warner, B., O'Donoghue, D. and Fairall, A.P. 1981, *M.N.R.A.S. 196*, 705.
Warner, B., Cropper, M.S., Fairall, A.P. and Allen, S. 1982, *M.N.R.A.S.* in press.
Watts, D.J., Greenhill, J.G. and Thomas, R.M. 1980, *M.N.R.A.S. 191*, 25P.
Watson, M.G., Sherrington, M.R. and Jameson, R.F. 1978, *M.N.R.A.S. 184*, 79P.
White, N.E. and Marshall, F.E. 1980, *I.A.U. Circ.* No. 3514.
White, N.E. and Marshall, F.E. 1981, *Ap. J. 249*, L25.
White, N.E., Mason, K.O. and Sanford, P.W. 1977, *Nature 267*, 229.
Wickramasinghe, D.T. and Visvanathan, N. 1980, *M.N.R.A.S. 191*, 589.
Wickramasinghe, D.T., Stobie, R.S. and Bessell, M.S. 1982, preprint.
Williams, G. and Johns, M. 1980, *I.A.U. Circ.* No. 3519.

DISCUSSION FOLLOWING B. WARNER'S TALK

TAPIA: The orbital period of all intermediate polars seems to be larger than the orbital period of the polars, do you have anything to say about that?

WARNER: I don't have any evolutionary model to suggest. As an observer all I can say is that this is what we are given. You will notice there is a strong concentration of these periods around 3.5 hours.

CHANMUGAM: If they have long orbital periods that does not rule out the fact that they could have strong fields, because the Alfven radius may not extend to the companion and the white dwarf could be spinning rapidly. Have you done optical polarimetry on these objects?

WARNER: The upper limits on the circular polarimetry are all around about 0.06% or so.

TAPIA: The time resolution that we can achieve is of the order of a minute and I think that we would notice anything for those that have 20-13 min polarimetric period, but we haven't. If the polarimetric modulation is faster than that, we may have do do fast polarimetry.

ROBINSON: Can I ask how you explain the simultaneous presence of two periods in WZ Sge?

WARNER: At the moment I don't have a full explanation of any of the very short period non coherent ones.

ROBINSON: But WZ Sagittae is coherent.

WARNER: Yes. What I was trying to draw attention to there, is that there is a sequence, as you look at the DQ Her stars which are coherent through UX UMa which is not coherent (but is not bad over a night) and through the very poorly coherent ones in the dwarf novae and if you believe the model of DQ Her then I think you almost have to accept the same model for UX UMa and by extrapolation to all the dwarf novae and I would have to include WZ Sagittae, but I don't have a clear picture.

LAMB: I would like to comment on the previous exchange. From analysis of the spin-up behaviour of these sources we can definitely say that the long period DQ Her stars as a group have field strengths that are about of factor of 10 less than the AM Her stars. There is no reason I know why that necessarily has to be so, unless the strength of the magnetic field has an effect on the evolution of the binary. So I expect that we should eventually find some long period DQ Her stars with fields comparable to the AM Her stars, just as we do have one short period system, EX Hydrae, with a period which is significantly less than that of AM Her and which is not synchronously rotating. That is because it also has a magnetic field which is about 10 times less than the other AM Her stars.

COWLEY: In the case of the systems like V2051 it is not straight forward to tell from spectroscopy whether the eclipse is of the star or a region very close to the star and the hot spot, the phasing in spectroscopy would be remarkably different.

WARNER: Yes, we have no spectra. The history of this is that Angel discovered it as an eclipsing binary about 1977, but nothing was published and then Howard Bond at the Rochester meeting said that Angel's period of 96 min was wrong and that it was really 90 min and again nothing was published, finally last year I started observing it and as far as I know this is the first detailed photometric study, but I have no spectra of it.

TAPIA: We see striking spectral similarities between the intermediate polars and the polars, isn't that evidence that the emission lines are coming from a region where the magnetic field is not important or that it is weak?

LAMB: I am not sure that that is the case. The magnetic field probably does not produce measurable Zeeman splitting of the emission lines, but it almost certainly does channel the accretion flow.

ENERGY DISTRIBUTION OF HARD X-RAY EMITTING CATACLYSMIC VARIABLES

Martine Mouchet
European Southern Observatory
8046 Garching bei München - FRG

ABSTRACT

We present the energy distribution of three hard X-ray sources H2252-035, 4U1849-31 and 2A0526-328 identified with intermediate polars. The ultraviolet and optical continuum can be well described by the standard disc model but this requires rather large discs and high accretion rates. Other possible contributions to the energy distribution are estimated. The predicted high accretion rates as well as the amplitude of the optical pulsations are in disagreement with the observed X-ray luminosity suggesting the existence of an unseen soft X-ray or extreme UV component.

1. INTRODUCTION

Among the hard X-ray sources recently identified with cataclysmic variables, three of them: H2252-035 (AO Psc), 4U1849-31 (V1223 Sgr) and 2A0526-328 (TV Col) exhibit a class of common properties. They were optically identified on the basis of their spectra showing blue continuum and strong hydrogen and helium emission lines. They differ from normal dwarf novae, being in a persistent high state without recorded outbursts.

These sources are multiperiodic systems. H2252-035 shows three optical periodicities: 3.59 hours, 805s and 859s (Patterson and Price, 1981). The longest one is interpreted as the orbital period. The 805s period being also present in the X-ray flux is likely to be associated with the rotation of the compact star. The second pulsation is the beat period between 805s and 3.59h and can be explained by the reprocessing of X-rays on a disc bulge or on the secondary in a prograde system. 4U1849-31 shows an optical pulsation (P = 794s) similar to those observed in H2252-035 but up to now no orbital period is known[1] (Steiner et al., 1981). 2A0526-328 displays two photometric periods, 5.2h and 4.02 days (Motch, 1981). A spectroscopic period of 5.5h has been discovered by Hutchings et al. (1981). The 4.02 day period appears to be the beat-

M. Livio and G. Shaviv (eds.), Cataclysmic Variables and Related Objects, 173–180.

ing period of the two short ones. Essentially two models have been proposed: Hutchings et al. (1981) identify the spectroscopic period with the orbital one, the 5.2h photometric period corresponding to the rotation of the white dwarf. Watts et al (1981) assume that the orbital period is the 4 days period and the spectroscopic one is the period of the rotation of the white dwarf. This last model requires a retrograde orbit.

These sources differ from the AM Her type systems in several aspects: no soft X-rays are detected and they do not exhibit polarisation, the systems are non synchronized and the corotation radius is probably smaller than the Roche lobe of the compact star allowing the formation of an accretion disc. We have attempted to test the presence of such a disc from the energy distribution in ultraviolet and optical.

2. OBSERVATIONS

The ultraviolet observations were made with the IUE satellite using the low dispersion mode and the large aperture. They were collected by ourselves or from the data bank of the ESA station. They were reduced using the revised calibration (ESA newsletters no. 11). The ultraviolet and optical data of H2252-035 used here were already published by Hassall et al. (1981). A first presentation of the ultraviolet spectrum of 4U1849-328 is in Bonnet-Bidaud et al. (1982). We added in this study three other spectra (two at short wavelengths and one at long wavelengths) obtained in January 81. The optical UBVRI colours are those given in Bonnet-Bidaud et al. (1982). Ultraviolet observations of 2A0526-328 were presented by Coe and Wickramasinghe (1981) and Mouchet et al. (1981). We present here average ultraviolet fluxes obtained from 8 short wavelength spectra and 5 long wavelength ones.

3. ACCRETION DISC FIT TO THE ENERGY DISTRIBUTION

The three sources exhibit a strong UV continuum with emission lines superimposed on it. Indication of an interstellar reddening is present in the average long wavelength (LWR) spectra around 2200Å. The colour excess was determined by fitting the LWR spectra with a power law reddened according to the near galactic extinction law given by Seaton (1979). The best values for E_{bv} are respectively 0.11, 0.16 and 0.06 for H2252-035, 4U1849-31 and 2A0526-328. The result for H2252-035 is in disagreement with the evaluation given by Hassall et al. (1981) who neglected reddening.

A standard optically thick disc model described by Bath et al. (1980), and references therein, was fitted to the UV and optical data (including IR for H2252-035). The shape of the disc continuum is determined by two parameters: a characteristic temperature T_* and the ratio R_0/R_1 of the outer and the inner radii of the disc, while the normalization factor is related to the distance, the inner radius and the inclination angle of the system.

The best fits determined by a minimum χ^2 routine are plotted in Figure 1. The corresponding parameters are given in Table 1. We did not use optical data of 2A0526-328 because of their great variability. The fitted disc from the UV data leads to a B-V colour slightly redder than observed. Assuming that the disc fills about 90% of the Roche lobe of the primary we can compute the inner radius, then the accretion rate and the luminosity of the disc are derived from T_*. An orbital period of 5.5h was assumed for 2A0526-328. The inner disc radius of 4U1849-31 must be small (~5 10^8 cm) in order to be compatible to an orbital period of a few hours. The deduced parameters are given in Table 1. As no eclipse is observed in these systems a reasonable upper limit for i is 70°.

A standard disc model can reasonably well fit the energy distribution of the three sources. The required temperatures are high implying a strong accretion rate and the inner radius of the disc is found to be very close to the white dwarf. This leads to a disc luminosity greater than the hard X-ray luminosity and would indicate the presence of an unseen soft X-ray or extreme UV component. Moreover, the resulting distance of 2A0526-328 is definitively too large.

Source	T_*(K)	R_0/R_1	$R_0(10^9$cm)	$R_1(10^9$cm)	d(pc)/$\sqrt{\cos i}$	$\dot{M}(10^{16}gs^{-1})$	L_D/L_X(2-10keV) (d = 300 pc)
H2252-035	140 000	75	30	0.4	460	9	44
4U1849-31	170 000	82	41[1]	0.5	625	37	50
2A0526-328	200 000	46	55	1.2	3650	978	1264

[1] Assuming an orbital period of 4h.

Table 1. Accretion disc fits parameters and deduced parameters

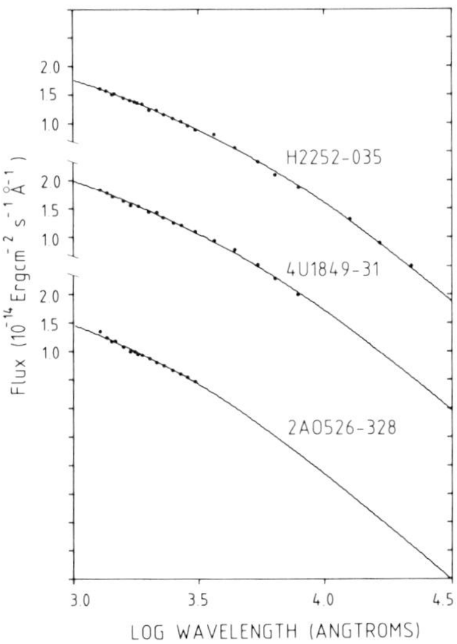

Figure 1: Energy distribution of H2252-035, 4U1849-31 and 2A0526-328. Drawn lines are the best fitted disc spectra with parameters given in Table 1.

4. OTHER POSSIBLE EMITTING COMPONENTS

The disc parameters derived in the previous section may be modified if the ultraviolet and optical fluxes are not entirely due to the accretion disc. In this paragraph we discuss other possible contributing emission regions.

4.1 The Secondary Star

Low mass red dwarf secondaries are expected to contribute at long wavelengths yielding an overestimation of the R_0/R_1 radii ratio. Assuming that the secondary is a main sequence star we can compute the contribution of this star in H2252-035 and 2A0526-328 using the relations given by Robinson (1976) and Warner (1976). In H2252-035 the secondary has a negligeable contribution in V but accounts for half of the observed K flux at a distance of 300 pc. The secondary in 2A0526-328 would contribute to 1/10 of the V flux at 500 pc for an orbital period of 5.5h. IR observations are required to determine the nature of this secondary more accurately. Though no precise estimate of the R_0/R_1 ratio can be made due to the unknown distance, a decrease of this ratio would be compatible with the presence of a magnetized white dwarf which prevents the disc from extending down to this object. Nevertheless, changes in the optical-IR part leave unchanged the temperature estimates.

4.2 Pulsating Light

A detailed study of the energy distribution of the two optical pulsations of H2252-035 is presented in Motch and Pakull (1981). They found a different colour behaviour for the direct pulsation (at the X-ray period) and for the reprocessed one.

905s Pulsations (X-ray Period). The energy spectrum of the pulsated light at this period is in λ^{-4} consistent with the Rayleigh Jeans tail of a hot blackbody possibly emitted by a polar cap on the white dwarf or by the inner edge of the disc. This contribution could be due to steady nuclear burning of accreted matter in the envelope of the degenerate dwarf or to reemitted thermalized bremsstrahlung. Assuming that this blackbody accounts for less than 20% at 1 keV (no soft X-ray having been detected) and considering that it represents 12% of the flux at 3250Å (Motch and Pakull, 1981) we derived a maximum temperature of about 4.5 10^5K for an X-ray pulsed fraction of 40% to 100%. The corresponding emitting areas are respectively 18% to 45% of the surface of the white dwarf for a typical radius of 10^9 cm and a distance of 300 pc. Considering a typical temperature of 10^5K the corresponding blackbody luminosity is about 5 10^{33} $(d/100pc)^2$ $ergs^{-1}$ consistent with the disc luminosity inferred from the parameters given above (1.5 10^{34} $ergs^{-1}$). Substracting this 10^5K blackbody from the observed energy spectrum yields a new disc model fit of similar quality with a characteristic temperature of 65000K and a smaller ratio R_0/R_1 of 32. These new parameters do not affect significantly the accretion rate and the disc luminosity which remains in disagreement with the X-ray observed luminosity.

959s Pulsations. These optical pulsations probably arise from the reprocessing of the X-ray beam on a bulge or in the atmosphere of the secondary. The theoretical energy distribution of such a reprocessed flux is not well established, so that this contribution cannot be easily removed. We can reasonably estimate from the optical observations that about 10% of the flux in the 1200-7000Å range is due to heating. The studies of the low mass X-ray binaries have shown that 1% to 5% of the X-ray flux is reprocessed (Van Paradijs, 1982). These two assumptions lead to an expected X-ray luminosity ranging from 10^{34} to $5\ 10^{34}$ $ergs^{-1}$ for a distance of 300 pc. This value is consistent with the strong X-ray luminosity predicted from the disc model. This conclusion is also valid for 4U1849-31.

4.3 Hot Spot

Although the hot spot is a feature often observed in cataclysmic variables, such a signature is absent in the light curves of H2252-035 and 4U1849-31, possibly due to a moderate inclination angle or to a large extension of the disc. We cannot however exclude the presence of a bright spot in 2A0526-328 as it was shown in the study of UV spectra near the photometric maximum (Mouchet et al, 1981).

4.4 Influence of the Magnetic Field

The influence of the compact object magnetic field on the disc must be taken into account. Ghosh and Lamb (1979) showed that this magnetic field modifies the temperature distribution of a standard disc in a region whose size is essentially a function of the rotation of the degenerate star and which can extend up to the outer parts of the disc for reasonable rotational velocities. This new distribution in radial temperature can lead to a very different energy spectrum in shape as well as in absolute flux (for more details see A. Bianchini (these proceedings)). Such an effect can be expected for the three sources in which an indication of an anisotropic X-ray emission is suggested.

5. VARIABILITY

The study of the energy distribution variability especially in the UV range is a key for understanding the nature of the different spectral contributions.

5.1 4U1849-31

Strong variations in the intensities of the four short wavelength IUE spectra are observed without any significant change in their slopes. An increase of 23% occurred between the two spectra taken in May 81 at 1.5h interval while the V magnitude deduced from the fine error sensor remained constant. This suggests the presence of at least two emitting regions. The study of spectral variations with the orbital period will certainly be valuable.

5.2 2A0526-328

The large number of available ultraviolet spectra made possible the study of the UV continuum variations with the different periods. The fluxes around 1770Å and 2700Å are plotted in Figure 2 versus the three phases using the ephemerides given by Hutchings et al. (1981). We have corrected the 5.2h photometric phase from the variation of the maximum of the optical light curve with the 4.02 days period (Motch 1981). No correlation of the UV flux is observed with both photometric periods but a regular variation is present versus the spectroscopic one showing a maximum around phase 0.6. This behaviour is consistent with the gradual decrease of the amplitude of the photometric 5.2h period with decreasing wavelength (B and U passbands) and with the presence of a modulation with the 5.5h spectroscopic period in the U band only, showing the same phasing as in the UV (Motch, private communication). None of the proposed models gives a clear explanation of this result.

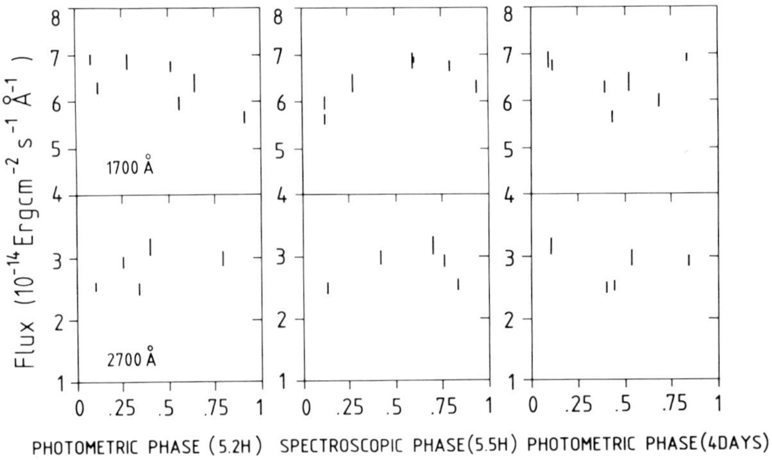

Figure 2: Ultraviolet fluxes of 2A0526-328 averaged in 50Å interval around 1770Å (upper part) and 2700Å (lower part) plotted versus the three optical phases.

CONCLUSION

Though a standard accretion disc can well fit the energy distribution of the three sources presented here, the presence of other emitting components is suggested both from optical and ultraviolet variability. The magnetic field of the compact star may also change the disc continuum. The amplitude of the optical pulsations and the evaluated disc parameters both lead to a predicted high energy luminosity larger than the hard observed X-ray luminosity.

ACKNOWLEDGEMENTS

I would like to thank C. Motch and J.M. Bonnet-Bidaud with whom this work has been done. I also thank A. Bianchini and M. Livio for stimulating discussions. I am grateful to Miss P. Bristow for the typing of the manuscript. M.M. acknowledges an E.S.O. fellowship.

NOTE

1. A possible orbital period of 3.22h is proposed by B. Warner (see these proceedings).

REFERENCES

Bonnet-Bidaud, J.M., Mouchet, M., Motch, C., 1982, Astron. Astrophys., 112, p.355
Coe, M.J., Wickramasinghe, D.J., 1981, Nature, 290, p.119
Ghosh, P., Lamb, F.K., 1979, Astrophys. J., 232, p.259
Hassall, B.J.M., Pringle, J.E., Ward, M.J., Whelan, J.A.J., Mayo, S.K., Echevarria, J., Jones, D.H.P., Wallis, R.E., Allen, D.A., Hyland, A.R., 1981, M.N.R.A.S., 197, p.275
Hutchings, J.B., Crampton, D., Cowley, A.P., Thorstensen, J.R., Charles, P.A., 1981, Astrophys. J., 249, p.680
Motch, C., 1981, Astron. Astrophys., 100, p.277
Motch, C., Pakull, M.W., 1981, Astron. Astrophys., 101, L9
Mouchet, M., Bonnet-Bidaud, J.M., Ilovaisky, S.A., Chevalier, C., 1981, Astron. Astrophys. 102, p.31
Patterson, J., Price, C.M., 1981, Astrophys. J., 243, L83
Robinson, E.L., 1976, Ann. Rev. Astron. Astrophys. 14, p.119
Seaton, M.J., 1979, M.N.R.A.S., 187, 73p
Steiner, J.E., Schwartz, D.A., Jablonski, F.J., Busko, I.C.,
Watson, M.G., Pye, J.P., McHardy, I.M., 1981, Astrophys. J., 249, L21
Van Paradijs, J., 1982, in "Accretion-driven Stellar X-ray Sources", in press
Warner, B., 1976, IAU Symposium 73, p.85
Watts, D.J., Greenhill, J.G., Hill, P.W., Thomas, R.M., 1982, M.N.R.A.S., 200, 1039

DISCUSSION FOLLOWING M. MOUCHET'S TALK

BIANCHINI: Do you think that the Ghosh and Lamb temperature distribution is a better explanation for what we observe, so that we need smaller inner radii and smaller accretion rates?

MOUCHET: The problem is that the standard disk model, without magnetic fields, already fits well, but we need to include other contributions as I have described, so it is difficult to conclude whether the Ghosh and Lamb model fits better.

LANGER: Do you think that you have seen, between the optical, UV and X-ray emission, most of the luminosity that you refer to as present in the system or do you actually think that there is a fair bit that is missing, that you have not seen?

MOUCHET: We need to observe maybe in very soft X-rays, because we don't observe soft X-rays, or in extreme UV, you must have energy there, but it is not observed.

COLUMN ACCRETION ON TO WHITE DWARFS

A. R. King
Astronomy Department, University of Leicester,
Leicester LE1 7RH, England.

The problem of column accretion on to white dwarfs in the AM Her and DQ Her systems is reviewed. Particular attention is paid to recent progress in explaining the large soft X-ray - EUV fluxes observed in these systems in terms of nonlocal electron energy transport into the white dwarf photosphere.

1. INTRODUCTION

In the AM Her and DQ Her systems ('polars' and 'intermediate polars' respectively) the magnetic field of the white dwarf is thought to be strong enough to channel the accretion flow from the secondary star radially on to the magnetic polecaps in a column-like configuration (Fig. 1). Because most of the accretion energy must be released near the white dwarf surface an understanding of column accretion is basic to any discussion of these systems. Moreover, from a theoretical point of view an accretion column represents almost the simplest accretion problem one can pose, being essentially one-dimensional. We might therefore hope for a more rapid progress in our understanding here than in the study of more complicated accretion flows. For these reasons the study of column accretion on to white dwarfs has received growing attention in recent years.

Figure 1 gives a basic picture of an accretion column (not to scale). The cool, freely-falling (and thus highly supersonic) accretion flow is channelled by the magnetic field on to a small fraction f ($\lesssim$ one percent) of the white dwarf surface area. For accretion rates relevant to the AM Her and DQ Her systems the accreting ions cannot penetrate to the white dwarf photosphere before being stopped in the atmosphere. Thus there must be some kind of standing shock above the polecaps in which the kinetic energy of the accretion flow (residing mostly in the ions) is thermalized. The precise nature of this shock and the 'settling' flow beneath it are the main things one hopes to learn from a study of column structure; given these one can for example predict where in the electromagnetic spectrum the accretion luminosity will ultimately be radiated.

M. Livio and G. Shaviv (eds.), Cataclysmic Variables and Related Objects, 181–197.

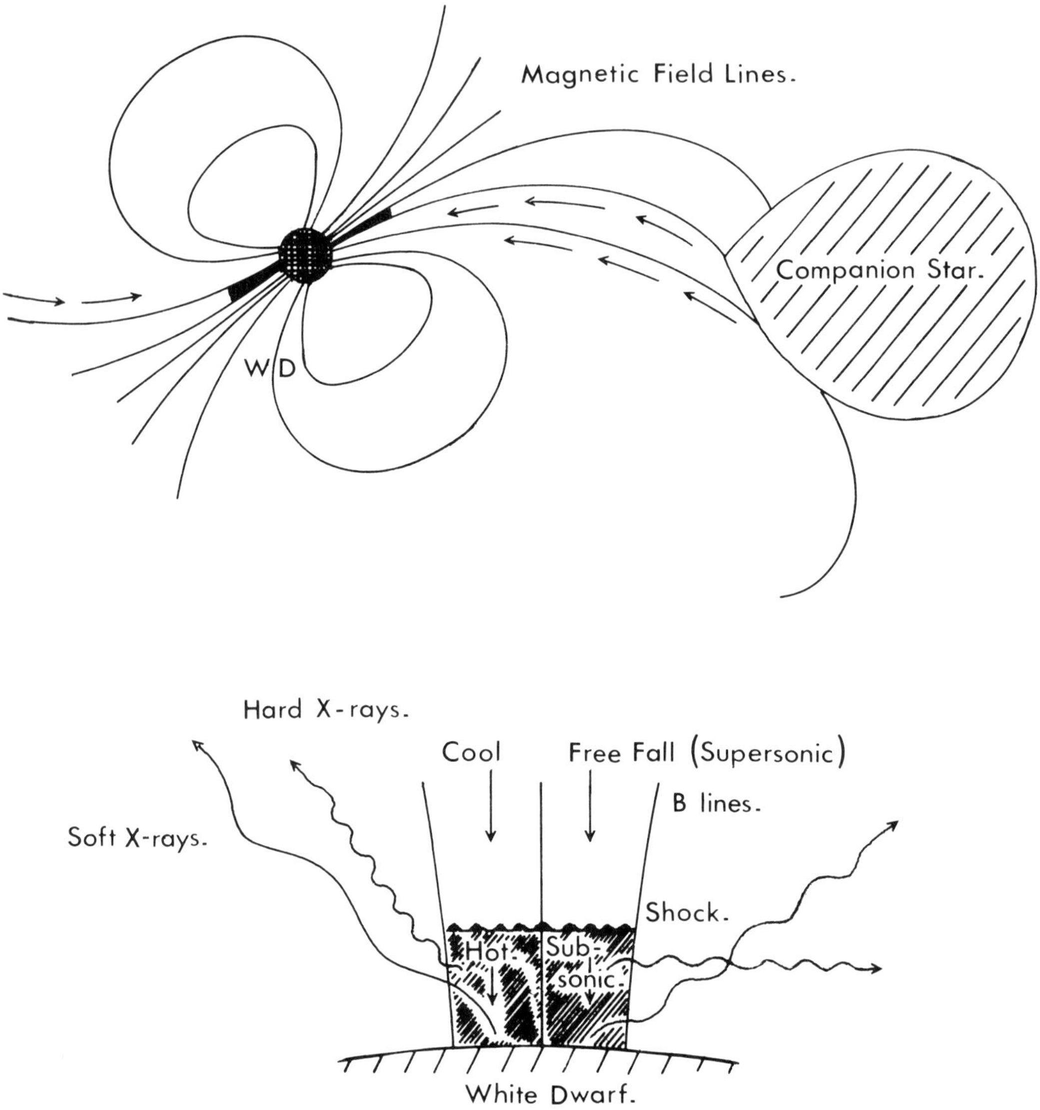

Figure 1. Accretion columns in an AM Her system. In a DQ Her ('intermediate polar') system the upper picture would be replaced by one in which the accretion flow from the companion star initially orbits the white dwarf in a disc, the disc being disrupted by the white dwarf magnetic field at some radius, whereupon the matter flows down field-lines to the magnetic poles. The two types of system are similar in the flow near the polecaps shown in the lower picture (not to scale).

In a steady state the accretion energy must be removed from the post-shock flow at the same rate at which it is supplied: it is the manner in which this removal is carried out, for example by direct radiation or particle transport processes, which determines the column structure. Note that this leaves open the possibility of distinct self-consistent column solutions for the same white dwarf parameters and accretion rate. We shall see examples of this in what follows, i.e. radiation-dominated column structures in which transport processes are negligible, and vice versa. The question of which solution actually occurs in a given case is then presumably decided either by stability considerations or by the history of the system. Langer *et al.* (1981, 1982) and Chevalier and Imamura (1982) have discussed the stability of radiative columns.

2. THE SOFT X-RAY PROBLEM

Historically the first type of solution to be discussed was the radiative one, in which the postshock flow cools by radiating the accretion luminosity directly. Since in this case the gas is characterized by the adiabatic shock temperature

$$T_s = \frac{3}{8}\frac{GM\mu m_p}{kR} = 3.7 \times 10^8 \text{ K } M_1 R_9^{-1} \tag{1}$$

(where $M = M_1 M_\odot$, $R = 10^9 R_9$ cm are the white dwarf mass and radius, and the other symbols have their usual meanings) most of this 'primary' emission is in the form of bremsstrahlung X-rays in the 10 - 50 keV region.[1] Because most of the AM Her stars are hard X-ray sources this was seen as encouraging, and a large number of models, of varying degrees of elaboration, were constructed on this basis both for spherical accretion (Hoshi, 1973; Aizu, 1973; Katz, 1977; Lamb and Masters, 1979; Kylafis and Lamb, 1979; Imamura *et al.*, 1979, Wada *et al.*, 1980) and, more relevantly for the AM Her and DQ Her stars, for column accretion (Fabian *et al.*, 1976; King and Lasota, 1979).[2] However, all such models were found wanting when soft X-ray and UV observations of AM Herculis became available.

For radiative columns with realistic accretion rates the shock height $D \ll$ white dwarf radius R, so that one half of the primary hard X-radiation from the shocked gas is intercepted by the white dwarf surface. Of this some fraction a_X is elastically scattered, while the remaining X-rays are absorbed by the white dwarf envelope and re-emitted as blackbody radiation with characteristic temperature

$$T_b = 1.2 \times 10^5 \text{ K } \dot{M}_{16}^{\frac{1}{4}} f_{-2}^{-\frac{1}{4}} M_1^{\frac{1}{4}} R_9^{-\frac{3}{4}} f_{soft}^{\frac{1}{4}} \tag{2}$$

Here $\dot{M}_{16}$ is the accretion rate in units of 10^{16} g s^{-1} $\simeq 1.6 \times 10^{-10}$ $M_\odot$ y^{-1}, $f_{-2} = 10^2 f$ and $f_{soft} = \frac{1}{2}(1-a_X)$ is the fraction of the total accretion luminosity.

$$L_{acc} = 1.33 \times 10^{33} \dot{M}_{16} M_1 R_9^{-1} \text{ erg s}^{-1} \tag{3}$$

which is re-emitted. Hence radiative column models predict that the luminosities L_X in hard X-rays (characterized by T_s) and L_{soft} in soft X-rays - EUV (characterized by T_b) should be in the ratio

$$\frac{L_X}{L_{soft}} = \frac{\frac{1}{2}(1 + a_X)}{\frac{1}{2}(1 - a_X)} \sim 2 \tag{4}$$

for a typical albedo $a_X \sim 0.3$ (Felsteiner and Opher, 1976). Early soft X-ray measurements already showed a spectrum rising steeply towards low energies, which if interpreted as the Wien tail of a blackbody spectrum violated (4) by large factors. Recognizing this King and Lasota (1979) attempted to salvage the existing radiative column models by adopting a non-blackbody interpretation of the soft X-ray spectrum. However the door was firmly shut on all such attempts by the phase-dependent UV observations of Raymond *et al.* (1979) using the IUE satellite. These showed that the component of the 3000 Å - 1200 Å flux eclipsed at the same binary phase as the hard and soft X-rays had a Rayleigh-Jeans distribution, with flux levels consistent with a single blackbody interpretation of soft X-rays and UV. The situation is shown in Fig. 2, where recent observations of Fabbiano *et al.* (1981) and Rothschild *et al.* (1981) are plotted. Clearly the total luminosity in the 'soft' component is very uncertain: the characteristic temperature T_b means that most of this component is hidden by interstellar absorption in the 13.6 eV - 100 eV range. From Fig. 2 we have

$$\left.\frac{L_X}{L_{soft}}\right|_{observed} \lesssim 0.1 - \text{few} \times 10^{-3} \tag{5}$$

The value 0.1 results from a very conservative interpretation between 10 and 100 eV in Fig. 2, and the small value from the blackbody interpretation. The large discrepancy between (4) and (5) implies that for AM Her simple radiative column models fail to account for rather more than 90% of the luminosity. Wherever sufficiently good data exist for other sources to make possible a comparison with theory a similar discrepancy exists. Further, the X-ray reprocessing which seems to be required to explain the beat periodicities observed in DQ Her systems like H2252-037 also calls for large unseen fluxes, presumably in soft X-rays, exceeding the hard X-ray flux by factors $\sim 10^2$ (e.g. Hassall *et al.*, 1981).

3. REMEDIES

There have been three main attempts to explain the large soft X-ray excesses of these systems exemplified by (5).

(i) Steady Nuclear Burning

The first approach is to assume that the radiative column models described above are broadly correct, but that the accreted material under-

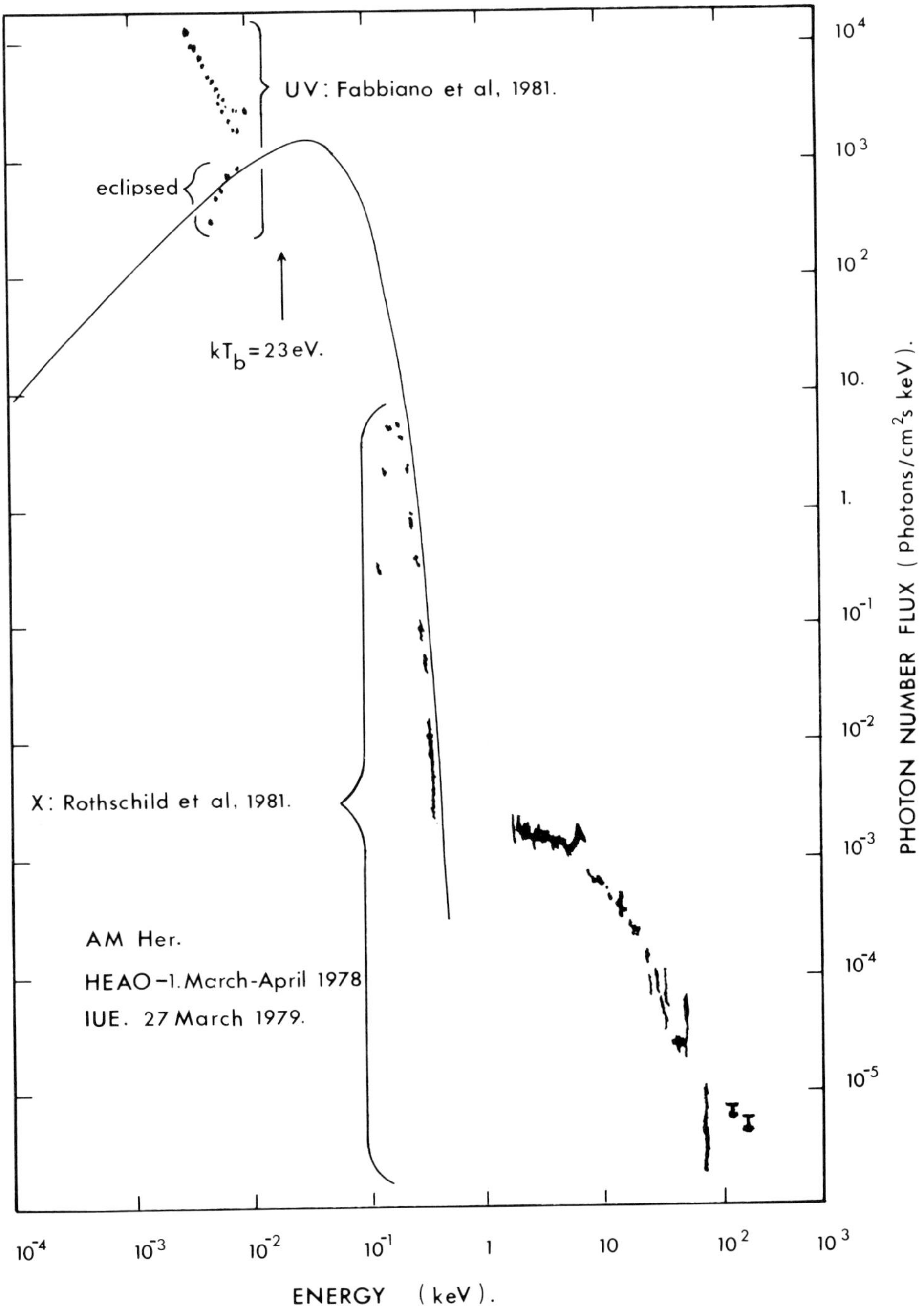

Figure 2. UV, soft and hard X-ray observations of AM Her (Fabbiano *et al.*, 1981; Rothschild *et al.*, 1981) showing the very large soft X-ray - EUV component of the eclipsed flux. The solid curve shows a 23 eV blackbody. The region between 13.6 eV and $\sim$ 0.1KeV is inaccessible to observation because of interstellar absorption.

goes steady nuclear burning at the base of the column, providing a large blackbody component. The main difficulty here is that steady (as opposed to runaway) burning does not seem to be possible at the required luminosities. The calculations of Papaloizou *et al.* (1982) imply an upper limit

$$L_{burn} \lesssim 2 \times 10^{32} \text{ f erg s}^{-1}$$

which is far too low to give the blackbody luminosity ($\sim 10^{34}$ erg s^{-1}, Fabbiano *et al.*, 1981) inferred for AM Her even if f = 1, which is very unlikely. There is a possible further stable burning regime if the accretion rate per unit surface area can be made to exceed the equivalent of $10^{-7}\ M_\odot\ y^{-1}$ over the whole white dwarf surface (e.g. Paczynski and Zytkow, 1978), implying a lower limit

$$L_{burn} \gtrsim 4 \times 10^{37} \text{ f erg s}^{-1} ,$$

which is much too high for the AM Her and DQ Her systems unless f is considerably smaller than currently believed.

(ii) Unsteady Flow

In a recent preprint, Kuijpers and Pringle (1982) have pointed out that if the accretion flow is sufficiently inhomogeneous and unsteady the bulk of the accreting matter could penetrate to large optical depths in the white dwarf envelope before releasing its energy, thus giving a large blackbody flux. At the time of writing it is unclear whether the required (high) degree of inhomogeneity (due to cooling instabilities in the accretion flow) can be realized.

(iii) Non-Radiative Transport into the White Dwarf Photosphere

In this approach one notes that the prediction of similar hard and soft X-ray fluxes (4) arises because of the assumption that all the losses from the shocked column are radiative. If most of the energy from this region could be transported under the white dwarf photosphere before being radiated, a ratio L_X/L_{soft} closer to the observed value (5) would be predicted. We follow here the simple analysis of King and Lasota (1980), who investigated the possibility mentioned by Fabian *et al.* (1976) that column solutions might exist which were dominated by electron thermal conduction. Such a solution requires a conductive flux

$$F_{cond} \sim L_{acc}/4\pi R^2 f \qquad (6)$$

into the white dwarf. In the usual diffusion approximation, F_{cond} is related to the temperature gradient dT/dz in the column by the thermal conductivity K:

$$F_{cond} = K \frac{dT}{dz} \quad . \qquad (7)$$

For the ionized postshock gas of density N and temperature T

$$K \sim \frac{Nk^2T}{m_e v_e} \lambda_{ei} \quad .$$

Here k is Boltzmann's constant, m_e is the electron mass, $v_e = (kT/m_e)^{\frac{1}{2}}$ and λ_{ei} is the mean free path. (This yields the familiar expression K = constant x $T^{5/2}$ when evaluated.) Also at the shock we must have the approximate equality (from energy conservation)

$$L_{acc}/4\pi R^2 f \sim NvkT \quad ,$$

where $v \sim (kT/m_p)^{\frac{1}{2}}$ is the postshock gas velocity (m_p = proton mass). To estimate the shock height D for this type of column structure we set $dT/dz \sim T/D$. Using all these relations in (6) yields

$$D \sim (m_p/m_e)^{\frac{1}{2}} \lambda_{ei} = \lambda_{eq} \tag{8}$$

Here λ_{eq} is the typical lengthscale over which the ions and electrons of the shocked flow establish equipartition. Since this is much less than the radiative shock height $D_{rad} \sim vt_{ff}$, with t_{ff} the free-free cooling time of the gas, this crude picture is self-consistent in that radiative losses from the column are small; a ratio L_X/L_{soft} of the observed order (eq. (5)) is predicted. However equation (8) means that two of the assumptions made in constructing this picture are questionable: first, since $D \sim \lambda_{eq}$, it is not clear that equipartition between ions and electrons is established, so that a full two-fluid treatment is required in which the electron and ion temperatures are allowed to differ. Second, since D is so small the mean free paths of electrons in the high energy tail of the Maxwellian distribution can exceed it: that is, the high energy electrons behind the shock are not 'localized' since there is a large difference in temperature over a mean free path. In particular this means that the diffusion approximation (7) for the conductive flux is poor, since this explicitly depends on being able to represent the temperature difference over an electron mean free path as $\Delta T \sim \lambda_{ei}\, dT/dz$. We shall see later (Section 5) how these defects can be overcome by a full two-fluid treatment taking account of the energy transport by nonlocal electrons.

4. THE GLOBAL BEHAVIOUR OF ONE-FLUID ACCRETION FLOWS

Before proceeding to the two-fluid description, it is worth taking a closer look at one-fluid accretion flows. There are several reasons for this. First, any two-fluid flow will eventually achieve equipartition: after all it must join on to a slowly 'sinking' white dwarf envelope solution at large optical depths, where equipartition must certainly hold. This is seen for example in the radiative (spherical) solution of Imamura *et al.* (1979), where because of the very high white dwarf mass assumed (1.4 $M_\odot$) Compton cooling of the postshock electrons by the black-body photon flux keeps them initially cooler than the ions, equipartition being achieved for $T \lesssim 0.3\ T_S$. For lower-mass white dwarfs Compton cooling is much less important (Frank *et al.*, 1982, eq. (13)): for $M < 0.5\ M_\odot$ it is less than 4% of free-free losses even near the shock, so that radiative columns for moderate-mass white dwarfs are effectively in equipartition throughout. Finally one would like to understand the relation of the radiative and conduction-dominated flows to each other: as these are solutions of the same set of equations the distinction must be specified by the boundary conditions.

Figure 3 (taken from Frank *et al.*, (1982) shows the global structure of all column flows in which equipartition holds and thermal conduction can be represented by the diffusion approximation (7). The variables θ (= T/T_S) and ϕ are dimensionless measures of the temperature and conductive flux at any point in the column: a column flow in which the matter is in equipartition throughout starts near $\theta = 1$ ($T = T_S$) at the shock and moves to very small values of $\theta \simeq T_b/T_S \sim 10^{-3}$ at the base of the column. A two-fluid solution reaching equipartition at $T < T_S$ starts at a smaller value of θ and again moves to $\theta \simeq T_b/T_S$. Positive values of ϕ represent a heat flow into the white dwarf; for $\phi < 0$ the flow is reversed. It is immediately apparent from the figure that all equipartition flows can be divided into just two families, labelled types 1 and 2; they are separated by a critical solution ($\phi = \phi_{crit}$ on the figure) which is the unique one satisfying $\phi = 0$ at $\theta = 0$. It can be shown (Frank *et al.*, 1982) that all column solutions have negligible optical depth so that optically thin radiative cooling may be assumed throughout.

Let us examine the type 2 solutions first. It is a simple matter to show from the equations describing the flow that the temperature T must have a minimum wherever the conductive flux F_{cond} vanishes. Thus the type 2 solution curves pass through $\phi = 0$ at their minimum value of θ and bend away to increasing values of θ with $\phi < 0$. (It is for this reason that a non-zero conductive flux is required behind the shock, so that by energy conservation T is slightly less than T_S - see Frank *et al.*, 1982 for details.) Because we want an accretion flow which cools to temperatures $T_b \sim 10^{-3}\ T_S$ we are only interested in solution curves which approach very close to the ϕ-axis ($\theta \simeq 0$). From the inset it can be seen that the type 2 solutions which do this run very near to the critical solution. Near the point $\phi = 0$ the type 2 solutions can be matched to a slowly sinking atmosphere in radiative equilibrium, so that the $\phi < 0$ part of the solution curve is not realized. A unique equipartition solution is then given by choosing the matching temperature so that the photospheric luminosity of the atmosphere ($\simeq$ blackbody luminosity at T_b) is just that supplied by X-ray heating from the column above. (Clearly conduction supplies nothing to the atmosphere as $\phi = 0$ near $T = T_b$.) Since the resulting solution runs so close to the critical solution and always has a small value of ϕ, this type of cooling flow is for many practical purposes indistinguishable from radiative flows where conduction is entirely neglected and the (singular) boundary conditions $v = 0$, $T = 0$ are adopted at the base of the flow (e.g. Aizu, 1973).

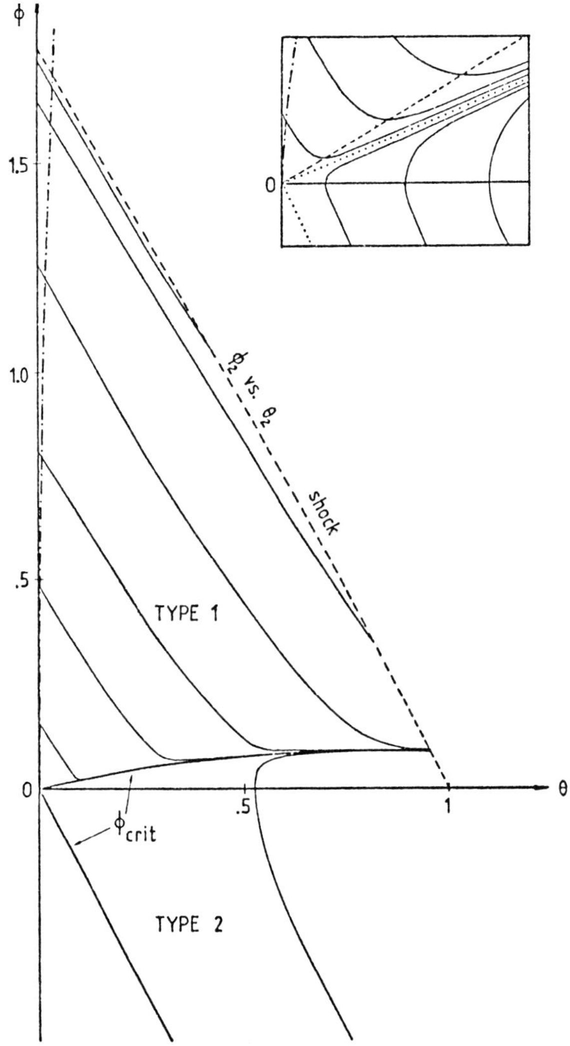

Figure 3. The global behaviour of all equipartition column solutions (Frank *et al.*, 1982). Each solution is characterized by the relation between its conductive flux and temperature, here represented by the dimensionless quantities ϕ and θ (= T/T_S). The solutions divide into two families (types 1 and 2) described in the text, corresponding to conductive- and radiative-dominated solutions respectively. The two families are separated on the ϕ - θ plane by the critical solution (labelled 'ϕ_{crit}') which is the unique one satisfying $\theta = 0$ at $\phi = 0$. The inset shows an expanded (not to scale) version of the neighbourhood of the origin. In all type 1 solutions the conductive flux ultimately 'saturates' (dot-dash curve) and the diffusion approximation breaks down (see text). The locus of minima of the type 1 solutions (dashed curve on inset) gives Aizu's (1973) solution in which conduction is entirely neglected; this runs very close to the critical solution (dotted).

The global behaviour of the type 1 solutions is quite different. As can be seen from Fig. 3 all of these reach low values of θ. All of them have minima in ϕ very close to the critical solution (this is best seen on the inset); as θ decreases further ϕ increases monotonically. Because $\phi \propto T^{5/2}\, dT/dz$ and T is decreasing, this implies that the temperature gradient dT/dz becomes very large - so large that on the dot-dash curve close to the ϕ-axis the conduction 'saturates'. This means that the temperature scale length $L = T/|dT/dz|$ has become comparable with the mean free path of the average electron, so that the diffusion approximation (7) for the conductive flux breaks down. This is just the behaviour we discovered in our crude analysis of conduction-dominated solutions in the last section. Indeed the resemblance becomes closer when one realizes that for consistency the saturation must occur at a temperature such that nonlocal electrons can transport the accretion energy under the white dwarf photosphere. This implies saturation at temperatures comparable to T_s (see below) so again it is unclear that equipartition would have been achieved.

5. TWO-FLUID ACCRETION COLUMNS WITH NONLOCAL ELECTRON ENERGY TRANSPORT

The discussion of the last two sections shows that to construct column models in which nonradiative energy transport is dominant a two-fluid treatment with nonlocal electrons is required. In this section I shall describe an approach to this problem developed by Dr. J. Frank and myself.

Let us consider first just how nonlocal the energy-transporting electrons are required to be. In order to penetrate the photosphere we require that the lengthscale λ_E over which a nonlocal electron deposits its energy in the white dwarf atmosphere should exceed the average photon mean free path $\lambda_{ph} \simeq (\kappa_b\rho_b)^{-1}$, where κ_b, ρ_b are the (Rosseland) opacity and density of the atmospheric plasma (i.e. we require the electron to penetrate to optical depth $\tau_E = \kappa_b\rho_b\lambda_E > 1$). Because for electrons with velocities $v_e \gg (kT_b/m_e)^{\frac{1}{2}}$ the deflection time t_D to electron-electron collisions is shorter than the energy exchange time t_E the electron will random-walk a distance $\lambda_E = (t_E t_D)^{\frac{1}{2}}\, v_e$ into the white dwarf. Using standard expressions to relate t_E, t_D to v_e, T_b and ρ_b, and recalling that the photospheric gas pressure $\rho_b kT_b/\mu m_p$ must approximately equal the ram pressure of the accretion flow, one finds that λ_E will exceed λ_{ph} (given by Kramers' opacity) for electron velocities

$$v_e \gtrsim v_{min} = 10^{10}\ \mathrm{cm\ s^{-1}}\ (\dot{M}_{16}/f_{-2})^{1/20}\ .\ M_1^{3/20}\ R_9^{-5/20} \tag{9}$$

(Frank *et al.*, 1982). A very similar numerical value for v_{min} results when electron scattering is more appropriate than Kramers', with a similar extremely weak dependence on white dwarf parameters and accretion rate (this arises because $\lambda_E \propto v_e^5$). If we were to require an electron temperature T_e such that the average electron had $v_e > v_{min}$ we would find $T_e \simeq 0.5\ T_s$; this is the result referred to at the end of Section 4. Our simple considerations of Section 3 suggest that the height of the shocked column will be $D \lesssim \lambda_{eq}$. It is easy to check that electrons with $v_e > v_{min}$ have mean free paths exceeding D, so they are indeed nonlocal.

The main practical obstacle remaining is the question of how to represent the energy transport by the nonlocal electrons. We note that as far as the electron gas in the column is concerned this is a loss process like, for example, a radiative loss process (although it is much more efficient). Indeed one can think of the column as being 'optically thin' to these electrons, so that energy 'leaks' into the photosphere directly. Analogous situations are encountered in plasma physics and stellar dynamics; if the timescale for the energetic particles to escape the region of interest is short compared to the timescale to refill by relaxation (collisions) the Maxwellian tail of the local distribution function the standard recipe is the so-called 'loss-cone' approximation. In our case the condition for the loss-cone approximation to hold is that the time of flight down the column for the escaping electrons should be short compared to the relaxation time for refilling the Maxwellian tail. Thus the loss-cone approximation will describe the 'leak' process well provided there exists a second minimum electron velocity v'_{min} such that

$$t_{flight}(v'_{min}) = \frac{D}{v'_{min}} < t_{tail}(v'_{min}) \tag{10}$$

where $t_{tail}(v'_{min})$ is the relaxation time for $v > v'_{min}$. If this holds the volume loss rate to the electron gas due to the leak process can be written

$$j_{leak} = \frac{\xi P_e}{t_{tail}} \quad . \tag{11}$$

Here $P_e = N_e k T_e$ is the electron pressure: with $v = \max(v_{min}, v'_{min})$, ξ is the fraction of the local electron distribution with $v_e > v$ directed towards the white dwarf, and t_{tail} the relaxation time for $v_e > v$. Note that (10) must be used as a consistency requirement: having found a column structure using (11) we must insert the calculated value of D to check that (10) holds, so that the structure is self-consistent.

The results of a full two-fluid calculation using (11) are shown in Figure 4. As usual for plasma shocks (see e.g. Zeldovich and Raizer, 1966) there is an ion shock where the ion temperature rises abruptly to a value $\sim 2T_s$ (since the electrons are not yet taking their share of the thermalized kinetic energy). The electron gas before the ion shock is 'preheated' by classical (diffusive) conduction from the shocked region. Behind the ion shock the ions attempt to heat the electrons collisionally to bring about equipartition. This is opposed by the very efficient loss of energy from the electron gas into the white dwarf photosphere given by (11). These two terms, the heating of the electrons by the ions ($\propto N_e^2 T_e^{-3/2} (T_i - T_e)$), and the leak term ($\propto \xi N_e^2 T_e$, eq. (11))

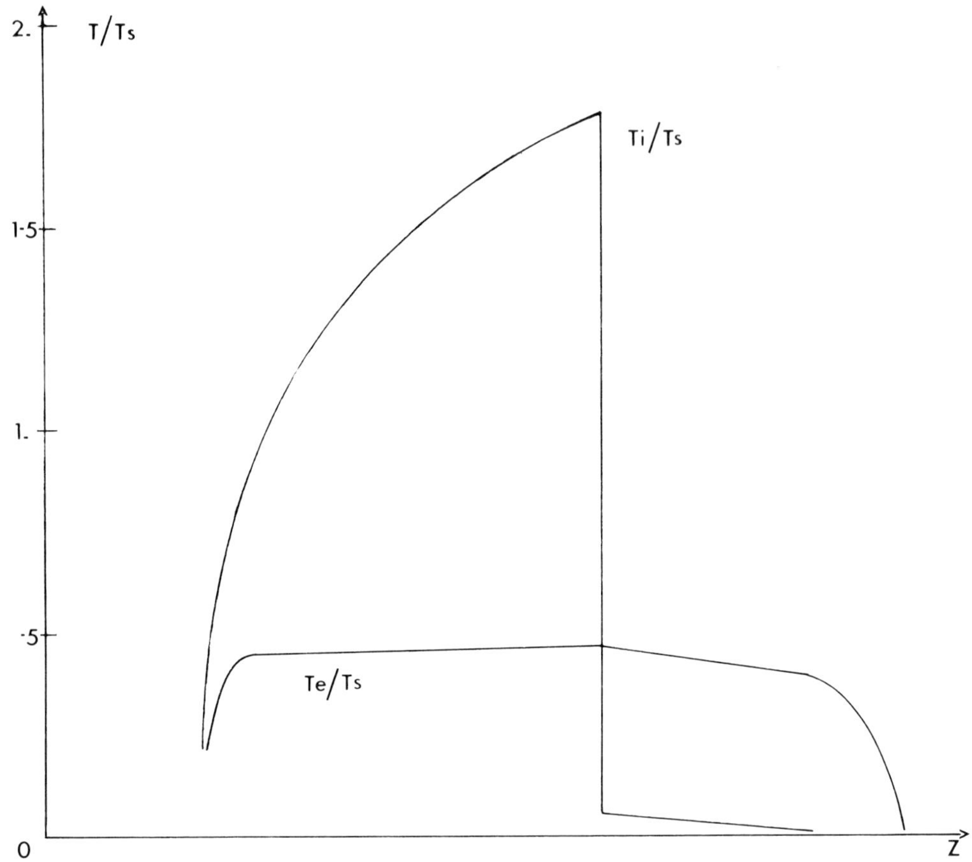

Figure 4. Electron and ion temperature profiles T_e, T_i in the accretion column model of Frank and King (1982) which is dominated by nonlocal electron energy transport. T_e, T_i are plotted as functions of vertical distance z in units of the adiabatic shock temperature T_s. The distance from the ion shock (vertical section of T_i curve) to the equipartition point ($T_i = T_e$) is of order 10^4 cm: there is very little hard X-ray emission from this region, most of the energy being transported directly under the white dwarf photosphere by energetic electrons.

dominate the electron energy equation behind the ion shock. The almost constant electron temperature profile there results from the great sensitivity of the fraction ξ in j_{leak} to T_e: although T_i gradually decreases behind the ion shock as the electrons drain energy from the ions, only a very small decrease in T_e produces a change in ξ sufficient to accommodate the change in the ion heating rate ($\propto (T_i - T_e)$). This behaviour ceases once $T_i - T_e$ is so small that ξ enters a parameter regime where it is no longer so sensitive to T_e: this gives the sharp decrease in T_e as equipartition ($T_i = T_e$) is approached at the left of the figure. For the parameters used to produce Fig. 3 the electron temperature is almost constant at $\sim 0.5\ T_s$ behind the ion shock, dropping abruptly to $\sim 0.25\ T_s$ at equipartition. Since at this point $T_i = T_e \simeq 0.25\ T_s$, three-quarters of the accretion energy has been removed from the post-shock gas at this stage. Almost all of this energy has been transported by the high-energy electrons into the white dwarf photosphere, whence it will be radiated as blackbody emission. The length of the two-fluid region is so short ($\sim 10^4$ cm) that direct radiation by hard X-rays accounts for only a few parts in 10^3 of the losses. The situation depicted in Fig. 3 is thus smaller in scale by a factor $\sim 10^3$ (for the same accretion rate) than radiative cases where a very high assumed white dwarf mass and resultant Compton cooling cause two-fluid behaviour, despite a superficial resemblance of the temperature profiles (e.g. Imamura *et al.*, 1979, with $M = 1.4\ M_\odot$). Further, this very small shock height implies that the condition (10) for the validity of the loss-cone approximation is satisfied.

Once equipartition is reached ($T_i = T_e \simeq 0.25\ T_s$) the gas has still to release $\sim \frac{1}{4}$ of the accretion energy before joining on to the white dwarf envelope. Thus it must follow one of the one-fluid solution curves of Fig. 3. It is not a straightforward matter to tell which type (1 or 2) of solution we have here, since this is extremely sensitive to the temperature gradient the flow 'inherits' from the two-fluid region: note in Fig. 3 how close to the critical solution *all* solution curves which reach $\theta \sim 10^{-3}$ run for $\theta \sim 0.25$, despite their entirely different behaviours for very small θ. The simplest possibility is that the gas cools radiatively for $T < 0.25\ T_s$: this emission at $\sim 0.25\ T_s$ would dominate the observed hard X-rays, and by a similar argument to that giving (4), produce a ratio

$$\frac{L_X}{L_{soft}} \sim 0.16 \qquad (12)$$

The other extreme possibility is a type 1 solution below $0.25\ T_s$: then the X-ray spectrum would be a mixture of $0.5\ T_s$ and $0.25\ T_s$ emission, with a very small ratio

$$\frac{L_X}{L_{soft}} \sim \text{few} \times 10^{-3} \qquad (13)$$

Clearly the estimates (12) and (13) show that we have made some progress towards accounting for the observational estimate (5) of this ratio. A further desirable consequence of our treatment is the lower hard X-ray temperature; our expected range (using (1))

$$kT_X \sim (8 - 16)\ \text{keV}\ M_1\ R_9^{-1} \qquad (14)$$

fits more comfortably with hard X-ray observations than the one-fluid estimate (1), which predicts $kT_X \sim 64$ keV for a $1\ M_\odot$ white dwarf.

ACKNOWLEDGEMENTS

Much of the work reported here has been carried out in collaboration with Drs. J. Frank and J.P. Lasota, with the support of the SERC. I thank the Royal Society for travel support and Dr. Frank for valuable help in the preparation of this review.

NOTES

1. Thermal cyclotron radiation, despite providing the classic observational property - phase-dependent optical polarization - of the AM Her stars, cannot be an important loss mechanism for columns producing the luminosities ($\gtrsim 10^{33}$ erg s^{-1}) observed for these systems. Cyclotron-dominated columns might be observationally important if one of the magnetic poles in an AM Her system receives a rather low accretion rate but radiates predominantly in the optical-near IR, the other pole providing most of the luminosity, or if there are systems with considerably higher magnetic fields than those (few x 10^7 G) so far found for the AM Her systems. See King and Lasota (1979) for discussion. Note that some early papers considerably overestimated the importance of cyclotron emission.

2. Spherical models can be rescaled to describe the postshock region of a column flow, but differ greatly from columns in the transfer of outgoing radiation and the ionization structure above the shock: in spherical models a photon released near the white dwarf surface must traverse the entire accretion flow before escaping, while all such photons eventually escape from the *sides* of a column. For the luminosities ($\lesssim 10^{34}$ erg s^{-1}) characteristic of the AM Her and DQ Her stars optical depth effects such as Compton degradation of the hard X-ray spectrum are negligible (King and Lasota, 1979).

REFERENCES

Aizu, K.: 1973, Prog. Theor. Phys. 49, p.1184.
Chevalier, R.A., and Imamura, J.N.: 1982, preprint.
Fabbiano, G., Hartmann, L., Raymond, J., Steiner, J., and Branduardi-Raymont, G.: 1981, Astrophys. J. 243, p.911.
Fabian, A.C., Pringle, J.E., and Rees, M.J.: 1976, Mon.Not.R.astr.Soc. 175, p.43.
Felsteiner, J., and Opher, R.: 1976, Astron. Astrophys. 46, p.189.
Frank, J., and King, A.R.: 1982, in preparation.
Frank, J., King, A.R., and Lasota, J.P.: 1982, Mon.Not.R.astr.Soc., in press.
Hassall, B.J.M., Pringle, J.E., Ward, M.J., Whelan, J.A.J., Mayo, S.K., Echevarria, J., Jones, D.H.P., Wallis, R.E., Allen, D.A., Hyland, A.R.: 1981, Mon.Not.R.astr.Soc. 197, p.275.
Hoshi, R.: 1973, Prog. Theor. Phys. 49, p.776.
Imamura, J.N., Durisen, R.H., Lamb, D.Q., and Weast, G.J.: 1979, in

'White Dwarfs and Variable Degenerate Stars', Proc. IAU Colloq. 53.
Katz, J.I.: 1977, Astrophys. J. 215, p.265.
King, A.R., and Lasota, J.P.: 1979, Mon.Not.R.astr.Soc. 188, p.653.
King, A.R., and Lasota, J.P.: 1980, Mon.Not.R.astr.Soc. 191, p.721.
Kuijpers, J., and Pringle, J.E.: 1982, preprint.
Kylafis, N.D., and Lamb, D.Q.: 1979, Astrophys. J. Lett. 228, p.L105.
Lamb, D.Q., and Masters, A.R.: 1979, Astrophys. J. Lett. 234, p.L117.
Langer, S.H., Chanmugam, G., and Shaviv, G.: 1981, Astrophys. J. Lett. 245, p.L23.
Langer, S.H., Chanmugam, G., and Shaviv, G.: 1982, Astrophys. J. 258, p.289.
Paczynski, B.E., and Zytkow, A.N.: 1978, Astrophys. J. 222, p.604.
Papaloizou, J., Pringle, J.E., and Macdonald, J.: 1982, Mon.Not.R.astr. Soc. 108, p.215.
Raymond, J.C., Black, J.H., Davis, R.J., Dupree, A.K., Gursky, H., and Hartmann, L.: 1979, Astrophys. J. Lett. 230, P.L95.
Rothschild, R.E., Gruber, D.E., Knight, F.K., Matteson, J.L., Nolan, P.L., Swank, J.H., Holt, S.S., Serlemitsos, P.J., Mason, K.O., and Tuohy, I.R.: 1981, Astrophys. J. 250, p723.
Wada, T., Shimizu, A., Suzuki, M., Kato, M., and Hoshi, R.: 1981, Prog. Theor. Phys., in press.
Zeldovich, Ya. B., and Raizer, Yu. P.: 1966, 'Physics of Shock Waves and High Temperature Hydrodynamic Phenomena' Vols. I and II (New York: Academic Press).

DISCUSSION FOLLOWING A. KING'S TALK

WILLIAMS: To what extent does this effect depend upon the presence of a magnetic field, that is, will this also serve to cut down the hard to soft X-rays in the non polars?

KING: First, it does not depend on magnetic fields in the sense that there were no magnetic fields in the calculation, but it does, in the sense that the flow was radial.

LAMB: Let me remark that the kind of one-fluid solutions you have talked about were discussed by Imamura and Durisen in a paper presented at the AAS meeting in Austin in 1978. They encountered the same problems with saturation of the conductive flux and so on. Following that meeting, I and Weast initiated two-fluid calculations. Several examples of our two-fluid calculations were published in the proceedings of IAU Colloquium No. 53 at Rochester in 1979. Since then, we have calculated an extensive series of two-fluid models for a wide range of conditions.

KING: I don't think they are quire the same. The effects that your group considered were primarily those of Comptonization which has the effect of keeping the temperatures different, but an important difference is that the losses to your gas were radiative.

LAMB: What I want to emphasize is that two-fluid calculations are necessary. When we did them, we found that the conductive flux was not important in conducting energies down to the region near the stellar surface. The diagram that you have shown is for one-fluid calculations; you can draw certain conclusions from it, but it does not apply to two-fluid calculations.

KING: It applies when the temperatures have equalized.

LAMB: No, it absolutely does not. For example in that figure you have an arbitrary choice of the temperature gradient at the shock and therefore the initial conductive flux; but when you do a two-fluid calculation, the initial conductive flux is specified.

KING: But when the temperatures equalize you have one fluid calculations. The gradient of the temperature can be anything you like, this diagram covers the whole parameter space.

LAMB: I would also like to say that suprathermal electrons cannot deposit energy deep in the photosphere in white dwarfs because the Coulomb cross section is sufficiently large that the ions must shock. Initially, the ions contain all the energy and they must transfer their energy to the electrons before it can go anywhere.

KING: The ions shock there. The difference is that your columns are very long, this column is very short, because the losses are not radiative.

LANGER: In your models here, you cannot possibly have a shock height which is less than the electron-ion equipartition distance. Essentially you have picked that point in the electron velocity distribution where the mean free path of an electron is equal to somewhat more than the distance to the photosphere, as the point where your tail cuts off and at that point the velocity will drop below a Maxwellian. You then compute the net flux of energy into that portion of the distribution driven by the electron relaxation processes, as the amount of energy density that would originally be beyond that cut off, divided by the relaxation time. How was that relaxation time computed and what is its relationship to some attempt to calculate, using the Fokker-Planck equation the energy flux across that cut-off point?

KING: To take the last point first, to do a Fokker-Planck treatment of this, is going to be difficult because you got spatial gradients, so it is the full Boltzmann. The answer to the first part, if I understand you correctly, is that it is the standard approximation used in stellar dynamics, for example.

CRAMPTON: Tapia showed us that, at least he thought, that in all of the AM Her stars the poles were inclined at a small angle to the line of sight. Can you tell us from the models whether you expect the X-rays only to be seen in a small cone angle? Because nearly all of the AM Her stars are found by X-rays.

KING: Well, the point is that the base of these flows is where the pressure in the white dwarf envelope equals the ram pressure and that is not necessarily near where the surface is. So you have a shock that probably lies very close to the surface, most of the X-rays will have to come out through the sides, they certainly cannot go up the column. So you would probably expect to see low energy absorption.

LAMB: There do exist Einstein OGS observations of AM Her which were nearly simultaneous with IUE observations and which require a blackbody temperature of 40-45 eV. The OGS observations do not tolerate a blackbody temperature of less than 35 eV. In order to join the soft X-rays to the UV using the Rayleigh-Jeans tail of a blackbody spectrum, one has to have, instead, a temperature of 27 eV. Taking a blackbody temperature of 40-45 eV does not give a large ratio for L_X/L_{total}, so in the case of AM Her it may be that the ratio is only 2-4 rather than 10 or 100. I just wanted to point this out, because it has been a long standing statement that you can join the soft X-rays and the UV, but it actually does not seem to be true.

TIME DEPENDENCE IN ACCRETION ONTO MAGNETIC WHITE DWARFS

Steven H. Langer
Departments of Physics and Astronomy, University of Illinois

G. Chanmugam
Department of Physics and Astronomy, Lousiana State Univ.

G. Shaviv
Faculty of Physics, Israel Institute of Technology

I. INTRODUCTION

In this talk we consider cataclysmic variable systems containing a white dwarf with a strong magnetic field. These include systems like AM Her (see, e.g., Chiappetti, Tanzi, and Treves 1981) in which the white dwarf rotates at the orbital period and systems such as AE Aquari in which the white dwarf rotates much faster than synchronously (see Patterson et al. 1980). The magnetic field in all of these systems is strong enough to disrupt the accretion disk at a point far above the surface of the white dwarf and may prevent the formation of a disk altogether. We will present theoretical models for the temperature, density, and velocity structure of the accretion flow in the region near the surface of the white dwarf where the kinetic energy of the flow is thermalized and radiated in the form of X-rays and ultraviolet radiation. This information is required to calculate accurate model spectra, and the results also have immediate consequences for the interpretation of observations.

The flow is characterized by a strong standoff shock that forms in the accretion column and thermalizes the kinetic energy of the infalling material. Below the shock, the hot gas cools by radiating and then settles onto the white dwarf. In an earlier paper (Langer, Chanmugam, and Shaviv 1981), we showed that the postshock flow is thermally unstable. This instability leads to periodic oscillations in the height of the standoff shock, and in the luminosity and the spectral temperature of the X-rays. We have recently studied the effects of varying the system parameters on the properties of the oscillation (Langer, Chanmugam, and Shaviv 1982).

Our model includes the effects of the gravitational field, heat

M. Livio and G. Shaviv (eds.), Cataclysmic Variables and Related Objects, 199–205.

conduction, and radiative cooling by bremsstrahlung radiation. We ignore, for now, the effects of electron cyclotron line radiation (see, e.g., Chanmugam and Wagner 1979), Compton scattering, nuclear burning of the accreted matter, non-hydrodynamic flow (see King 1983), and possible differences between the electron and ion temperatures. We solve the three time-dependent, non-linear differential equations for the transport of mass, momentum, and energy numerically, using an implicit, finite-difference method. We obtain the solution on a discrete grid of radius points and follow its evolution with time. At the upper boundary of our grid, the matter is cool and in free fall. At the lower boundary, which is placed well inside the photosphere of the white dwarf, the mass flux is zero. The models start with a smooth, but arbitrary, initial density, temperature and velocity profile.

II. RESULTS

The models have as free parameters the mass, M, and radius, R, of the white dwarf, the accretion rate, $\dot{M}$, and the area on the surface of the white dwarf over which accretion occurs, A. In figure 1 we plot the pressure, velocity, and density as functions of the height above the surface of the white dwarf for a model in which $M = M_\odot$, $R = 10^9$ cm, $\dot{M} = 2.5 \times 10^{15}$ g s^{-1}, and $A = 10^{16}$ cm^2. The shock is located at $r = 1.22R$; it has been spread across several mesh points by the artificial viscosity in the numerical model. The sharp jump in the density and pressure at $r = 1.01R$ marks the point where the accreting matter rapidly cools to the photospheric temperature and the hydrostatic, isothermal atmosphere of the white dwarf begins. The radiative cooling is cutoff at the photospheric temperature to simulate the approach to LTE. Figure 2 shows the shock height for the same model as a function of time. The first cycle is affected by the initial conditions, but after that the shock height undergoes a periodic limit cycle with repetitive properties and no indication of damping. The shock height ranges from just above the photosphere to a value somewhat less than twice the shock height found by solving the static flow equations. The results may be understood by noting that on dimensional grounds the shock height is roughly the product of the radiative cooling time of the postshock gas and the postshock velocity, while the period is a few times the radiative cooling time.

The results for this particular model are quite interesting, and have led us to examine the behavior of the limit cycle as a function of the system parameters. For the models we have considered, the cooling is dominated by bremsstrahlung and heat conduction is unimportant. Under these conditions the solutions of the dimensionless hydrodynamic equations depend on a single parameter, which is a combination of M, R, $\dot{M}$, and A. In figure 3 the maximum shock height during the cycle and the period of the limit cycle are shown as functions of this parameter. The points are taken from models with a variety of white dwarf masses and radii, and with a range of mass accretion rates. The points all lie on a single curve, as the equations predict.

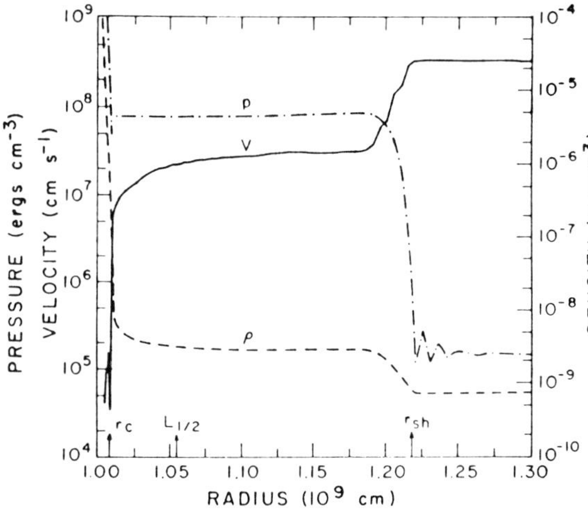

Fig. 1: The density, pressure and velocity as functions of radius for $M = 0.5\ M_\odot$, $R = 10^9$ cm, $\dot{M} = 2.5 \times 10^{15}$ g s^{-1} and $A = 10^{16}$ cm^2. Half of the accretion luminosity is radiated above the point marked $L_{1/2}$.

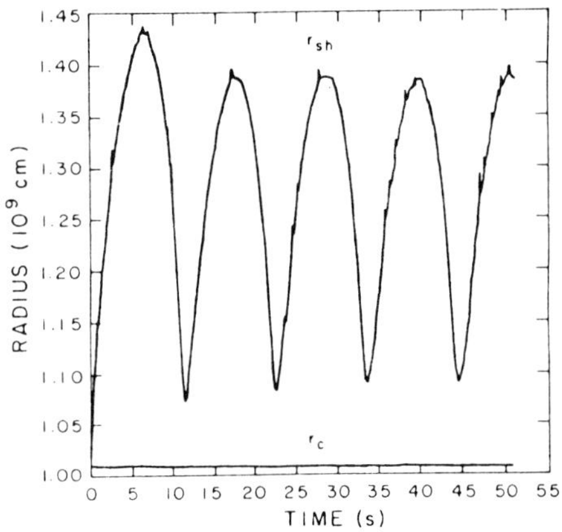

Fig. 2: The shock height, r_{sh}, and the photospheric radius, r_c, as functions of time for the model of Figure 1.

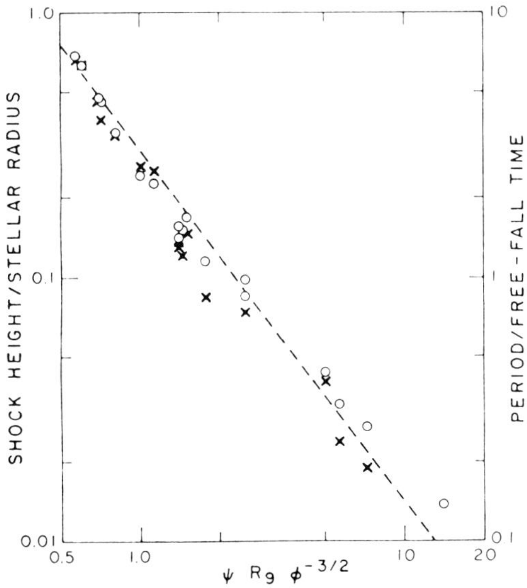

Fig. 3: The shock height in units of R and the period in units of the free-fall time are shown as functions of $(\dot{M}/A)\ R\ (M/R)^{-3/2}$ in dimensionless units. The scatter in the points is due to numerical uncertainty. [Figures taken from the Astrophysical Journal 258, 289 (1982).]

III. DISCUSSION

The accretion flow is characterized by a stand-off shock with a region dominated by radiative cooling below it. The shock height tries to adjust itself so that the flow time from the shock to the surface is equal to the time required to cool from the postshock temperature to the photospheric temperature. The cooling gas exhibits a thermal instability which instead forces the shock to move in a periodic limit cycle. The varying amount of hot material between the shock and the photosphere leads to a periodically varying X-ray luminosity. Because there is a range of gas temperatures behind the shock at any one time, the X-ray spectrum is not a pure thermal bremsstrahlung spectrum. There are thus several potentially observable effects in the X-rays from the system. In a real system the behavior is apt to be more complicated than indicated for several reasons. The matter probably becomes attached to the magnetic field lines in a non-uniform manner so that the mass flux varies across the accretion column. The flow is only weakly coupled across magnetic field lines, so the accretion column should consist of several independent flux tubes, each with its own mass flux. The resulting X-ray flux will not be strictly periodic, but there should be a bump in its power spectrum at the frequency corresponding to the average mass flux.

The hard X-ray flux from these systems is so weak that it is difficult to detect variability at periods on the order of one second. However, the same variations will also occur in any other radiation emitted in the accretion column and in any radiation resulting from the reprocessing of radiation emitted in the column. This suggests that it might be interesting to examine not only X-ray data, but also optical radiation for evidence of this oscillation. This has been done for AN UMa, E1405-451, and AM Her (Middleditch 1982). Both E1405-451 and AN UMa show a bump in their power spectrum at a period of about 1.7 seconds. The data for AM Her failed to reveal any evidence for periodicity near this period. Our model predicts periods that are of the order of one second for reasonable accretion rates and polar cap areas, in agreement with these observations.

We have recently run a few models which include electron cyclotron line emission. If the magnetic field is strong enough, or the accretion rate is low enough, the cyclotron lines dominate the radiative cooling, and the flow becomes stable. The characteristic magnetic field at which the flow stabilizes is roughly 2×10^7 Gauss. This is comparable to the values measured for AM Her (Schmidt, Stockman, and Margon 1981) and VV Pup (Visvanathan and Wickramasinghe, 1979). This suggests that some of the AM Her systems will exhibit oscillations and others will not, depending on the magnetic field. In addition, it is possible that the oscillations will appear and disappear in a single system as the accretion rate varies. It should be emphasized that the results on the cyclotron cooling are preliminary, and that the identification of the feature in the power spectrum with the limit cycle must still be confirmed. However, the study of these features can provide new information about the AM Her system and seems worthy of further effort, both observationally and theoretically.

REFERENCES

Chanmugam, G., and Wagner, R. L. 1979, Ap. J. 232, 895.

Chiappetti, L., Tanzi, E. G., and Treves, A. 1980, Space Sci. Rev., 27, 3.

King, A. R. 1983, "Column Accretion on White Dwarfs," (this volume).

Langer, S. H., Chanmugam, G., and Shaviv, G. 1981, Ap. J. (Letters), 245, L23.

Langer, S. H., Chanmugam, G., and Shaviv, G. 1982, Ap. J., 258, 289.

Middleditch, J. 1982, Ap. J. (Letters), 257, L71.

Patterson, J., Branch, D., Chincarini, G., and Robinson, E. L. 1980, Ap. J. (Letters), 240, L133.

Schmidt, G. D., Stockman, H. S., and Margon, B. 1981, Ap. J. (Letters), 243, L157.

Visvanathan, N., and Wichramasinghe, D. T. 1979, Nature, 281, 47.

DISCUSSION FOLLOWING S. LANGER'S TALK

KING: I have a question about your lower boundary condition, you have a sort of box there, that the matter falls into, it is not allowed to go into the star, does that not mean that your whole structure gradually rises out of the star as matter piles up?

LANGER: Yes, it is true that it would rise upward, but if you look at the specific volume of material at the temperature where we cut it off, at a few 10^5°K, it is pretty small and it goes nowhere. As a matter of fact, you can see that the photospheric radius in the curve is not doing anything discernible on the timescale of this calculation.

KING: Have you tried playing around with that boundary condition to see if it effects your results?

LANGER: We have not played with the boundary condition, but I can't possibly see how it could, the boundary is buried deep inside by far more matter than the accreting mass can in anyway influence, we can see that we are not getting any response to those oscillations down at the point where we cut it off.

KING: But the point I am driving at is that you have not really matched that flow to the atmosphere of the white dwarf.

LANGER: We have not matched that flow to the white dwarf only because we have taken a rather faky means of handling optical thickness effects at the cut off. That was a computational decision.

KING: This instabilities you are talking about are global in nature, so it is very important that your boundary condition is really the physical one.

LANGER: There is one thing I can say on that, it is a comparison between two numerical solutions, neither of which is entirely physical. In the first paper which we had in Ap. J. Letters over a year ago, we had a different treatment for what happened when the matter got too cool to cope with the cooling time. In that case, the periods and the shock heights that we got, are consistent within the numerical noise with what we get now with this entirely different treatment.

KING: You get the same period because that is the cooling time, I don't think that is telling you anything, I think you are going to get that regardless of what you did at the bottom.

LANGER: There is no requirement that I know of that requires the thing to oscillate and that is what we are interested in, is there some physical process that would damp the oscillations. It is hard to see how you can do that at the photosphere, where you don't have the energy to do anything to what is up at the shock.

LAMB: John Middleditch's data is optical data whereas what you are talking about here are X-ray oscillations. Reprocessing of the hard and/or soft X-rays would therefore be necessary. Tuohy, Lamb, Garmire and Mason (1978) do discuss quasi-periodicities in the soft X-ray flux, but at a few tenths of a second, but to my knowledge those oscillations were not detected in the hard X-ray component. The number of counts in the hard X-rays is much smaller than in the soft X-rays and therefore it may be that if the amplitude is small, it would be below detector threshold. Also, it seems premature to be sure that these kinds of oscillations can take place for two reasons, besides the possible stabilizing effect of cyclotron emission in a strong field. One is that the region right below the shock, whether you have nuclear burning or not, is going to be predominantly cooled by Compton cooling from the photospheric black body flux. Such cooling depends only on the electron density in the column and is greater if the electron temperature is higher. I would be surprised if this does not have a stabilizing influence on the structure. The second point is that conduction transfers energy from just below the shock to further down, and again I would intuitively expect that this would have a stabilizing influence on the shock.

LANGER: We have included the standard $K\frac{dT}{dr}$ type heat flux in our equations in the latest runs, when we revised cooling law, we did not have it in the earlier ones. We did not see any difference, we did not see it damping out, the heat flux never got to be a large fraction of the accretion luminosity for the models that we considered. If you had a shock that was very close to the surface, perhaps that would break down.

LAMB: But these calculations are only one fluid calculations and the effect of the conductivity within the emission region is completely different if you actually allow the conductivity to be treated in a two fluid situation.

LANGER: Which requires that you have, say, Compton cooling to cause a difference between the electron and ion temperatures. In some situations where Compton cooling was very important, the electron and ion temperatures were different, I cannot claim that this works, I do not know what happens then.

LAMB: What about just the Compton cooling from the black body flux itself?

LANGER: If we had put in the observed black body flux in AM Her, I think that would have been enough to do something. Depending upon the shock height and the geometry we would have had different optical depths and it would depend upon the optical depth to Compton scattering. We have not included that strong black body component, we want to put it in.

CIRCULAR POLARIMETRY OF AM HERCULIS

V. Piirola, O. Vilhu and I. Tuominen
University of Helsinki
Observatory and Astrophysics Laboratory

ABSTRACT

Circular polarimetry in the red and simultaneous photometric observations in the UBVRI bands during the period June 1-3, 1981, are discussed. The peak value of negative circular polarization $P_V \sim -15$ % is stronger than observed in 1976-79. Variations in the shape of the polarization and light curves occur from night to night. The positive crossover and reversal of the sign of the circular polarization are only marginal. A probable explanation of the short term variations seems to be the changing shape and position of the accretion columns with respect to the magnetic axis.

1. INTRODUCTION

In the present paper we give a preliminary report of circular polarimetry and simultaneous UBVRI photometry of AM Herculis on three nights in June 1981.

The binary system AM Herculis shows a strong circular polarization varying with the $3^h\!.1$ orbital period (Tapia, 1977). Both the polarization and light variations are largest in the near infrared and decrease steeply towards the ultraviolet. The general interpretation for this behaviour is that circular polarization is produced by cyclotron radiation from thermal electrons near the magnetic poles of a highly magnetized ($B \sim 10^7$ G) white dwarf accreting material from a close companion star (see e.g. Kruszewski, 1978, Chanmugam and Dulk, 1981).

2. OBSERVATIONS AND RESULTS

Our observations were carried out with the 1.25 m telescope of the Crimean Astrophysical Observatory using the double image chopping polarimeter of the Observatory of the University of Helsinki. Dichroic filters were used to split the light into five spectral regions. The re-

M. Livio and G. Shaviv (eds.), Cataclysmic Variables and Related Objects, 207–210.

sulting passbands are close to the standard UBVRI-system. For circular polarization observations an achromatic quarter-wave retarder was rotated with 90° steps in front of the plane parallel calcite beam splitter. Each polarization observation consists of two 40 s integrations with rotation of the retarder between the integrations.

The observations of the circular polarization in the red colour on the nights June 2 and 3, 1981 are displayed in Figure 1. Standard error bars ($\pm \sigma$) computed from photon statistics are given. The peak value of the negative circular polarization $P_V \sim -15$ % is stronger than observed in 1976-79. The polarization approaches zero when the direction of the magnetic field is perpendicular to the line of sight (about 22^h UT on June 3). In contrary to the earlier observations the polarization does not clearly change the sign. The few positive values may be due to observational scatter. This indicates changes in the effective direction of the magnetic field since 1979. It has been suggested (Bailey and Axon, 1981) that the long term variations in polarization, especially the different widths of polarization reversals in 1976 and 1979, are due to precession of the axis of rotation of the white dwarf about the binary axis. However, the duration of the phase interval where circular polarization remains close to zero (near 22^h UT on June 3) is clearly different on the two nights in Figure 1, indicating that other mechanisms than precession can considerably change the polarization curve on time scale of days. Probable explanation of the short term variations seems to be the changing shape and position of the accretion columns with respect to the magnetic axis.

For comparison the total flux in the I band is given below the polarization curves in Figure 1. Both the polarization and photometric maxima and minima occur approximately at the same phase suggesting that both are of the same origin, i.e. due to cyclotron radiation. The light curves show considerable variation and flarelike activity is present. The wavelength dependence of light curves from infrared to ultraviolet can be seen in Figure 2. The variability decreases rapidly towards the blue and the ultraviolet.

The proposed models cannot explain very well the details of the light curves. Especially the changes in the shape of the light curves are larger than in the polarization curves, suggesting that a part of the radiation from the accretion column is not due to cyclotron radiation.

Acknowledgements. We are grateful to Prof. A.A. Boyarchuk of Crimean Astrophysical Observatory for the opportunity to use the 125 cm telescope in connection with our polarimeter. We are also grateful to Drs. N.M Shakhovskoi and Y.S. Efimov for valuable help during the observations and the modification of the equipment.

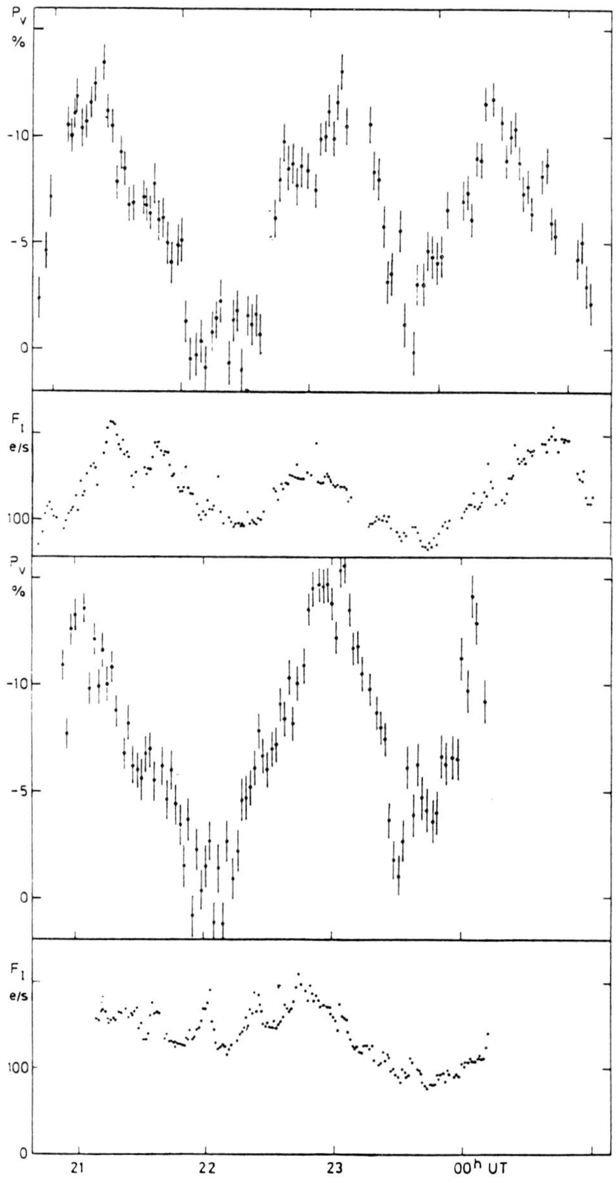

Figure 1. Circular polarization P_V (%) and infrared flux F_I (photoelectrons/s) of AM Herculis on June 2/3 (upper part) and June 3/4, 1981

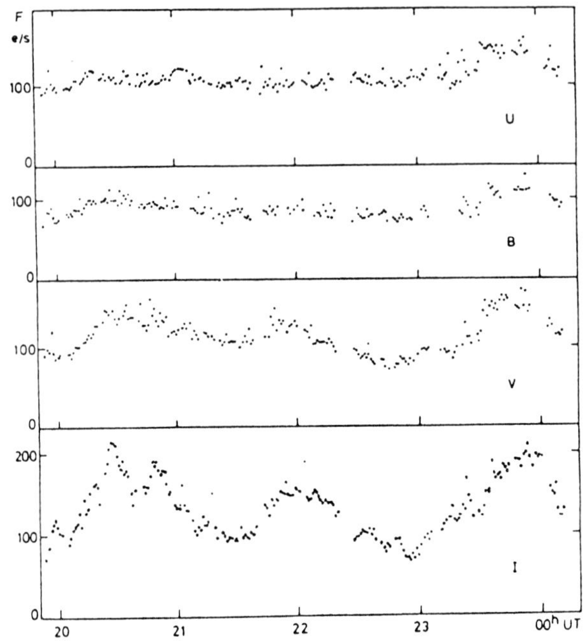

Figure 2. Light curves of AM Herculis on June 2/3, 1981, in the U,B,V and I regions. Each dot represents a 40 s integration. Intensity is expressed in units of photoelectrons/s.

REFERENCES

Bailey, J. and Axon, D.J.: 1981, M.N.R.A.S. 194, 187.
Chanmugam, G. and Dulk, G.A.: 1981, Astrophys. J. 244, 569.
Kruszewski, A.: 1978, in Nonstationary Evolution of Close Binaries, ed. A.N. Zytkow, Warsaw, p. 55.
Tapia, S.: 1977, Astrophys. J. 212, L125.

POLARIMETRY OF SYMBIOTIC STARS

V. Piirola
University of Helsinki
Observatory and Astrophysics Laboratory

ABSTRACT

Five symbiotic stars have been observed for linear polarization (UBVRI) in September 1981. Three systems, CH Cyg, CI Cyg and AG Peg show intrinsic polarization while in the case of Z And and AX Per the observed polarization seems to be mostly of interstellar origin. The position angle of polarization of CI Cyg and AG Peg rotates strongly vs. wavelength, as observed also for CH Cyg in 1977-80. The polarization of CH Cyg has decreased since May 1980, especially in the I, R and U bands, so that the maximum polarization is now in the blue ($P_B \sim 0.3$ %). Probably we are monitoring the formation, growth and disappearance of dust particles in the atmosphere of this star. Two related systems, PU Vul (Nova Vul 1979) and R Aql (Mira) have polarization behaviour rather similar to that of symbiotic stars which suggests that the M type giant present in these systems is responsible for most of the intrinsic polarization.

1. INTRODUCTION

The peculiar wavelength dependence of the polarization of CH Cyg and the time variability observed during the recent active phase in 1977-81 (Piirola, 1982) indicate the presence of a complicated scattering envelope in this system. Both the density, particle size distribution and scattering geometry of the envelope are varying on time scale of weeks. We report further polarization observations of CH Cyg and four other symbiotic stars.

2. OBSERVATIONS AND DISCUSSION

The observations were made during the period September 10-25, 1981 with the 1.25 m photometric telescope of Crimean Astrophysical Observatory using a multichannel version of the double image chopping photopolarimeter of the University of Helsinki. The observations were made

M. Livio and G. Shaviv (eds.), Cataclysmic Variables and Related Objects, 211–216.

simultaneously in five colours using dichroic filters to split the light into five spectral regions. The resulting passbands are close to the standard UBVRI system, with effective wavelengths 0.37, 0.44, 0.53, 0.69 and 0.83 microns, respectively.

The polarization of CH Cygni (Figure 1) has decreased since May 1980, especially in R, I and U bands, so that the maximum polarization is now in the blue ($P_B \sim 0.3$ %). Also the position angle has rotated vs. time from $\Theta_U \sim 170^o$ in 1980 to $\Theta_U \sim 70^o$ in 1981. Accordingly, changes in particle size distribution and scattering geometry have occurred. In particular, the large dust particles (a $\sim$ 1 μm) responsible for the polarization component increasing towards the I band (Piirola, 1982) have disappeared. The drop in ultraviolet polarization on the other hand could be explained by growth of the average small particle size from less than 0.05 μm to 0.08-0.10 μm.

In CI Cyg and AG Peg the position angle rotates strongly vs. wavelength (Figure 2) which indicates that the polarization cannot be entirely interstellar. In AG Peg the position angle is approximately constant from I to V but then clearly changes to B and U. This seems to coincide with the wavelength ($\lambda_o \sim 4000$ Å given by Viotti, 1982) where the spectrum of the hot source emerges. For wavelengths $\lambda > \lambda_o$ the cool spectrum dominates. In CI Cyg the rotation starts already in V band, consistent with the somewhat larger value of $\lambda_o \sim 5000$ Å.

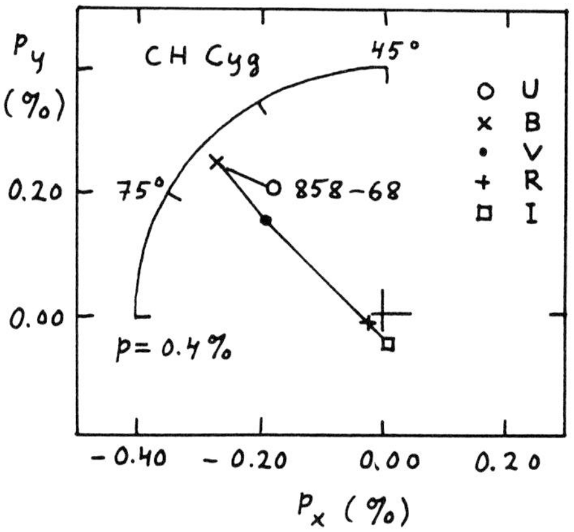

Figure 1. Polarization observations of CH Cygni in September -81 plotted in polar coordinates. The degree of polarization has maximum in the blue ($P_B \sim 0.35$ %, $\Theta_B \sim 70^o$) and decreases steeply towards the red.

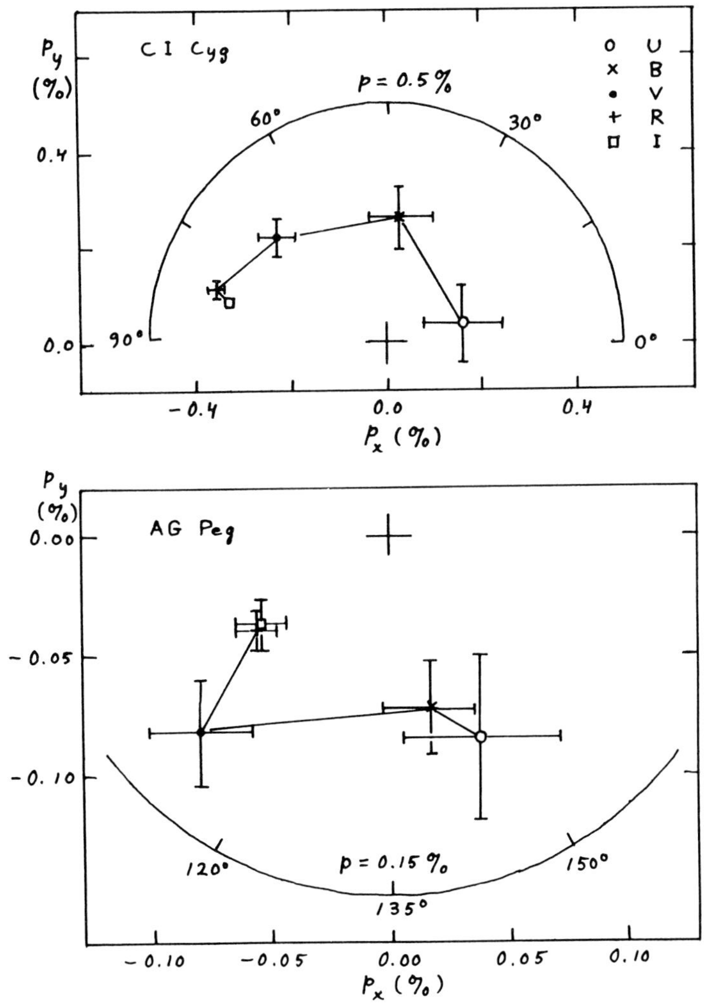

Figure 2. Polarization observations of CI Cyg and AG Peg show strong rotation of position angle as a function of wavelength which indicates the presence of intrinsic (circumstellar) polarization.

The wavelength dependence of polarization in Z And ($P_V \sim 1.2$ %, $\Theta_V \sim 53^o$) and AX Per ($P_V \sim 1.3$ %, $\Theta_V \sim 95^o$) approaches that of interstellar polarization and more observations are needed to reveal possible time variations and consequently intrinsic polarization.

For comparison, Figure 3 shows polarization diagrams of related objects, namely PU Vul (Nova Vul 1979) and a Mira-type star R Aql, obtained in Sept. -81 (R Aql was too bright to be observed in R and I): The approximately similar polarization behaviour points to the possibility that the common feature, the presence of an M-type giant may be the main source of polarization. The degree of polarization usually increases towards the ultraviolet nearly proportional to λ^{-4} which requires either small particle size or Rayleigh scattering by gas molecules. The correlation between infrared excess near 10 μm and polarization in many late type giants (see e.g. Zellner and Serkowski, 1972) supports dust particles as the main source of polarization in these stars. Further infrared photometry of symbiotic stars is needed to establish the possible correlation between infrared excess and polarization.

The contribution of the hot companion, e.g. a white dwarf and accretion disk to the polarization depends on the relative fluxes from the WD and the M giant and the existence of scattering agents in the vicinity of the white dwarf. The near λ^{-4} dependence of polarization of CH Cyg during time intervals when high additional hot continuum is present would suggest that the light from the hot companion is effectively scattered by the dust particles in the M giant envelope and/or by free electrons in the disk, but more quantitative modelling is still required to decide which one is dominant.

We are grateful to Prof. A.A. Boyarchuk of Crimean Astrophysical Observatory for the opportunity to use the 125 cm telescope in connection with our polarimeter. Thanks are also due to Drs. N.M. Shakhovskoi and Y.S. Efimov for valuable help during the observations.

REFERENCES

Piirola, V.: 1982, in The Nature of Symbiotic Stars, eds. M. Friedjung and R. Viotti, IAU Coll. 95, 139

Viotti, R.: 1982, private communication

Zellner, B. and Serkowski, K.: 1972, P.A.S.P. 84, 619

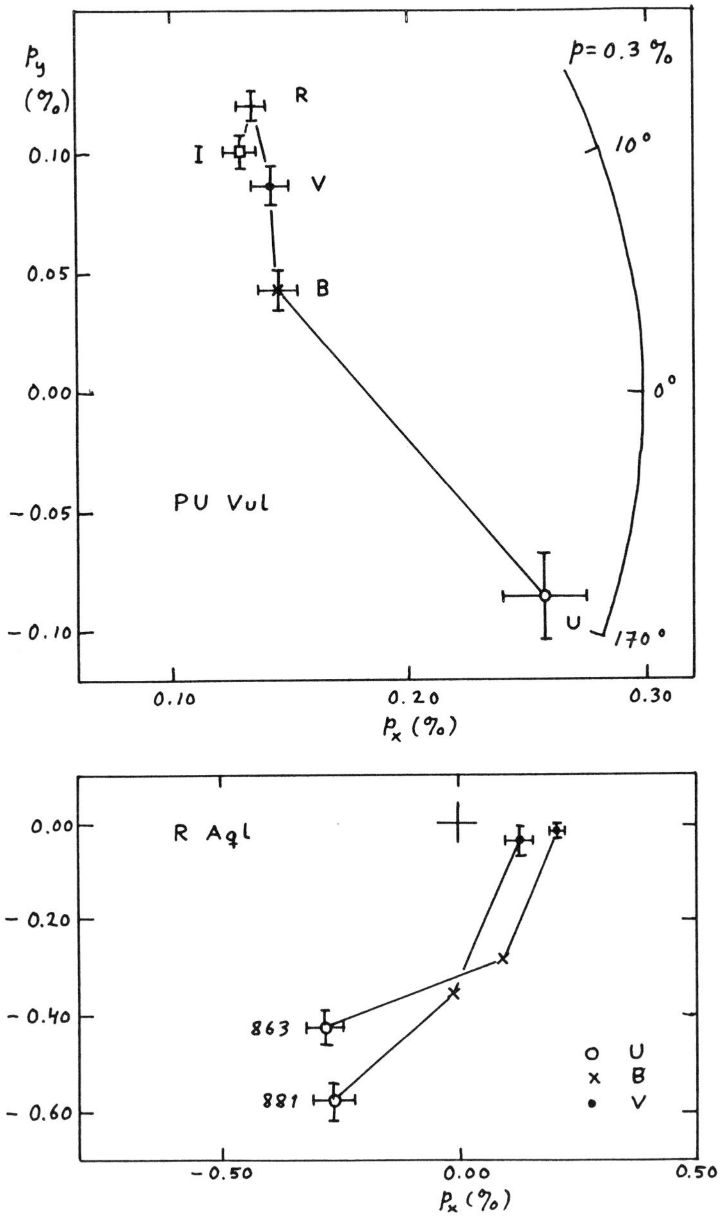

Figure 3. Polarization observations of two systems related to symbiotic stars, PU Vul (Nova Vul 1979) and R Aql (Mira) in September -81 plotted in polar coordinates.

DISCUSSION FOLLOWING V. PIIROLA'S TALK

MITCHELL: Do you see any pulsations in the red component and if so is it possible to correlate those with the inferred changes in the grain size characteristics of CH Cyg?

PIIROLA: So far I cannot tell about the pulsations from these observations.

MATTEI: Could you tell me when you did the CI Cyg and Z And observations? Is the detection of polarization phase dependent in the light curve?

PIIROLA: I started these observations of CI Cyg in September 1981 and observed it also in June 1982.

MATTEI: So it was out of its eclipse.

PIIROLA: I observed Z And in September 1981.

MATTEI: At its minimum phase then. So they were in different phases in their light curve.

EVANS: Can you say anything about the composition and the distribution of the dust?

PIIROLA: The recent observations of CH Cyg show that the average dust particle size has increased, because the maximum was in the blue and not in UV. You can fit grain models to the wavelength dependence and quite accurately fit the average size of the grains. It is impossible to get the geometry of the scattering matter without eclipses.

MITCHELL: Can you put in any numbers on the grain size?

PIIROLA: It was something like 0.1μ.

TAPIA: First, I am happy that you have demonstrated that very large changes in the polarization in AM Her occur not only in my observations. Now, I would like to ask, were the observations of CH Cyg motivated by any particular reason?

PIIROLA: Yes, I was informed that an active phase started.

POLARIZED RADIATION FROM AM HERCULIS STARS

P. E. Barrett and G. Chanmugam
Department of Physics and Astronomy
Louisiana State University
Baton Rouge, Louisiana 70803-4001

ABSTRACT

Calculations of polarized cyclotron radiation from magnetized hot plasmas are modified to include the effects of collisions and photon scattering. These effects produce a general reduction in both the linear and circular polarization. The background unpolarized flux which had previously been invoked, in an ad hoc manner for the collisionless case, is no longer required to bring agreement with observations. The most striking effect is at small viewing angles with the magnetic field, where the fractional circular polarization is reduced from nearly 100% to close to 0%. The calculations also show that the shape of the linear polarization pulse is essentially unchanged.

1. INTRODUCTION

AM Herculis stars are binary systems comprised of a magnetic white dwarf (magnetic field $\approx$ few $\times 10^7$ gauss; see e.g. Chiappetti, Tanzi, and Treves, 1980, Chanmugam and Dulk, 1981) and a lower-main-sequence star. The strong magnetic field controls the motion of the plasma in a sphere containing both stars. As a result both stars are in synchronous orbit and the accreted matter is funnelled directly onto the surface of the white dwarf at one or both of the magnetic poles. It is in this region that the X-ray, UV, and optical fluxes originate. Earlier calculations showed that the observed large fractional polarization ($\sim$ 10%) is emitted at cyclotron harmonic number $\approx$ 5 to 10 from the cooler ($\sim$ 1keV) region of the plasma column just above the shock-heated region near the polar cap (Chanmugam and Dulk, 1981, c.f. Meggitt and Wickramasinghe, 1982). These earlier calculations assumed a collisionless plasma in LTE emitting cyclotron radiation. In this paper calculations of the cyclotron absorption coefficients are modified to include absorption of radiation from electron-ion collisions (inverse bremsstrahlung) and Thomson scattering. These modifications give better agreement between the calculated and the observed polarization.

M. Livio and G. Shaviv (eds.), Cataclysmic Variables and Related Objects, 217–221.

2. CALCULATIONS

The absorption coefficients are calculated using the method developed by Pavlov et al (1980). The method involves calculating the dielectric tensor, at a given frequency ω, which is expanded in a series of terms each of which corresponds to the contribution from cyclotron harmonic number s (see e.g. Stix 1962). The effects of collisions and photon scattering are incorporated by including an imaginary term in the argument of the plasma dispersion function $W(z_s)$. Hence $z_s = (\omega - s\omega_c)/\nu_D + i\nu/\nu_D$, where ω is the frequency of the radiation, ω_c the cyclotron frequency, ω_D the Doppler frequency and ν the effective collision frequency. Here $\nu = \nu_c + \nu_r$ where ν_c is the electron-ion collision frequency and ν_r the radiative damping frequency due to Thomson scattering. The case $\nu = 0$ corresponds to the collisionless plasma. The absorption coefficients μ_o, μ_x for the ordinary and extraordinary modes are then deduced from the imaginary part of the complex index of refraction. The equations of radiative transfer are then solved for the case of large Faraday rotation (Ramaty 1969).

3. RESULTS

Numerical calculations were done at cyclotron harmonic numbers 4,6 and 8 for a tenuous plasma slab at a temperature of 1keV ($\simeq 10^7$K) and width $L=10^8$ cm. The plasma slab was set parallel to the stellar magnetic field ($B=3\times10^7$ gauss). The dimensionless parameter $\Lambda \equiv \omega_p^2 L/\omega_c c = 10^8$ where ω_p is the plasma frequency, so that the electron number density is 10^{16} cm^{-3}. Figures 1a and 2a are curves of the absorption coefficients versus $\cos\alpha$ where α is the angle between the line of sight and the magnetic field, for harmonic numbers 4 and 6, respectively. The corresponding linear and circular polarization curves are shown in Figs. 1b and 2b. From Figs. 1a and 2a we note for the collisionless case that $\mu_{o,x}$ is largest for $\alpha \approx \pi/2$ and decreases strongly with decreasing α. Furthermore, $\mu_{o,x}$ decreases strongly as the cyclotron harmonic number increases. The effects of collisions on $\mu_{o,x}$ are not as strongly dependent on α or the harmonic number and hence become more important at high frequencies and small α. This is clearly demonstrated in Figs. 1a and 2a. For harmonic numbers $\lesssim 6$, the collisionless approximation is adequate in determining the cyclotron absorption coefficients except at small α. The absorption coefficients at the higher harmonics ($\omega/\omega_c \gtrsim 7$) become angle independent and have approximately the same value for the two modes.

The principal effect due to the inclusion of collisions is an overall reduction in both the linear and circular polarization which is brought about by the increased opacity. The most striking effect is for the circular polarization at small viewing angles. For the collisionless case, the circular polarization approaches 100% whereas when collisions are included it approaches 0%.

At the fourth harmonic the plasma is optically thick for $\cos\alpha \lesssim$

0.7 and the polarization becomes negligible in this region. For small viewing angles ($\cos\alpha \gtrsim 0.7$), the plasma becomes optically thin and there is a circular polarization pulse reaching a peak of 40%. For harmonic numbers 6 and 8 the plasma is optically thin for most α and there is significant circular polarization which reaches values of about 35% and 15% respectively. The sharp linear polarization pulse at α close to $\pi/2$ is not noticeable for increasing frequency until the sixth harmonic where it has a maximum pulse height of 50% polarization. Collisional effects do not change the shape of the linear polarization pulse: only the maximum pulse height decreases.

4. CONCLUSIONS

The principal result of our inclusion of collisional effects is that both the linear and circular polarizations are reduced. Furthermore, at small viewing angles the fractional circular polarization approaches 0% whereas in the collisionless case it approaches 100%. This is most encouraging since observations of AM Her (Tapia 1977) indicate that the circular polarization approaches zero at small viewing angles. For the collisionless case, Chanmugam and Dulk (1981) postulated an unexplained background flux to dilute the polarization so that it agreed with observations. This postulate is no longer necessary.

Our results confirm the earlier results (Chanmugam and Dulk 1981) that for systems with fields $\gtrsim 5 \times 10^7$ gauss the polarization shifts into the UV making their detection difficult-since UV polarimetry is not developed. However such systems are likely to be spectacular UV sources because of high harmonic, self-absorbed cyclotron emission from the shock (Lamb and Masters 1979). Observations of 7 cataclysmic variables with the IUE have not revealed the strong UV emission expected from systems with strong fields (Bond and Chanmugam 1982). This suggests that there is a paucity of systems with strong magnetic fields ($\gtrsim 5 \times 10^7$ gauss).

We thank Dr. Pavlov for valuable correspondence regarding his paper. This research was supported by NSF grant AST-8025250.

REFERENCES

Bond, H. E. and Chanmugam, G.: 1982 in Y. Kondo, J. M. Mead and R. D. Chapman (eds.), *Advances in Ultraviolet Astronomy: Four Years of IUE Research*, pp. 530-533, NASA, Washington D. C.

Chanmugam, G. and Dulk, G. A.: 1981, Astrophys. J. 244, pp. 569-578.

Chiappetti, L., Tanzi, E. G., and Treves, A.: 1980, Space Sci. Rev. 27, pp. 3-33.

Lamb, D. Q. and Masters, A. R.: 1979, Astrophys. J. (Letters) 234, pp. L117-L122.

Meggitt, S. M. A. and Wickramasinghe, D. T.: 1982, Monthly Notices Roy. Astron. Soc. 198, pp. 71-82.

Pavlov, G. G., Mitrofanov, I. G. and Shibanov, Yu. A.: 1980, Astrophys. Space Sc. 73, pp. 63-82.
Ramaty, R.: 1969, Astrophys. J. 158, pp. 753-770.
Stix, T. H.: 1962, *Theory of Plasma Waves* McGraw-Hill, New York.
Tapia, S.: 1977, Astrophys. J. (Letters) 212, pp. L125-L129.

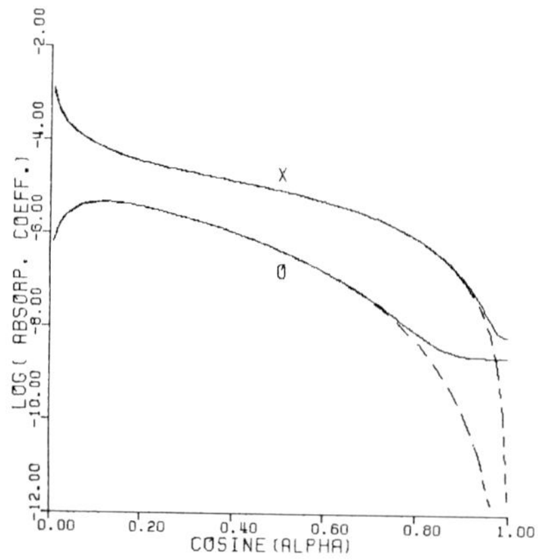

Fig. 1a

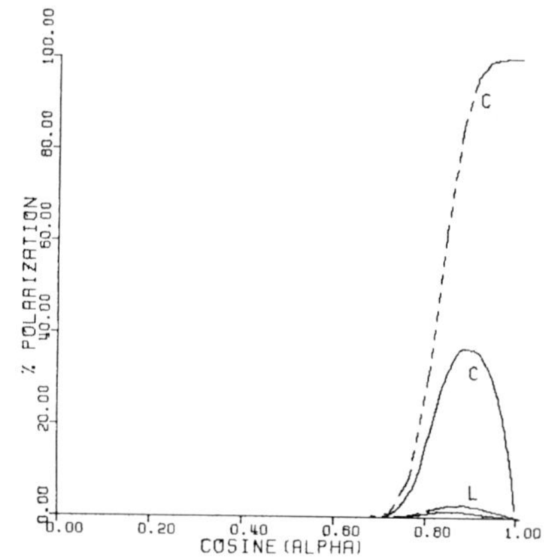

Fig. 1b

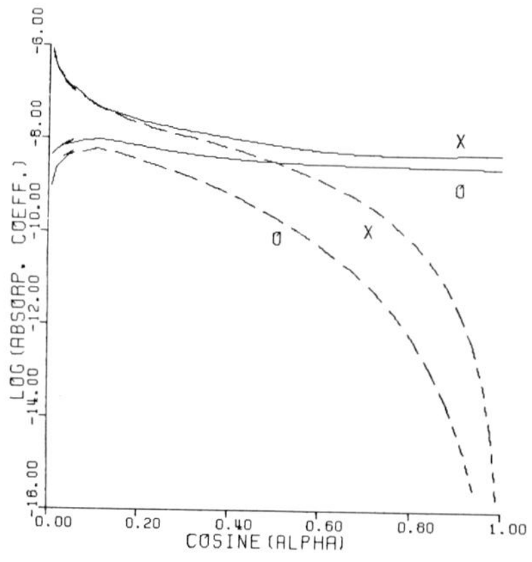

Fig. 2a

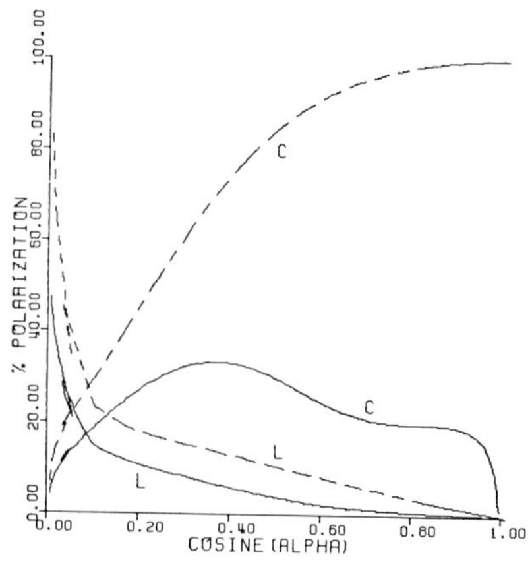

Fig. 2b

Figures 1a and 2a are curves of absorption coefficient versus angle for harmonic number 4 and 6 respectively. Ordinary and extraordinary modes are labeled O and X. Dashed lines correspond to the collisionless case. Figures 1b and 2b are curves of polarization versus angle for corresponding harmonic number. Linear and circular polarization are denoted by L and C. The plasma slab has a temperature of 1 keV and width of 10^8 cm.

DISCUSSION FOLLOWING G. CHANMUGAM'S TALK

TAPIA: I wanted to point out that in the collisional case the linear polarization seems to be significantly different from zero at very high angles with respect to the magnetic field. Is that possible?

CHANMUGAM: These are qualitative numbers, you shouldn't compare these immediately with observations.

LAMB: I think that the observations pose a challenge to present theoretical calculations, to explain how one can get a linear polarization pulse that is so broad. We are talking about a pulse that is something like 70 or 80 degrees wide.

CHANMUGAM: So this you think, is in the wrong direction.

LAMB: Yes.

CHANMUGAM: But the level has come down and that's good.

LAMB: I also want to remark that in our calculations the percentage of circular polarization does reach a maximum when you are looking down the field line but the polarized flux drops, because you have less optical depth through the emission region. Therefore, while the collisions do give you a turnover and can possibly account for the secondary minimum, our calculations give a fan beam and would naturally give a minimum when you look down the field lines, provided there is a constant background source of light. So I think it is probably not necessary to include these effects to get rather good agreement with the observations.

CHANMUGAM: That's what I said earlier. But the point is that in the previous calculations you get a very high polarization and you cut that down somehow, with some artificial background source.

LAMB: But the actual polarized flux is dropping, so if you have a constant background source of any kind you will get the right behaviour. Regarding your search for polarized UV emitters, I want to mention that whether the cooling is cyclotron dominated or not depends not only on the field strength, but also on the accretion rate. Therefore one should not look only for stars with high magnetic fields. If the field is a few x 10^7 Gauss, but the accretion rate is rather low, and therefore the density in the emission region is low. This would also tend to move the polarized cyclotron emission into the UV.

CHANMUGAM: The problem is that if you look at low accretion rates then it is difficult to observe them, the luminosity will be much lower.

LAMB: That depends, because the parameters of the system may change. For example, in AM Her stars the field is high enough but the stream is very small in cross sectional area, so the density is high and it is hard to get into the cyclotron dominated regime. On the other hand, in the long period DQ Her stars the field is weaker, but the disk comes in close and therefore the number of field lines that are accessible to the accreted matter is large and cover, probably, a substantial fraction of the stellar surface. So you can have the same overall accretion rate, but because the area is so much larger the density would be smaller and one would move into the cyclotron dominated regime.

CHANMUGAM: Yes, that's true.

AM HERCULIS BINARIES: PARTICLE ACCELERATION, RADIO EMISSION AND SYNCHRONIZATION*

G. Chanmugam
Department of Physics and Astronomy, Louisiana State University
Baton Rouge, Louisiana 70803-4001 and
G. A. Dulk
Department of Astro-Geophysics, University of Colorado, Boulder, Colorado 80309

ABSTRACT

It has been suggested that the recently discovered radio emission from AM Her arises as a result of gyrosynchrotron radiation from electrons at energies ≈400 keV in the magnetosphere of the white dwarf. However, no mechanism for producing such energetic electrons was discussed. In this paper, we argue that small departures from synchronous rotation can cause the companion star to act as a unipolar inductor. This leads to high voltages being produced across the companion star, which provides the necessary acceleration mechanism. This also implies that if the magnetic white dwarf was formed with a rapid rotation, synchronization would be achieved on a time scale ~10^4 yr.

1. INTRODUCTION

The AM Herculis binaries are a subclass of cataclysmic variables and are distinguished by the strong magnetic field $\gtrsim$ few x 10^7 gauss of the white dwarf (Schmidt et al 1981, Chanmugam and Dulk 1981). The field is sufficiently strong to prevent formation of an accretion disk and the companion is likely to be inside the magnetosphere of the white dwarf. Furthermore all of these systems are believed to be rotating synchronously. The prototype of these systems AM Her, in addition to emitting soft and hard X-rays, UV radiation, infrared radiation and polarized light (see the review by Chiappetti et al 1980) has recently been discovered to be a radio source (Chanmugam and Dulk 1982). The observed flux density of 0.67±.052 mJy at 4.9 GHz implies a radio luminosity ~ 10^{25} erg s^{-1} assuming a distance of 100 pc. If the projected area of the emitting region is πr^2 the brightness temperature of the radio emission is $T_b \approx 2.7 \times 10^9$ r_{11}^{-2} K where $r_{11} = r/10^{11}$ cm. This is probably too high to be explained by thermal bremsstrahlung. Instead a self-consistent model in which the radio emission arises as a result of gyro-synchrotron emission at a cyclotron number ≈ 30 to 50 has been proposed (Chanmugam and Dulk 1982). The emission probably occurs in the magnetosphere of the white dwarf at a radial distance ≈7-9 x 10^{10} cm from the white dwarf, with the average

M. Livio and G. Shaviv (eds.), Cataclysmic Variables and Related Objects, 223–227.

energy of the electrons ≈ 400 - 300 keV. However, the source of the energetic electrons is unknown.

The purpose of this paper is to show that departures from synchronous rotation (or nutations) could lead to the companion star acting as an unipolar inductor. This could lead to large potential differences being set up across the companion star, which can provide the necessary acceleration mechanism. We also show that if the white dwarf is formed spinning rapidly and the system acts as a unipolar inductor, synchronization can take place on a timescale ~ 10^4 yr. This is much shorter than the timescale ~ 10^{10} yr obtained if ohmic dissipation from the magnetic interaction of the two stars brings about synchronization (Joss et al 1979).

2. ACCELERATION MECHANISM

AM Her is analogous to the Jupiter-Io system in that the companion star (which corresponds to Io) is inside the magnetosphere of the white dwarf (shich corresponds to Jupiter). However, there is an important difference: the AM Her system rotates synchronously whereas the Jupiter-Io system does not. As the magnetic flux tubes of Jupiter sweep by Io, large potential differences are set up across the flux tube connecting Io to Jupiter, leading to acceleration of electrons (Goldreich and Lynden-Bell 1969). Here we consider the possibility that there are small departures from synchronous rotation in AM Her which gives rise to large potential differences being set up across the companion star.

The surface magnetic field of the white dwarf is $B_{WD} \simeq 2 \times 10^7$ gauss (Schmidt et al 1981). If this field is dipolar the magnetic field at the companion is $B = 2 \times 10^7 \, d_9^{-3}$ gauss, assuming the radius of the white dwarf to be 10^9 cm, and the distance between the two stars to be d. The induced electric field seen by the companion is then (Goldreich and Lynden-Bell 1969, Michel 1979) $\underset{\sim}{E} = -\underset{\sim}{V} \times \underset{\sim}{B}/c$, where $\underset{\sim}{V} = \underset{\sim}{V}_{orb} - \underset{\sim}{V}_{corot}$, with $\underset{\sim}{V}_{orb}$ the orbital velocity of the companion star and $\underset{\sim}{V}_{corot}$ the velocity of the plasma corotating with the white dwarf as it passes the companion. Let $\underset{\sim}{\xi}$ be the angular velocity characterizing the departure from synchronism. Hence

$$E \simeq 2 \times 10^8 \, \xi \, d_9^{-2} \text{ Volt cm}^{-1}.$$

If the companion has a diameter D (≈ 6×10^{10} cm), the potential difference across it is

$$\Phi \simeq 2 \times 10^{18} \, \xi \, d_9^{-2} \, D_{10} \text{ Volts}.$$

Thus with d_9=70 to obtain Φ~1MeV only a very small departure from synchronism is needed, i.e. $\xi \sim 10^{-9}$ rad s^{-1}. This potential difference acts across the flux tube from the white dwarf which encompasses the companion and drives an electric current from the companion to the white dwarf along half the surface of the flux tube. The current then crosses the flux tube in the atmosphere of the white dwarf near the polar cap and returns to the companion along the other half surface of the flux tube.

The circuit is completed across the companion.

In order to ensure that the large potential differences could be set up, it is necessary to show that a d.c. circuit can be established. This requires estimates of the electrical conductivities in the circuit. The most important of these is the Pedersen conductivity in the atmosphere of the white dwarf at the polar cap: $\sigma_p \approx 1.3 \times 10^8$ s^{-1}, since the temperature is $\sim 10^5$ K (Chiappetti et al 1980) and the electron number density is $\sim 10^{17}$ cm^{-3}. Hence the height integrated resistance of the atomosphere is $R_{WD} \approx 8 \times 10^{-14}$ in Gaussian cgs units or $\approx$ 0.1 ohm. This is much higher than the resistance of the portion of the circuit through the atmosphere of the companion or the flux tube. Hence if $\xi \lesssim 10^{-3}$ rad s^{-1}, the electric current flowing in the circuit is not large enough to create a magnetic field which would perturb the flow past the companion and hence reduce the electric field substantially (Dessler and Hill 1979). Another requirement for the d.c. circuit model to be valid, is that the flux tube cannot slip by more than D during the time T_A it takes for an Alfven wave to travel from the companion to the white dwarf and back. We estimate T_A to be $\sim 10^3$ s. This condition is satisfied because of the lower resistance of the companion (Dermott 1970). Hence, although there are uncertainties in our estimates of the resistance it is very likely that a d.c. circuit can be established. If so, electrons can be accelerated to energies $\sim$1MeV provided $\xi \sim 10^{-9}$ rad s^{-1}; upon diffusing into the magnetosphere they are a possible source of the observed radio emission.

3. SYNCHRONIZATION

During the formation of the white dwarf, mass loss can result in magnetic braking which would reduce ξ to a value $\sim 10^{-3}$ rad s^{-1} in about 10^4 yr (Brecher and Chanmugam 1978). The unipolar inductor model should then be applicable and since the power loss rate is Φ^2/R_{WD} the equation for ξ is given by

$$I \, d\xi/dt = -(dBD/c)^2 \, \xi/R_{WD},$$

where $I \sim 10^{50}$ gm cm^2 is the moment of inertia of the white dwarf. Hence ξ decays exponentially on a timescale $\sim 10^4$ yr. Once the value of ξ is reduced to $\sim 10^{-8}$ rad s^{-1}, the ohmic dissipation mechanism becomes more efficient in synchronizing the rotation. This is because $d\,\xi/dt \propto \xi^{-\frac{1}{2}}$ for that mechanism whereas $d\xi/dt \propto \xi$ for the unipolar inductor. It is noteworthy that Joss et al (1979) mentioned that the unipolar inductor may be more important than the ohmic dissipation mechanism but did not give any details (see also Lamb and Lamb (1979)).

4. CONCLUSIONS

Observations indicate that AM Her undergoes synchronous rotation. However, accretion torques would tend to change the rotation period of the white dwarf, and nutations could exist if the orbit is slightly eccentric. A small departure from synchronous rotation with $\xi \sim 10^{-9}$ rad s^{-1} is adequate to create large potential differences which could accelerate the electrons to energies $\gtrsim$ 400 keV, as required for the gyrosynchrotron radio-emission. Note that ξ is much less than the orbital angular velocity of 5.6×10^{-4} rad s^{-1} and hence not easily detectable. Departures from synchronism will tend to be damped either as a result of electromagnetic torques due to the current flowing through the companion or as a result of ohmic dissipation in the atmosphere of the companion; further investigation is needed to determine what value of ξ is necessary.

A consequence of our analysis is that if the white dwarf was formed spinning rapidly, extremely high potential differences would be set up and could result in ultra-high energy synchrotron emission. A possibly more significant consequence is that the large electromagnetic torque acting on the white dwarf can synchronize its rotation on a timescale which is much shorter than given by the ohmic dissipation model of Joss et al (1979).

This research was supported by the National Science Foundation under grant NSF-AST-8025250 to Louisiana State University and by NASA under grants NAGW-91 and NSG-7287 to the University of Colorado.

REFERENCES

Brecher, K. and Chanmugam, G.: 1978, Astrophys. J. 221, pp. 969-972.

Chanmugam, G. and Dulk, G. A. 1981, Astrophys. J., 244, pp. 569-578.
1982, Astrophys. J. (Letters) 255, pp. L107-110.

Chiappetti, L., Tanzi, E. G., and Treves, A. 1980, Space Sci. Rev., 27 pp. 3-33.

Dermott, S. F. 1970, Monthly Notices Roy. Astron. Soc. 149, pp. 35-44.

Goldreich, P. and Lynden-Bell, D. 1969, Astrophys. J., 156, pp. 59-78.

Joss, P. C., Katz, J. I., and Rappaport, S. A.: 1979, Astrophys. J. 230, pp. 176-183.

Lamb, D. Q., and Lamb, F. K. 1979: Bull. Am. Astr. Soc. 11, pp. 463.

Michel, F. C. 1979, Space Sci. Rev., 24, pp. 381-406.

Schmidt, G. D., Stockman, H. S., and Margon, B.: 1981, Astrophys. J. (Letters) 243, pp. L157-161.

DISCUSSION FOLLOWING G. CHANMUGAM'S TALK

ROBINSON: Is either the current flow or the electric field that you set up large enough to affect the accretion?

CHANMUGAN: This current I presume will flow on the surface of the flux tube, it is a skin effect so you can have accretion through the middle of the flux tube. So I don't think it will affect the mass flow because the energetics involved are much lower.

MEYER: Do you know how strongly the fields are swept back near the secondary when you couple with this resistance?

CHANMUGAN: I have not estimated that.

MEYER: It is an interesting question because you would not expect magnetic fields in this configuration to be swept back by much more than 45°, because usually then you get MHD instabilities which break the whole configuration, so you have there a hole in the strength for the coupling of magnetic fields. What determines the resistance in the polar cap?

CHANMUGAN: It is the ordinary conductivity of an ionized gas, but you have to divide it by $\omega_c^2\tau^2$, where ω_c is the cyclotron frequency and τ is the collision time.

KING: When you got it in synchronism, it is held there by the Joss, Rappaport mechanism. The dissipation required, where does that go, is it in the secondary?

CHANMUGAN: In that mechanism it is in the outer part, in the skin of the secondary.

KING: Your mechanism gets it into synchronism and Joss and Rappaport's is supposed to keep it there.

LAMB: Joss, Katz, Rappaport (1979) discuss two torques. One is a dipole-induced dipole torque which is dissipative, and which acts to synchronize the system. With the smaller magnetic fields now known to occur in the AM Her stars, this dissipative torque is too small to bring about synchronous rotation on the timescale required. Dr. Chanmugam and I are talking about new dissipative torques which can act to synchronize the system. Once the system is cynchronized, a second torque, due to the magnetostatic interaction between the stars, acts to maintain synchronous rotation. This torque is much stronger than the kind of torques we are talking about right now, and does not dissipate anything anywhere.

CHANMUGAN: I would like to make the comment that we have not detected radio emission from EF Eri, VV Pup, AN UMa.

SPIN-UP AND MAGNETIC FIELDS IN DQ HER STARS

D. Q. Lamb and J. Patterson
Harvard-Smithsonian Center for Astrophysics

ABSTRACT

We have carried out a detailed comparison of the spin-up behavior of the recently discovered faint long-period X-ray sources, using the accretion torque theory of Ghosh and Lamb. We confirm that these sources are degenerate dwarfs. We find that they have magnetic fields $B \sim 1 \times 10^6$ G, an order of magnitude smaller than those of AM Her stars. On this basis, we predict that they will show significant polarization in the infrared.

1. INTRODUCTION

Recently, several faint galactic X-ray sources have been found to pulse in both X-rays and optical light with periods $\gtrsim$ 1000 seconds. Table 1 lists six of these sources, together with the three previously known DQ Her stars with rotation periods < 100 seconds, and summarizes some of their properties. Figure 1 shows the distribution of rotation period versus binary period for these nine stars. Also shown are the ten AM Her stars and the nine pulsing neutron stars for which the binary period is known. There has been controversy about whether the long-period sources are degenerate dwarfs, like the AM Her stars, or are neutron stars. Their optical appearance and their X-ray to optical luminosity ratio of order unity suggest that they are degenerate dwarfs (Patterson and Price 1981), however their properties are otherwise indistinguishable from the long-period neutron star sources (White and Marshall 1981).

We have carried out a detailed comparison of the spin-up rate of these stars and that expected from the classic accretion torque theory of Ghosh and Lamb (1978, 1979a,b). In this paper we summarize our findings. The full details of our analysis and a more complete discussion of our results will be published elsewhere (Lamb and Patterson 1982).

2. SPIN-UP BEHAVIOR

Ghosh and Lamb (1978, 1979a,b; hereafter GL) have developed a magnetohyrodynamic theory of disk accretion onto magnetic stars. According to this theory (see also Lamb, Pethick, and Pines 1973), the accretion torque is the sum of a matter torque and a torque arising from stressed magnetic field lines that thread the disk. If the star is a slow rotator (in the sense $\omega_s \equiv \Omega_K(r_o)/\Omega \ll 1$, where $\Omega_K(r_o)$ is the Keplerian angular velocity at the

M. Livio and G. Shaviv (eds.), Cataclysmic Variables and Related Objects, 229–237.

TABLE 1

DQ Her Stars

STAR	P_b (Hours)	d (pc)[a]	P (s)	$\lvert\dot{P}\rvert^{-1}$	L_s (10^{30} ergs s^{-1})[b]	T_{bb} (eV)	L_h (10^{30} ergs s^{-1})[b]	T_{br} (keV)	REFERENCES
AE Aqr	9.88	84	33.076737	$> 2 \times 10^{13}$	5.4	...	...	...	1, 10, 11, 13
V533 Her	≃7	1000-1500	63.63307	$> 3 \times 10^{12}$	< 20-40	...	...	...	3, 9, 11
DQ Her	4.65	420	71.0653	1.2×10^{12}	< 2.4	...	...	...	3, 15
V1223 Sgr		400	794.380	$> 2 \times 10^{10}$	...	...	1600	≳ 10	17, 18
H2252-035	3.59	250	805.21	$> 5 \times 10^{9}$	...	...	500	> 20	12, 13, 20, 21
3A0729+103	3.75	[200]	913.48	$> 5 \times 10^{9}$			160	[10]	7
H2215-086	4.03	[200]	1254.5	$> 5 \times 10^{9}$	20	...	48 120	≳ 10 ≳ 10	16
Ex Hya	1.64	100	4021.62	2.2×10^{10}	14	570	60 100	≃ 5 ≃ 4.5	4, 5, 19
TV Col (= 3A0526-328)	5.49	[200]	18690.	$> 2 \times 10^{6}$			140	[10]	2, 6, 8

[a] Distances in brackets are assumed.

[b] Luminosities assume a distance of 200 pc if distance is unknown and include only observed flux.

(1) Bailey (1981)
(2) Charles et al. (1979)
(3) Cordova, Mason, and Nelson (1981)
(4) Cordova and Riegler (1979)
(5) Gilliland (1982)
(6) Hutchings et al. (1981)
(7) McHardy et al. (1982)
(8) Motch (1981)
(9) Patterson (1979a)
(10) Patterson (1979b)
(11) Patterson (1982a)
(12) Patterson (1982b)
(13) Patterson et al. (1980)
(14) Patterson and Price (1981)
(15) Patterson, Robinson, and Nather (1978)
(16) Patterson and Steiner (1982)
(17) Steiner (1981)
(18) Steiner et al. (1981)
(19) Swank (1980)
(20) Warner, O'Donoghue, and Fairall (1981)
(21) White and Marshall (1981)

inner radius r_o of the disk and Ω is the angular velocity of the star), the matter torque dominates and the star is spun up. On the other hand, if the star is a fast rotator ($\omega_s \simeq 1$), the torque due to the stressed magnetic field lines dominates and the star is spun down. The spin-up rate is given in general by

$$-\dot{P} = 4.09 \times 10^{-11}\, n(\omega_s)\, (P_3 L_{34}{}^{3/7})^2\, B_6{}^{2/7} (M/M_\odot)^{-3/7}\, (R/5 \times 10^8\ \mathrm{cm})^{12/7} I_{50}{}^{-1}\ \mathrm{s\ s^{-1}}, \quad (1)$$

where P_3 is the rotation period in units of 10^3 seconds, L_{34} is the luminosity in units of 10^{34} ergs s^{-1}, and M, R and I_{50} are mass, radius, and moment of inertia of the star, the last in units of 10^{50} g cm^2. In the remainder of this paper, we use the values of M, R, and I_{50} for a 1 $M_\odot$ degenerate dwarf. The dimensionless torque function $n(\omega_s)$ is given by GL (1979b; eq. 10), while the fastness parameter ω_s is given by

$$\omega_s = 0.114\, (P_3 L_{34}{}^{3/7})^{-1} B_6{}^{6/7}\, (M/M_\odot)^{-2/7} (R/5 \times 10^8\ \mathrm{cm})^{15/7}. \quad (2)$$

Letting the constants in the theory take their standard values, the accretion torque vanishes for $\omega_s = 0.35$. This defines an equilibrium state for the accreting star. Assuming that the accreting star is near this equilibrium state, the spin-up rate is

$$-\dot{P}_{eq} = 5.96 \times 10^{-11}\, (\omega_s/0.35)^{1/3}\, (P_3 L_{34}{}^{3/7})^{7/3}\, (M/M_\odot)^{-1/3}\, (R/5 \times 10^8\ \mathrm{cm}) I_{50}{}^{-1}\ \mathrm{s\ s^{-1}} \quad (3)$$

For somewhat larger values of ω_s, the centrifugal force exceeds the gravitational force at r_o and steady accretion is not possible; thus $\omega_s \sim 1$ also represents an upper limit to the fastness of the star. This translates into an upper bound on the stellar magnetic field,

$$B \lesssim 3.71 \times 10^6\, (\omega_s/0.35)^{7/6}\, (P_3 L_{34}{}^{3/7})^{7/6}\, (M/M_\odot)^{1/3}\, (R/5 \times 10^8)^{-5/2}\ \mathrm{G}, \quad (4)$$

and on the inner radius of the disk,

$$r_o/R \lesssim 14.9\, (\omega_s/0.35)^{2/3}\, P_3{}^{2/3}\, (M/M_\odot)^{1/3} (R/5 \times 10^8\ \mathrm{cm})^{-1}. \quad (5)$$

All the sources display large amplitude X-ray and/or optical pulsations, indicating that the magnetic field is strong enough to disrupt the disk and channel the flow of accreting matter toward the magnetic pole(s). This implies a lower bound on the stellar magnetic field,

$$B \gtrsim 2.24 \times 10^5\, (r_o/3R)^{7/4}\, L_{34}{}^{1/2}\, (M/M_\odot)^{-1/4} (R/5 \times 10^8\ \mathrm{cm})^{-3/4}\ \mathrm{G}. \quad (6)$$

Lower bounds on B and r_o/R can also be derived from the observed upper limit or measured value of $\dot{P}$, but are generally not very stringent because $\dot{P}$ depends so weakly on B (see eq. 1 above). Similar expressions hold for the pulsing neutron star X-ray sources, and when they are needed in the remainder of the paper, we use the values of M, R, and I for a 1.3 $M_\odot$ PPS star (Pandharipande, Pines, and Smith 1976).

In order to compare theory and observation, we need estimates of the accretion rates and luminosities of the sources. We have derived the accretion rates using the techniques described in Patterson (1982b) and Lamb and Patterson (1982), and have converted them to luminosities assuming a 1 $M_\odot$ star. The results are listed in the second and third columns of Table 2.

3. DISCUSSION

3.1. Nature of Long-Period Sources

Figure 2 shows the spin-up rates given by the theory of GL for degenerate dwarfs with magnetic fields $B = 10^4-10^8$ G (solid curves) and neutron stars with $B = 10^{10}-10^{14}$ G (dashed curves). Also plotted are the observed upper limits or measured value of the spin-up rates for the nine stars listed in Table 1 and the nine pulsing neutron star sources for which $\dot{P}$'s have been measured (cf. Rappaport and Joss 1977; GL 1979b, Table 2). The region of Figure 2 accessible to degenerate dwarfs lies inside the region accessible to neutron stars. Therefore, there is always a neutron star spin-up curve that passes through any given degenerate dwarf spin-up rate. However, these solutions always require that the star be an extremely fast rotator. Even Her X-1, for which $\omega_s \simeq 0.35$ and for which the accretion torque is 30 times less than the slow rotator value, is not nearly fast enough, as can be seen from Figure 2. Furthermore, the seven upper limits and two measured values of the spin-up rate for the nine stars listed in Table 1 are all consistent with the spin-up behavior expected for degenerate dwarfs that are not particularly fast rotators. This agreement supports the accretion torque theory of GL. We also conclude the the nine stars listed in Table 1 are indeed all degenerate dwarfs. Henceforth we shall refer to them as DQ Her stars.

Throughout the remainder of this paper we shall assume the validity of the GL accretion torque theory for the DQ Her stars. If we further assume that the stars lie near the equilibrium state corresponding to $\omega_s = 0.35$ described earlier, we can derive the equilibrium spin-up rate $-\dot{P}_{eq}$, the range of allowed stellar magnetic field, and an upper bound to the ratio of the inner radius of the disk and the stellar radius r_o/R, for each of the DQ Her stars listed in Table 1, using equations (3), (4) and (6), and (5), respectively. The results are given in Table 2.

The values of $-\dot{P}_{eq}$ agree well with the observed spin-up rates of DQ Her and EX Hya (compare column 5 of Table 1 and column 4 of Table 2). The values of $-\dot{P}_{eq}$ for the remaining sources constitute predictions, provided that the stars are near equilibrium. They suggest that the spin-up rates of most of the sources will soon be measurable, but that those of AE Aqr and TV Col may not be measurable for some time to come. On the other hand, if the stars are not near equilibrium, they must have experienced episodes of spin-down as well as spin-up in the past, since their spin-up time scales ($T_s \equiv P\,|\dot{P}|^{-1}$) are all much less than the evolutionary time scales of their binary systems (cf. the discussions by GL 1979b and Elsner, Ghosh, and Lamb 1980 of the analogous situation in the long period neutron star X-ray sources).

3.2. Magnetic Field Strengths

The range of stellar magnetic field strengths allowed by the spin-up and pulsing behavior of the DQ Her stars is given in column 5 of Table 2. In the cases of the three short-period sources, the maximum and minimum allowed field strengths coincide (in the cases of DQ Her and V533 Her, they are even slightly inconsistent). Assuming the stars to be very slowly rotating, rather than nearly in equilibrium, would make the minimum and maximum field strengths more inconsistent. We conclude that these stars are moderately fast rotators near equilibrium, and predict that they have magnetic fields $B \sim 6 \times 10^5$ G. For the long-period sources, a range of magnetic fields is consistent with their spin-up and pulsing behavior. The allowed magnetic fields are $B \sim 1 \times 10^6$ G, an order of magnitude smaller than those of AM Her stars. On this basis, we predict that the long-period sources will show significant polarization in the infrared, while the short-period ones will show

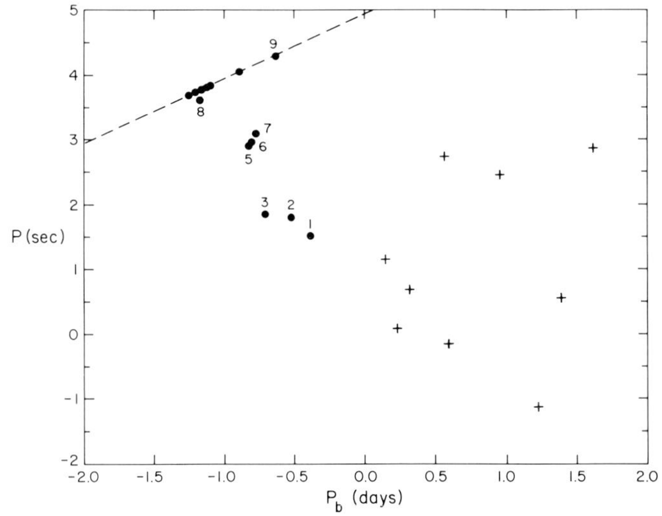

Fig. 1--Distribution of rotation period P versus binary period P_b for the nine stars in Table 1 (the numbers label them by their position in the table). Also shown are the ten AM Her stars and the nine pulsing neutron star X-ray sources for which the binary period is known.

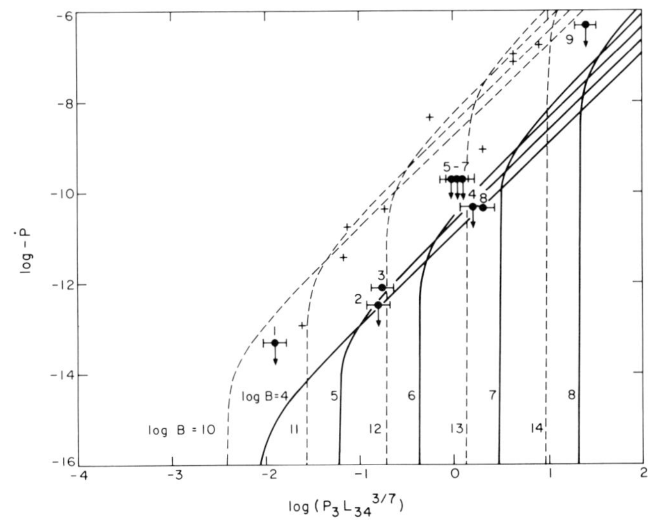

Fig. 2--Curves of the spin-up rate for degenerate dwarfs (solid lines) and for neutron stars (dashed lines). Also plotted are the observed upper limits or measured values of $\dot{P}$ for the nine stars in Table 1 (again labeled by their position in the table) and the nine pulsing neutron star sources for which $\dot{P}$'s have been measured.

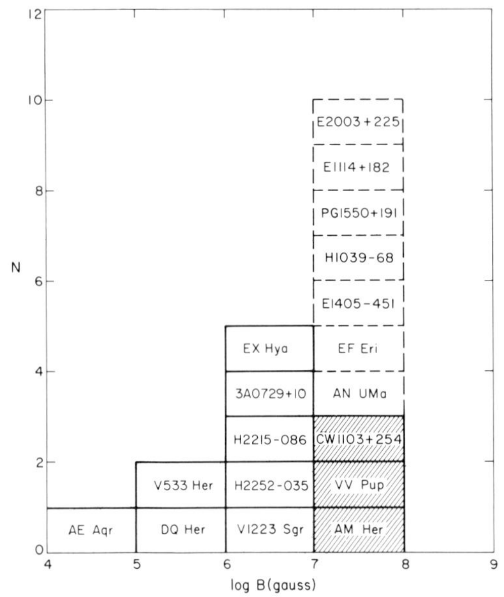

Fig. 3--Histogram of the stellar magnetic field strengths of the known magnetic cataclysmic variables. The shaded boxes are AM Her stars for which the magnetic field has been directly measured; the dashed boxes are the remaining AM Her stars (assumed to have similar fields). The unshaded boxes are DQ Her stars; their fields have been inferred from their spin-up and pulsing behavior.

TABLE 2

Derived Properties of DQ Her Stars

STAR	$\dot{M}$ ($10^{-9} M_\odot$ yr^{-1})	L (10^{34} erg s^{-1})	$\lvert \dot{P}_{eq} \rvert^{-1}$	B (gauss)[a]	r_o/R[a]
AE Aqr	0.06	0.1	7×10^{14}	$\simeq 6 \times 10^4$	$\simeq 3$
V533 Her	5	8	2×10^{12}	$\simeq 6 \times 10^5$	$\simeq 3$
DQ Her	5	8	1×10^{12}	$\simeq 6 \times 10^5$	$\simeq 3$
V1223 Sqr	3	5	8×10^9	$5 \times 10^5 - 5 \times 10^6$	$\lesssim 10$
H2252-035	2	2	2×10^{10}	$3 \times 10^5 - 3 \times 10^6$	$\lesssim 10$
3A0729+103	1	2	1×10^{10}	$3 \times 10^5 - 4 \times 10^6$	$\lesssim 10$
H2215-086	0.3	0.5	3×10^{10}	$1 \times 10^5 - 3 \times 10^6$	$\lesssim 20$
EX Hya	0.1	0.2	4×10^9	$9 \times 10^4 - 6 \times 10^6$	$\lesssim 30$
TV Col	1	2	1×10^7	$3 \times 10^5 - 1 \times 10^8$	$\lesssim 90$

[a]The larger value or the equality applies if the star is near equilibrium, i.e., if the accretion torque is nearly zero.

polarization only at longer wavelengths. Finally, we note that the fact that EX Hya does not rotate sychronously with its binary period, although the period is shorter than those of many AM Her stars, is now easily understood as due to the fact that it has a magnetic field at least an order of magnitude smaller than the fields in AM Her stars.

The upper bound on the ratio r_0/R is given in column 6 of Table 2. It is a function only of the rotation period of the star (see eq. 5) and increases as the period increases. If the star is near equilibrium, the equality applies; otherwise, the ratio must be smaller. The ratio is only ~ 3 for the three short-period sources, indicating that the disk extends down nearly to the stellar surface (note, however, that $r_0/R \propto L^{-2/7}$, so that an increase in luminosity of a factor of ~ 50 is still required to crush the magnetosphere to the stellar surface). For the long-period sources, the ratio is larger, although not as large as the value ~ 200 typical of pulsing neutron star X-ray sources. In fact, the emission lines from the disks in these sources should all be rather broad (velocity profiles $\gtrsim 1000$ km s^{-1}. However, if the emission line widths seen in TV Col (Hutchings et al. 1981) are due to Keplerian velocity broadening, the large inferred velocity ($\simeq 3000$ km s^{-1}) suggests that r_0/R is not very large and that this source is far from equilibrium. This result and the lack of observed optical polarization (Charles et al. 1979) indicates that the magnetic field in TV Col is closer to the minimum value $\simeq 3 \times 10^5$ G than to the maximum of $\simeq 1 \times 10^8$ G.

Figure 3 shows a histogram of the stellar magnetic field strengths of the known magnetic cataclysmic variables (AM Her stars and DQ Her stars). The stars in the interval 10^7-10^8 G are all AM Her stars; the shaded stars are those in which the magnetic field has been directly measured, while the remainder have been assumed to have similar field strengths on the basis of their similar X-ray, UV, and optical spectra, and the fact that they are all phase-locked to their binary periods. The stars in the unshaded boxes drawn with solid lines are all DQ Her stars; their magnetic fields have been inferred from their spin-up and pulsing behavior, as described above. The observational selection effects affecting this histogram are obviously severe; for example, the identification of AM Her stars is relatively easier. Thus, it is not clear at present what to make of the distribution of magnetic fields; however, the situation may inprove as an X-ray-selected sample of stars becomes possible.

4. CONCLUSIONS

We have carried out a detailed comparison of the spin-up rate of the recently discovered faint long-period X-ray sources and that expected from the classic accretion torque theory of Ghosh and Lamb (1978, 1979a,b). Our analysis supports the theory and confirms that the long-period sources are indeed degenerate dwarfs. These stars have rotation periods appropriate to their luminosities and magnetic fields. We therefore believe that they and the DQ Her stars should be regarded as members of a single class, as are the short- and long-period pulsing neutron star X-ray sources.

Our analysis also shows that the long-period sources have magnetic fields $B \sim 1 \times 10^6$ G, an order of magnitude smaller than those of AM Her stars, and that the three previously known short-period DQ Her stars have still smaller fields. On this basis, we predict that the long-period sources will show significant polarization in the infrared, while the short-period ones will show polarization only at longer wavelengths.

This research was supported in part by NASA grant NAGW 246. One of us (DQL) also gratefully acknowledges support from the Fluid Research Fund of the Smithsonian Institution.

REFERENCES

Bailey, J.: 1981, Mon. Not. R. Astron. Soc., 197, 31.
Charles, P. A., Thorstensen, J., Bowyer, S., and Middleditch, J.: 1979, Astrophys. J. Letters, 231, L131.
Cordova, F. A., Mason, K. O., and Nelson, J. E.: 1981, Astrophys. J., 245, 609.
Cordova, F. A., and Riegler, G. R.: 1979, Mon. Not. R. Astron. Soc., 188, 103.
Elsner, R. F., Ghosh, P., and Lamb, F. K.: 1980, Astrophys. J. Letters, 241, L155.
Ghosh, P., and Lamb, F. K.: 1978b, Astrophys. J. Letters, 223, L83.
Ghosh, P., and Lamb, F. K.: 1979a, Astrophys. J., 232, 259.
Ghosh, P., and Lamb, F. K.: 1979b, Astrophys. J., 234, 296.
Gilliland, R. L. 1982, Astrophys. J., in press.
Hutchings, J. B., Crampton, D., Cowley, A. P., Thorstensen, J. R., Charles, P. A.: 1981, Astrophys. J., 249, 680.
Lamb, D. Q., and Patterson, J.: 1982, submitted to Astrophys. J.
Lamb, F. K., Pethick, C. J., and Pines, D.: 1973, Astrophys. J., 184, 271.
McHardy, I. M., Pye, J. P., Fairall, A. P., Warner, B., Allen, S., Cropper, M., and Ward, M. J.: 1982, IAU Circ. No. 3687.
Motch, C.: 1981, Astron. Astrophys. 100, 277 (1981).
Pandharipande, V. R., Pines, D., and Smith, R. A.: 1976, Astrophys. J., 208, 550.
Patterson, J.: 1979a, Astrophys. J. Letters, 233, L13.
Patterson, J.: 1979b, Astrophys. J., 234, 978.
Patterson, J.: 1982a, Astrophys. J., in press.
Patterson, J.: 1982b, Astrophys. J., in press.
Patterson, J., Branch, D., Chincarini, G., and Robinson, E. L.: 1980, Astrophys. J. Letters, 240, L133.
Patterson, J., and Price, C. M.: 1981, Astrophys. J. Letters, 243, L83.
Patterson, J., Robinson, E. L., and Nather, R. E.: 1978, Astrophys. J., 224, 570.
Patterson, J., and Steiner, J. E.: 1982, Astrophys. J. Letters, in press.
Rappaport, S., and Joss, P. C.: 1977, Nature, 266, 683.
Steiner, J. E.: 1981, private communication.
Steiner, J. E., Schwartz, D. A., Jablonski, F. J., Busko, I. C., Watson, M. G., Pye, J. P., and McHardy, I. M.: 1981, Astrophys. J. Letters, 249, L21.
Swank, J. H.: 1980, private communication.
Warner, B., O'Donoghue, D., and Fairall, A. P.: 1981, Mon. Not. R. Astron. Soc., 196, 705.
White, N. E., and Marshall, F. E.: 1981, Astrophys. J. Letters, 249, L25.

DISCUSSION FOLLOWING D. LAMB'S TALK

CHANMUGAM: I have a comment regarding the distribution of magnetic fields. These isolated white dwarfs have no magnetic fields less than $3x10^6$ Gauss or something like that and I think there are upper limits, down to 10^4 Gauss in isolated magnetic white dwarfs. This means that the field distribution here, since you are seeing a lot of systems with fields in between is quite different.

LAMB: I think it is premature to conclude that. To my knowledge the most careful work that has been done on the distribution of magnetic fields in isolated degenerate dwarfs is by Liebert, Borra, and Handstreet. Their results are consistent with a uniform distribution of magnetic field strengths.

CHANMUGAM: According to Angel et al., that is between $3x10^6$ and $3x10^8$, there is a cut off at about 10^6.

ROBINSON: AE Aqr has a very low accretion rate and its Alfven radius is still relatively close to the white dwarf, how much could you increase that accretion rate before the Alfven radius got squashed right down to the white dwarf?

LAMB: It goes like $L^{-2/7}$ so if you want to change this by a factor of 2 it would be a factor of 10 in accretion rate.

ROBINSON: So, as the star evolves and its accretion rate goes up or down, I am talking about secular variations, the magnetic field of the white dwarf can be important or absolutely negligible depending on the accretion rate.

LAMB: That's correct. But this is true only for low field strengths.

WILLIAMS: I find it curious that several of the intermediate polar objects are active novae whereas non of the AM Her type are novae and I wonder what the possible effects of magnetic fields on the nova phenomenon are.

LAMB: It could have effects, but they are poorly understood and I don't wish to try to address them here.

THE ORIGIN AND EVOLUTION OF CATACLYSMIC BINARIES*

P.P. Eggleton
Institute of Astronomy, Madingley Rd.
Cambridge CB3 OHA, U.K.

ABSTRACT

Some cataclysmic binaries may be products of Case C evolution of low mass stars (orbital period $\sim$ 1 yr; masses $\sim$ 1 - 4 $M_\odot$), involving a common envelope phase. Other mechanisms, probably involving late Case B and even early Case B, but with significant loss of angular momentum, may be necessary to account for some evolved binaries such as AA Dor or V Sge. Further angular momentum loss, probably by magnetic braking coupled with tidal friction, causes secular evolution in cataclysmic binaries. It is suggested that tidal friction may account for the shortage of cataclysmics with periods $\lesssim$ 1.3 hr; but this cutoff, as well as the gap in the period distribution between 2 and 3 hrs, is hard to explain and imposes more severe constraints on possible theories than is commonly acknowledged.

1. INTRODUCTION

Cataclysmic binaries were presumably young binaries once, but it has long been difficult to see clearly what kind of young system evolves into the typical close white dwarf/red dwarf configuration of a cataclysmic binary. Kraft (1963) suggested W UMa systems, whose angular momenta and galactic distribution are similar; but it seems more likely (Webbink 1976, Robertson and Eggleton 1977, Bopp and Rucinski 1981) that W UMa binaries evolve into single rapidly-rotating red subgiants. Current thinking, stimulated largely by Paczyński (1976), is that the binary was once wide enough to have contained a red supergiant, with the present white dwarf as its core. The main sequence secondary, after causing Case C (or very late Case B) mass transfer, became rapidly embroiled in the distended convective supergiant envelope, and spiralled inwards within a differentially rotating "common envelope". It is necessary to suppose that this spiralling-in

* Work supported in part by NSF grants AST 78-20123 and AST 78-20124

M. Livio and G. Shaviv (eds.), Cataclysmic Variables and Related Objects, 239–255.

process removed just enough angular momentum from the orbital motion (depositing it in the common envelope) to reduce the orbital period from ∿ 100 days to ∿ ½ day, and that the gravitational energy released in this process was efficiently directed into blowing off the common envelope before the period could become so short that the two cores might coalesce.

It seems likely, but perhaps not inevitable, that some kind of common envelope phase occurs if a binary starts with such a period that Case C mass transfer is to be expected. It is less clear, however, that what will emerge from the common envelope will be a cataclysmic binary, or something closely related. But the idea received strong support from the discovery of a short-period eclipsing binary (UU Sge, Miller *et al.* 1976) at the centre of a planetary nebula (Abell 63) shortly after Paczyński (1976) suggested this kind of evolution. UU Sge, however, is sufficiently wide that it will have to lose a significant further amount of angular momentum before it can become cataclysmic; this might be by a combination of magnetic braking (by stellar wind) and of tidal friction (Huang 1966, Eggleton 1976, Verbunt and Zwaan 1981).

In Section 2, possible evolutionary paths are discussed for systems of fairly low mass and long period. In Section 3, various systems are described which might be thought of as post-common-envelope, pre-cataclysmic binaries. In Section 4 I discuss the period gap in the distribution of cataclysmic binaries. This gap has very important implications for the evolutionary precursors of cataclysmics. Section 5 summarises the main conclusions.

2. EVOLUTIONARY PATHS FOR LONG-PERIOD LOW-MASS BINARIES

Figure 1 shows, very schematically, three evolutionary paths which might start from systems with initial periods ≳ 100 days, and with component masses sufficiently low that either star on its own would evolve to a white dwarf rather than a supernova, neutron star or black hole. An obvious fourth path, where the separation is too great for any significant interaction to happen, is omitted. It will, in the long run, save space if I use a shorthand to describe various types of system, thus:-

MM; D - Main sequence star + Main sequence star; Detached
MS; S - Main sequence star + Supergiant; Semidetached
SW; C - Supergiant + White dwarf; Common envelope.

The other acronyms can be inferred from these. The paths in Figure 1 are based on the assumption that any Roche lobe overflow from a red supergiant (i.e. Case C or possibly late Case B) will be followed by formation of a common envelope, with spiralling-in. If this assumption is not correct other possible paths would have to be considered, but they would also have to involve substantial angular momentum loss (hereinafter "AML") if they are to lead to cataclysmic binaries.

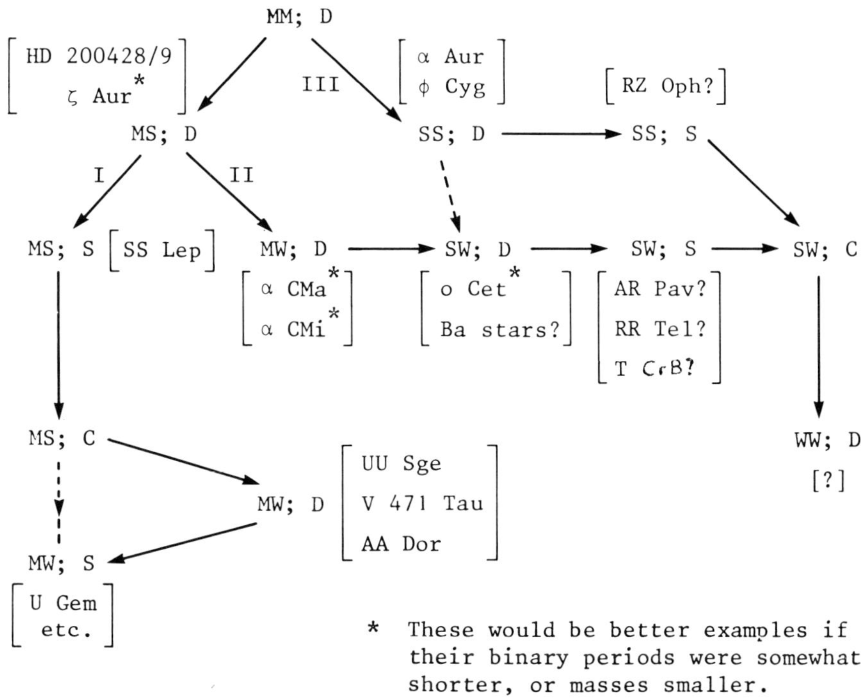

Figure 1. Schematic paths of binary evolution, for systems of low mass and fairly long period (P ∿ 100 - 1000d). Path I may apply to shorter periods within this range and Path II to longer periods. Path III may apply to systems of closely equal initial mass. Acronyms such as "MM;D" are explained in the text (beginning of Section 2). Possibly typical systems at various stages are indicated in brackets.

Path I refers to systems with periods of perhaps 100-300 days, systems which interact strongly when the initial primary becomes a supergiant. Shortly before this interaction we might have an MS;D system like ζ Aur, although this well-known binary is perhaps more massive, and has a longer period, than we would like. A better example would be HD200428/9 (KO III + A5 V, P = 113d, Griffin *et al.* 1976). Once the system becomes semidetached (MS;S) the convective character of the loser's envelope may ensure a rather drastic phase of Roche-lobe overflow followed by a common envelope phase (MS;C). What emerges will hopefully be something of the character of UU Sge (MW;D), which may after further AML become a cataclysmic binary (MW;S).

Path II starts with somewhat longer periods (maybe 300-1000d), so that the initial primary becomes a supergiant (MS;D) and then a white dwarf (MW;D) before significant interaction takes place. Then the companion evolves to a supergiant (SW;D), and some interaction begins to take place, initially between the white dwarf and a wind from the supergiant but later from Roche lobe overflow (SW;S). There are several types of observed system which are thought (Starrfield *et al.* 1976, Tutukov and Yungelson 1976, Paczyński and Rudak 1980) to consist of interacting white dwarf/red supergiant pairs,, i.e. symbiotics (like Z And), slow novae (like RR Tel) and recurrent novae (like T CrB). Some of these systems are possibly semidetached, but others are almost certainly detached, with wind rather than overflow providing the observed line-emitting gas. But once overflow begins it may become drastic, as on Path I, and be followed by a common envelope phase (SW;C). Presumably, by analogy with Path I, this leads to the blowing off of the envelope, leaving a close detached pair of fairly massive white dwarfs (WW;D). However, Sparks and Stecher (1972) suggested that the outcome of a rather similar scenario would be the coalescence of the two white dwarf cores, which could lead to a single core exceeding the Chandrasekhar limit and hence to a supernova. This seems at least as plausible an outcome. If it were correct, it rather suggests that at the analogous stage in Path I the main sequence companion would simply be disintegrated as it spiralled in very close to the white dwarf core. It is a major problem with the common envelope model to see why the envelope should dissipate at the convenient separation rather than too early or too late.

Path III refers to systems with nearly equal initial masses, so that both stars evolve into giants and supergiants at about the same time (SS;D). This might seem to require improbably nearly equal masses, but some systems (α Aur, ϕ Cyg) achieve this all the same. Even so, when the components begin to interact one will probably be appreciably larger than the other and become the loser in a Roche lobe overflow (SS;S), as appears to be the case for RZ Oph (Hiltner 1946). This could lead to a common envelope (SS;C), which would probably not be different in principle from the SW;C step in Path II, and so might also lead to a close pair of white dwarfs (WW;D). Alternatively, particularly if the period were a little greater, the more advanced supergiant may blow off its envelope as if it were single, to leave a white dwarf (SW;D), which also links up with Path II.

Figure 1 was drawn up partly on the basis of what might be expected theoretically, assuming the common envelope mechanism, and partly on the basis of what kinds of systems are observed. There seem to me to be at least three discrepancies between these two approaches, which may have to be resolved. On Path I, there are several known systems in the last two states (MW;D and MW;S) and yet a shortage, as far as I can tell, in the preliminary stage MS;S. This might only mean that the MS;S stage is very short-lived, whereas the two later stages can apparently be quite long-lived. However, on Path II one sees several types of system which have been attributed to the analogous SW;S phase (or perhaps to a slightly earlier phase when the supergiant is close to filling its lobe), and one would expect this phase to be about equally short-lived. On the other hand there are no known WW;D systems to belong to the end of Path II. Such systems would be hard to recognise, of course, but not necessarily harder than some of the close MW;D systems like UU Sge (see Section 3). This point makes me wonder if several symbiotic and related systems contain accreting main sequence stars rather than white dwarfs, the accretion process heating them up so that they somehow resemble hot blue subdwarfs.

A second point, reinforcing the first though independent of it, is that one might not expect Path II to be followed at all. For if the binary was close enough to interact during the SW;D stage, it was presumably already close enough to interact at the earlier MS;D stage, and so to follow Path I. This might be another reason for supposing that symbiotics etc. contain main sequence stars rather than white dwarfs; but it may also be that there is some AML when a supergiant in an MS;D binary blows off its envelope on the way to becoming a white dwarf (the main sequence component giving some of its orbital angular momentum to the proto-planetary nebula), so that the remaining system is somewhat closer in the SW;D stage than in the earlier MS;D stage. Since the separation is proportional to the square of the angular momentum, a modest amount of AML goes a fairly long way.

A third point derives from the binary RZ Oph (no. 700 in the catalogue of Batten *et al.* 1978), which appears to be a semidetached pair of supergiants (F3 Ib + K5 Ib, P = 262d). The system is surprisingly like a normal Algol (Hiltner 1946), despite considerably greater radii, and so seems to imply that Case C mass transfer need not always be drastic. However, the system presumably started with nearly equal masses; and since the more evolved star may have lost some mass by stellar wind before filling its Roche lobe, the primary may have already been the less massive component before it began to overflow. Hence its convective envelope may not have been faced with the usual difficulty of trying to fill a contracting Roche lobe while wanting to expand in response to mass loss.

An alternative picture of RZ Oph is that the hotter supergiant is actually a bloated accretion disc or ring around, say, a main sequence star. If so, the initial masses were presumably less closely equal, so it is the more surprising that the onset of mass transfer did not, apparently drive the system into a common envelope. The same conclusion appears to follow if the gainer is a bloated white dwarf rather than main sequence star. Generally, it seems possible that systems starting with equal or moderately unequal masses may avoid the common envelope phase if the primary suffers substantial single-star mass loss as a red supergiant before filling its lobe.

I shall argue in the next two sections that many of the low-mass main sequence components of cataclysmic binaries and UU-Sge-like systems must have been low-mass M dwarfs (or even black dwarfs) all along, and not just because they lost mass during and after the common envelope phase. This leads me to suppose that there should be a class of MS;D and MS;S binaries in which the main sequence component is an M dwarf.

Table 1

Name (Name 2)	Period (days)	Spectra	Binarity indicator*	Mass Function	Notes	Ref
NGC 2346	16.0	AV† + SDOB?	SB1	.007	a,b	1
BD -3°5357 (FF Aqr)	9.2	G8III† + SDOB	ecl,SB1	.019		2
NGC 1630 (CPD -26°389	8.21	? + SDO†	SB1	.19	a	3
Feige 24	4.23	MVe† + WDOBe	SB1	.13		4,5
PG 1155+492 (BE UMa)	2.29	? + SDOBe†	ell,SB1	.20		6,7
Case 1	0.67	M2Ve† + WDA	SB1	.11		8
HZ 22 (UX CVn)	0.57	? + SDB†	ell,SB1	.13		9,10
HZ 9	0.56	M4.5Ve† + WDA	SB1	.12	c	11
BD 16°516 (V471 Tau)	0.52	K2V† + WD	ecl,SB1	.18	c	12,13
V Sge	0.51	AVe? + SDOBe	ecl,SB2	2.8+0.75	b,d	14
Abell 46 (V477 Lyr)	0.47		ecl		a	15
Abell 63 (UU Sge)	0.47	KV + SDO	ecl		a	16,17
PG 1413+01 (GK Vir)	0.34	M2V + WDOB	ecl			18
LB 3459 (AA Dor)	0.26	WDe? + SDO†	ecl,SB1	.001	b	19,20,21
NGC 6826	0.24	? + SDOBe†	SB1		a	22
Abell 41	0.11		ell		a	23

* SB1, SB2 = single-lined, double-lined spectroscopic binary; ecl = eclipsing; ell = "ellipsoidal" light variations.

† the component with measured radial velocity amplitude (if SB1).

Notes
a. Nucleus of planetary nebula
b. Spectral type of one component conjectural (see text)
c. A member of the Hyades
d. Possible contact binary; m $\sin^3 i$, not mass function, in previous column.

References on next page.

Clearly such a companion would he hard to recognise next to an M supergiant; but if heated by accretion could it nevertheless resemble the kind of hot blue object seen in symbiotics, as much as would a more massive main sequence star or a white dwarf?

3. DETACHED BUT CLOSE BINARIES WITH A WHITE DWARF OR SUBDWARF COMPONENT

In Table 1 some information is listed for a collection of sixteen somewhat heterogeneous objects, whose only common factors are that (a) one component appears to be the highly evolved remnant of something which was presumably once a red giant or supergiant (b) they are detached (c) the period is sufficiently short (except perhaps for the first three objects) that it is hard to see how a giant, let alone supergiant, could have been contained in the system unless there has been considerable AML. Six of these objects are in planetary nebulae. These six might be thought to have just emerged from a common envelope, while the other ten may be somewhat older so that the nebula has dissipated. Do they all look like the expected products of common envelope evolution? My own view is that some do (e.g. V471 Tau) and some do not (e.g. AA Dor).

In NGC 2346, it is not clear (Mendez and Niemela 1981) that the second component of the 16d spectroscopic binary is the subdwarf OB star which excites the nebula: the system might conceivably be triple, with the hot subdwarf some substantial way away from the single-lined spectroscopic binary. Triple systems are not at all rare, of course, but Mendez and Niemela opt for the less complicated possibility. In AA Dor, the secondary is thought (Conti *et al.* 1981, Kudritzki *et al.* 1982) to be essentially a black dwarf of very low mass (perhaps $\sim 0.05\ M_\odot$, given the low mass function despite an inclination high enough to give eclipses). The black dwarf shines mainly by reflected light, to give the spurious appearance of a second hot subdwarf; it should be thought of as a failed main sequence star rather than as the white dwarf remnant of an evolved star. In V Sge, Herbig *et al.* (1965) suggested both components were hot stars below the main sequence. However, a re-analysis of their light-curve (Wilson and Eggleton, to be published)

---

1. Mendez & Niemela (1981)
2. Etzel *et al.* (1977)
3. Mendez & Niemela (1977)
4. Margon *et al.* (1976)
5. Thorstensen *et al.* (1978)
6. Margon *et al.* (1981)
7. Ferguson *et al.* (1981)
8. Lanning (1982)
9. Young *et al.* (1972)
10. Greenstein (1973)
11. Lanning & Pesch (1981)
12. Nelson & Young (1970)
13. Nelson & Young (1976)
14. Herbig *et al.* (1965)
15. Grauer & Bond (1981)
16. Miller *et al.* (1976)
17. Bond *et al.* (1978)
18. Green *et al.* (1978)
19. Kilkenny *et al.* (1978)
20. Conti *et al.* (1981)
21. Kudritzki *et al.* (1982)
22. Noskova (1980)
23. Grauer & Bond (1982)

suggests the possibility that (a) the system is in contact, rather than semi-detached, and (b) the secondary also shines mainly by reflected light, so that its true nature may be a fairly normal star of spectral type ∿ A V, roughly in accordance with the mass estimate of Herbig *et al.* (1965), and with the radius implied by Roche geometry. The primary has to be considerably more luminous than most of the other hot objects in Table 1, in order to be as hot as observed and yet to fill its Roche lobe, so it is probably a helium burning core with a hydrogen burning shell; it may resemble the expected remnant of a star of ∿ 5 $M_\odot$ which has lost its envelope during core helium burning. Note, however, that the minimum masses in Table 1 should be taken with some reserve, as explained by Herbig *et al.* (1965).

Possibly the hardest object to explain in terms of common envelope evolution is AA Dor, although a lot hinges on the mass estimates (Kudritzki *et al.* 1982) of ∿ 0.25 $M_\odot$ for the SDO component and ∿ 0.05 $M_\odot$ for the putative black dwarf. These estimates were based on eclipse geometry, as well as the mass function, and give fairly consistent radii if the SDO component still has a hydrogen burning shell. Such an object would be the remnant of a subgiant, rather than a giant or supergiant. Presumably the subgiant lost an envelope of ∿ 0.75 $M_\odot$. Could such a massive envelope have been ejected by the spiralling-in of such a small companion? There does not appear to be enough orbital energy available, even if this was directed with 100% efficiency into mass loss. It seems more likely, in this case, that the inherent tendency of a giant or subgiant to lose mass slowly, by stellar wind, was somehow enhanced by the presence of a small (but possibly close) companion, much as subgiant companions in RS CVn binaries show unusually strong chromospheric activity (Walter and Bowyer 1981). Indeed, some RS CVn's show evidence for the loss of ∿ 0.1 - 0.2 $M_\odot$ (Popper and Ulrich 1977) from the more evolved subgiant component. Perhaps in AA Dor the black dwarf spiralled in within the comparatively rarefied "envelope" of the wind emanating from the subgiant, rather than within the relatively dense "envelope" of the subgiant's outer regions. Note that it is only necessary for a subgiant at the base of the giant branch to have a wind of ∿ 10^{-9} $M_\odot$/yr, in order to end up as a low-mass white dwarf.

FF Aqr (Table 1) may have also been something like an RS CVn binary in the past. In fact, since the two components must have started with nearly equal masses (so that both could have been subgiants simultaneously before one lost its envelope to become a hot subdwarf), the resemblance to an RS CVn system may have been quite close. I find it almost impossible to believe that a system like FF Aqr containing a giant can have emerged from a common envelope phase. It seems much more probable to me that this system has simply suffered extensive mass loss from at least one component. Observations establishing the velocity curve of the subdwarf as well as the giant (and hence giving both masses) would be of enormous value in understanding not just this system but the overall problem of forming objects like those in Table 1. Given the small mass-function (Etzel *et al.* 1977) and the eclipses, it is hard to believe the subdwarf is more than $\sim 0.5\ M_\odot$, and it could be less if the G8III companion has also lost mass, as seems likely.

Four of the systems in Table 1 have recognisably M-dwarf components. Given that it is difficult, though not impossible, to see how an M dwarf will eject a supergiant envelope in the "common envelope" scenario, is it possible that all these M dwarfs are actually remains of originally more massive companions, the companion having also lost some of its envelope during the spiralling-in phase? There are two arguments against this. Firstly, it takes even more work to strip the outer layers off a main sequence star than to do the same service for a subgiant or giant (per unit mass lost). Secondly, if a G dwarf, say, has perhaps half its mass stripped away in the 10^3 - 10^4 years of a common envelope phase, it will emerge as a very hot subdwarf, and should take about as long to cool down as its white dwarf companion once the envelope has been ejected. In GK Vir and Feige 24 the companion is a very hot white dwarf, presumably not very old, and yet the companion is already a fairly normal M dwarf. In GK Vir, especially, the M dwarf is 7 magnitudes fainter than the hot white dwarf (Green *et al.* 1978) which does not suggest to me that they both emerged from the fiery furnace at the same time; not, at least, if the M dwarf lost half its mass in the furnace, although even if it conserved its mass I would find this extreme case rather hard to accept. The other two systems (Case 1, HZ 9) are less worrying in this respect, although Lanning and Pesch (1981) remark of HZ 9 "It is curious that after such an evolution [i.e. common envelope] the component stars should appear so ordinary".

V Sge (Table 1) is also awkward. Although its structure, let alone its evolutionary status, is not clear, it is not unlike the product expected from early Case B evolution, except that its period is much too short. Whereas early Case B should lead in many circumstances to a compact helium-burning star and a main sequence companion, the final period should be $\sim$ 100d, not $\sim$ 0.5d. This suggests to me simply that in some, though not all, Case B systems angular momentum and mass are not conserved. V Sge may be a system which has lost $\sim$ 50% of its mass and $\sim$ 80% of its angular momentum. It is interesting that the helium-burning component may, after helium exhaustion, become a C/O white dwarf without going through a further red giant phase (since helium stars with $\lesssim 0.9\ M_\odot$ do not expand much in the shell-burning phase, Paczyński 1971), and so the system may reasonably become cataclysmic without going through a further detached phase, and without having been through a common envelope phase.

A final comment about the systems of Table 1 is that, even though several are nuclei of planetary nebulae, it is not clear that the planetary nebula was produced by the same mechanism as the one which is thought to operate for single red giants (Abell and Goldreich 1966, Wood and Cahn 1977). For instance, the hydrodynamic mass transfer (Whyte and Eggleton 1980) that can be expected in close white dwarf/red dwarf systems, if the red dwarf is more massive than the white dwarf when mass transfer begins, might easily lead to substantial mass loss from the binary, and so to the formation of something not unlike a planetary nebula. The expected binary remnant would be a detached system of somewhat longer period than before, with the red dwarf substantially reduced in mass and the white dwarf not much altered. This might be rather like UU Sge; however, I would expect the "red" dwarf to be a very hot object for some considerable time, as it may not cool off much faster than a white dwarf.

4. THE PERIOD GAP AND PERIOD CUTOFF IN CATACLYSMIC BINARIES

It now seems clear that there is a real shortage of cataclysmic binaries in the period range 2 - 3 hrs (Warner 1976, Whyte and Eggleton 1980). I will assume this gap to be real without further qualification. There is also, though not quite so clearly, a cutoff at $\sim$ 1.3 hrs. There are $\sim$ 20 binaries known in each of the period ranges 1.3 - 2 and 3 - 4.5 hrs, and no more than 2 in the ranges 0.9 - 1.3 and 2 - 3 hrs together. It also seems clear that AML (whether by gravitational radiation or by magnetic braking) is the main cause of the long-term evolution of cataclysmics, although nuclear evolution may be a significant driving mechanism in a few systems like GK Per. AML can be expected to operate as well on MW;D systems (Table 1) as on MW;S systems of comparable period, and so MW;D systems must not be omitted in a discussion of the period distribution of cataclysmics. Figure 2, a plot of orbital period against secondary mass, shows in a highly schematic form some of the factors that must influence the period distribution. No less than 5 constraints seem to be imposed by the existence of the gap and cutoff:-

(i) MW;S systems with $P > 3$ hr cannot evolve steadily to periods < 3 hr; either they turn round at $\sim$ 3 hr, or they accelerate or decelerate their evolution strongly, or they cease to be cataclysmic

(ii) the same applies, *mutatis mutandis*, to MW;S systems with $P < 2$ hr

(iii) MW;D systems with periods short enough to be affected by AML must avoid secondary masses in the range $\sim 0.2 - 0.3\ M_\odot$

(iv) similarly they must avoid masses in the range $\sim 0.01 - 0.02\ M_\odot$

(v) secondaries in the mass range $0.02 - 0.15\ M_\odot$ cannot be in thermal equilibrium, even if detached, at periods less than $\sim$ 1.3 hrs.

These constraints are not always independent. If the answer to (i), (ii) is that systems speed up their evolution considerably on reaching the gap, then (iii), (iv) may not be necessary. But if on the contrary systems slow down, or cease to be cataclysmic (by becoming detached, Robinson *et al.* 1981), then (iii), (iv) are necessary.

Figure 2 shows lines corresponding to semidetached black dwarfs (BD) and main sequence (MS) stars in thermal equilibrium. For secondary masses $\lesssim 0.5\ M_\odot$ a departure from thermal equilibrium usually means

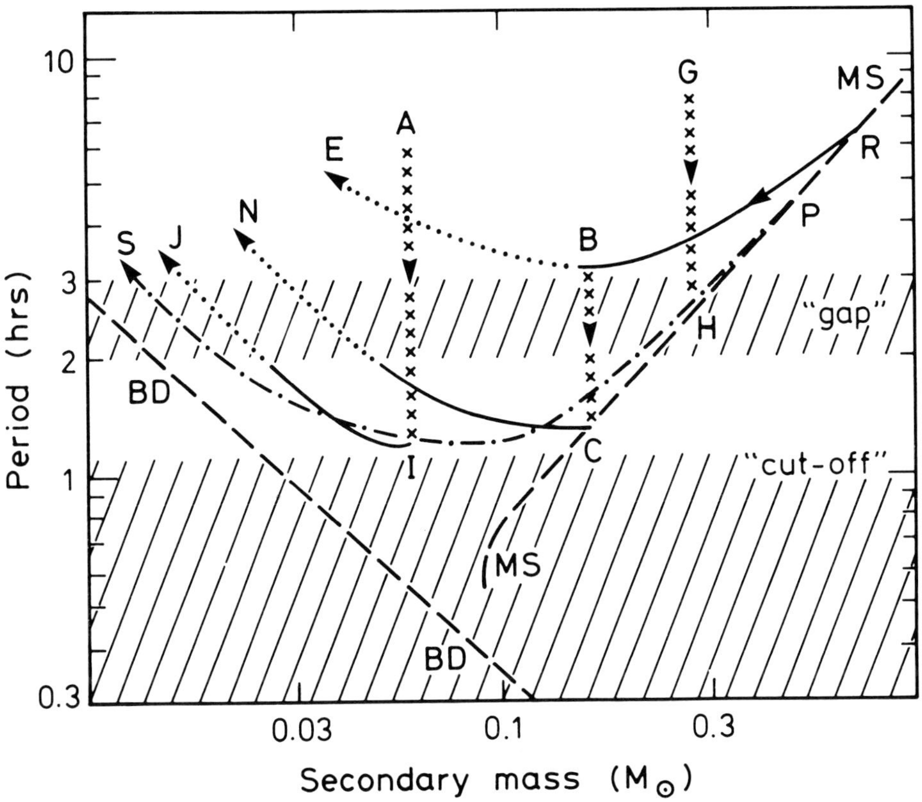

Figure 2. Schematic picture of the secular evolution of cataclysmic binaries. Period is plotted against mass of secondary. Systems containing semidetached "black" dwarfs or low-mass main sequence stars, in thermal equilibrium, would lie on the dashed lines BD and MS. The dash-dot line PS is the model of Paczyński and Sienkiewicz (1981), involving angular momentum loss by gravitational radiation only. The line RBCN is an interpretation of the suggestion of Robinson *et al.* (1981) that cataclysmic binaries become detached (crosses, BC) at periods of ∿ 3 hr. G and A are possible locations of the detached systems GK Vir and AA Dor (Table 1). The paths GH, AIJ, BE are speculative possibilities discussed in the text. Dotted lines represent faster semidetached evolution than solid lines.

lower densities, hence longer periods. Thus the BD and MS lines should be a lower envelope. Evolution by AML can lead to considerable departures from thermal equilibrium for low-mass secondaries, if the AML timescale is comparable to or shorter than the thermal timescale, which can be $\sim 10^{10}$ yrs in the region where the BD and MS lines approach each other. The path PS on Figure 2 is the path that would be followed if the secondary starts somewhere on the MS and evolves by gravitational radiation only (Faulkner *et al.* 1972, Paczyński and Sienkiewicz 1981). This might explain the cutoff, provided all systems started with $m_2 \gtrsim 0.15\ M_{\odot}$, but fails to explain the gap, although a second passage through the gap might be avoidable simply because evolution is so slow that no system may have got there yet. The path RBC corresponds roughly to the suggestion of Robinson *et al.* (1981), who noted that a small number of systems just above and just below the gap appear to switch off mass transfer, at least temporarily. It would appear to be necessary that the secondary be far from thermal equilibrium at B: otherwise a small amount of AML would make it semidetached again, at a period within the gap. Fairly rapid AML (perhaps by magnetic braking) might lead to a path like RB. Presumably when the system becomes semidetached again, at C, it will again depart from thermal equilibrium, and follow a path like CN. So another mechanism is needed to prevent it reappearing in the gap.

A detached system like AA Dor, indicated (very tentatively) as A on Figure 2, poses a further problem. If AML reduces its period on a reasonable timescale, it should become cataclysmic at a period of ~ 0.5 hr, well below the cutoff; any secondary mass in the range $\sim 0.03 - 0.15\ M_{\odot}$ could do this. I speculate that tidal friction, which is inevitably involved in AML by magnetic braking, may heat such a secondary sufficiently, once $P \lesssim 2$ hr, that it is out of thermal equilibrium and so fills its Roche lobe sooner than otherwise. It might then follow a path like AIJ in Figure 2. If there is an MW;D system at a point like G (and GK Vir may be near there, although the spectral type quoted in Table 1 suggests a slightly greater mass) it might be expected to become cataclysmic at H. Hence either there is a shortage of systems near G, or MW;S systems near H evolve much more rapidly than at either longer or shorter periods.

The following set of assumptions, while not put forward as a "theory", is intended to illustrate the complexity forced on any model which attempts to explain both the gap and the cutoff in terms of secular evolution of cataclysmics. The assumptions are

(a) two different mechanisms lead to the formation of MW;S (and also MW;D) systems; one mechanism produces secondaries in the mass range $m_2 \gtrsim 0.4\ M_{\odot}$, the other in the range $0.02 - 0.2\ M_{\odot}$.

(b) AML by magnetic braking and tidal friction operates on timescales $\sim 10^9$ yr, decreasing to $\sim 10^8$ yr for secondaries with $m_2 \lesssim 0.03\ M_{\odot}$. Tidal friction causes heating in the secondary, which may be its main source of luminous energy at low mass and/or short period.

These assumptions may ensure that

(i) systems starting from the main sequence above the gap follow a path in Figure 2 like RBE, but evolve rather rapidly on the section BE. If by some accident they become detached at a point like B, they might follow the sort of path RBCN implied by Robinson *et al.* (1981), but evolving progressively more rapidly on the section CN so that they spend most of their life below the gap;

(ii) Systems never start from either the MS or the BD curve near where they nearly join, because such a secondary will always be heated by tidal friction and so be out of thermal equilibrium. Such systems will therefore start cataclysmic life at points like C or I, perhaps having already had detached evolution from a point like A. The MW;S systems will follow paths like CN, IJ, but spend most time near points C, I respectively.

It is not difficult to fault such an elaborate scheme both for the implausibility of its assumptions and for the optimistic interpretation of their consequences. But can one do better? It is not going to be easy to explain both the gap and the cutoff; and there are many other observed features with which the explanation must also be compatible.

5. CONCLUSIONS

There are several difficulties in accounting for both the origin and the subsequent evolution of cataclysmic variables. These are interdependent: for if a theory of their origin predicts a smooth distribution of secondary masses from say 0.1 or 0.5 $M_{\odot}$ upwards, then evolution has a harder (but maybe not impossible) job of explaining the observed period distribution.

I feel that possibly three different mechanisms may lead to the formation of cataclysmics:-

1. a common envelope mechanism (Paczyński 1976), which might produce detached systems like V471 Tau, and semidetached systems like EM Cyg.

2. a mechanism where the outer layers of a *subgiant* can be stripped off, and by a companion of low mass; this might lead to detached systems like AA Dor, and semidetached systems like VW Hyi

3. non-conservative early Case B, in fairly massive binaries; this might lead to a system like V Sge, and then perhaps to a system like Sco X-1, supposing that the compact companion in Sco X-1 is a white dwarf rather than a neutron star.

Once an MW;D or MW;S system has been formed by whatever route, I believe that angular momentum loss by magnetic braking and tidal friction cannot be ignored, even though this may not make the overall picture any easier to understand. The tidal friction may be an important cause of thermal disequilibrium in short-period systems. Magnetic braking may introduce randomness into binary evolution, as it appears to do for single main-sequence B stars. Magnetic B stars often have low, or very low, angular momentum, but this is probably not going to affect their interior evolution very much. In contrast, the evolution of a *binary* depends enormously on the angular momentum, so the spread of possible evolutionary paths should be much greater.

I am indebted to Drs. C. Whyte, R. Wade, R. Webbink, M. Livio and E. van den Heuvel for helpful discussions; to Drs. I. Iben and J. Truran for their support at the University of Illinois where part of this work was carried out (NSF grants AST 78-20123 and AST 78-20124); to Dr. J. Faulkner for organising the Cataclysmic Binaries Workshop at Santa Cruz, where I learnt much that was useful; and to CECAM for supporting a Common Envelope Workshop, at Meudon, where further discussion was very helpful.

REFERENCES

Abell, G.O., and Goldreich, P. (1966). P.A.S.P., 78, 232.
Batten, A.H., Fletcher, J.M. and Mann, P.J. (1978). Pub. D.A.O., 15, 121.
Bond, H.E., Liller, W., Mannery, E.J. (1978). Ap.J., 223, 252.
Bopp, B.W., and Rucinski, S.M. (1981). IAU Symp. 93, 177. (ed. Sugimoto, Schramm, Lamb).
Conti, P.S., Dearborn, D. and Massey, P. (1981). Mon.Not.R.astr.Soc., 195, 165.
Eggleton, P.P. (1976). IUA Symp. 73, 209 (ed. Eggleton, Mitton, Whelan).
Etzel, P.B., Lanning, H.H., Patenaude, D.J., and Dworetsky, M.M. (1977). P.A.S.P., 89, 616.
Faulkner, J., Flannery, B.P. and Warner, B. (1972). Ap.J.Lett., 175, L79.
Ferguson, D.H., Liebert, J., Green, R.F., McGraw, J.T., Spinrad, H. (1981). Ap.J., 251, 205.
Grauer, A.D. and Bond, H.E. (1981). P.A.S.P., 93, 388.
Grauer, A.D. and Bond, H.E. (1982). IAU Circ. 3714.
Green, R.F., Richstone, D.O., and Schmidt, M. (1978). Ap.J., 224, 892.
Greenstein, J.L. (1973). Astron. Astrophys., 23, 1.
Griffin, R.F., Radford, G.A., Harmer, D. and Stickland, D.J. (1976). Observatory, 96, 153.
Herbig, G.H., Preston, G.W., Smak, J. and Paczyński, B. (1965). Ap.J., 141, 167.
Hiltner, W.A. (1946). Ap.J., 104, 396.
Huang, S.-S. (1966). Ann.d'Astr., 29, 331.
Kilkenny, D., Hilditch, R.W., and Penfold, J.E. (1978). Mon.Not.R.astr. Soc., 183, 523.
Kraft, R.P. (1963). Adv.Astron.Astrophys., 2, 43.
Kudritzki, R.P., Simon, K.P., Lynas-Gray, A.E., Kilkenny, D. and Hill, P.W. (1982). Astron. Astrophys., 106, 254.
Lanning, H.H. (1982). Ap.J., 253, 752.

Lanning, H.H., and Pesch, P. (1981). Ap.J., 244, 280.
Margon, B., Downes, R.A. and Katz, J.I. (1981). Nature, 293, 200.
Margon, B., Lampton, M., Bowyer, S., Stern, R., and Paresce, F. (1976). Ap.J.Lett., 210, L79.
Mendez, R.H. and Niemela, V.S. (1977). Mon.Not.R.astr.Soc., 178, 409.
Mendez, R.H. and Niemela, V.S. (1981). Ap.J., 250, 240.
Miller, J.S., Krzeminski, W., and Priedhorsky, W. (1976). IAU Circ. 2974.
Nelson, B., and Young, A. (1970). P.A.S.P., 82. 699.
Nelson, B., and Young, A. (1976). IAU Symp. 73, 141. (ed. Eggleton, Mitton, Whelan).
Noskova, R.I. (1980). Astron. Tsirk, 1128.
Paczyński, B. (1971). Acta Astr., 21, 1.
Paczyński, B. (1976). IAU Symp. 73, 75 (ed. Eggleton, Mitton, Whelan).
Paczyński, B. and Rudak, B. (1980). Astron. Astrophys., 82, 349.
Paczyński, B. and Sienkiewicz, R. (1981). Ap.J.Lett., 248, L27.
Popper, D.M. and Ulrich, R.K. (1977). Ap.J., 212, L131.
Robertson, J.A. and Eggleton, P.P. (1977). Mon.Not.R.astr.Soc., 179, 359.
Robinson, E.L., Barker, E.S., Cochran, A.L., Cochran, W.D., and Nather, R.E. (1981). Ap.J., 251, 611.
Sparks, W.M. and Stecher, T.P. (1972). Ap.J., 188, 149.
Starrfield, S., Sparks, W.M. and Truran, J. (1976). IAU Symp. 73, 155 (ed. Eggleton, Mitton, Whelan).
Thorstensen, J.R., Charles, P.A., Margon, B. and Bowyer, S. (1978). Ap.J., 223, 260.
Tutukov, A.V. and Yungelson, L.R. (1982). IAU Coll. 70, 283. (ed. Friedjung, Viotti).
Verbunt, F. and Zwaan, C. (1981). Astron. Astrophys., 100, L7.
Walter, F. and Bowyer, C.S. (1981). Ap.J., 245, 671.
Warner, B. (1976). IAU Symp., 73, 85 (ed. Eggleton, Mitton, Whelan).
Webbink, R.F. (1976). Ap.J., 209, 829.
Whyte, C.A. and Eggleton, P.P. (1980). Mon.Not.R.astr.Soc., 190, 801.
Wood, P.R. and Cahn, J.H. (1977). Ap.J., 211, 499.
Young, A., Nelson, B. and Mielbrecht, R. (1972). Ap.J., 174, 27.

DISCUSSION FOLLOWING P. EGGLETON'S TALK

BATH: The evidence that you have got white dwarfs in these SWS systems does not exist, it is purely a myth. There has been a lot of discussion of symbiotic star models, containing white dwarfs, with no observational evidence that these systems exist at all, and yet they have become accepted. I would argue that in fact the symbiotic stars where the blue component gets brighter in the optical than the giant, must contain main sequence stars, unless they are losing considerable amounts of mass and have a pseudo photosphere around the white dwarf. This is because the white dwarf is just too small ever to be hot enough (below the Eddington limit) to be bright enough in the optical, to compete with the giant.

EGGLETON: In the case of Z And at least, the ourbursts have been thought to have something to do with whether you have steady nuclear burning or unsteady nuclear burning. That is not something I personally believe.

BATH: Yes, but the point is that now we know that you can produce eruptions by accretion events. It is much easier to explain symbiotic star eruptions in many cases, by accretion events than by nuclear burning, that is particularly the case with Z And CI Cyg which is even a better example.

EGGLETON: I agree with you.

SHAVIV: The spiralling-in sotry, which was the basis of a large part of this kind of overall picture does not necessarily end at a close binary, they might really amalgamate into a single star so you have the problem how to stop such a successful process.

EGGLETON: Well, I wonder whether in fact that might not be the most common outcome of a common envelope situation and I only put it forward in the most tentative way, but it might be a reason for a shortage of moderate mass secondaries, around 0.25 $M_{\odot}$, maybe you need a more massive secondary than that to survive the common envelope process. I don't think that anybody has really claimed that the common envelope situation would inevitably end with a binary of the sort of period that we would like to see, which would be something of the order of half a day to a day. It might be ten days, or it might be one hundredth of a day, which effectively means that one has amalgamated in the other. Some of the systems that I have shown, do have periods of several days but it is not clear to me that they actually are products of common envelope evolution. Another thing one must bear in mind about the common envelope, is that we shouldn't equate all red super-giant envelopes with each other. There are red super-giants and there are orange super-giants which are very different from red super-giants in the fact that their envelopes are largely radiative, whereas the red super-giants would have envelopes which are largely convective and this can make an enormous difference to the way things happen. Also, if you have a red giant, the envelope is very much less favourable for being blown off in the process of the common envelope scenario. I don't find it difficult to imagine that you would blow off the envelope of an extreme red super-giant fairly easily, by some kind of common envelope mechanism, but I do find it difficult to imagine you can blow off the envelope of an ordinary red giant, because it is much more tightly bound. Yet, at least one of these objects on the list seems to be a rather low mass white dwarf.

SHAVIV: Bob Williams has introduced the idea that it might very well be that the secondaries are highly evolved, with a composition which is different from solar. What if in at least part of the binaries the companion is highly evolved?

EGGLETON: I certainly wouldn't be surprised if it weren't solar. However, very tentative as the data is, the secondaries do seem to fit in with being fairly normal main sequence stars. I can well believe they are enriched in He^3, for instance.

SHAVIV: So it does pose a serious problem.

EGGLETON: It does, indeed.

SUGIMOTO: I did not understand what is the essential physical difference between the common envelope evolution and the two other mechanisms you are proposing. What is the contents of the new mechanism?

EGGLETON: What I would like in the kind of mechanism which I refer to as being more or less a planetary nebular thing, is that the spiralling-in of the secondary would take place within the more dilute envelope that you might get if the envlope from the primary is already expanding on its way to becoming a planetary nebula envelope, rather than being in the comparatively high density environment of a red giant or even a red super-giant envelope. So I wouldlike the star to loose its envelope first and then the spiralling-in take place within that, rather than to have the spiralling-in take place and lead as a consequence, at a mugh later stage, when the binary has become much closer, to the blowing-off of the envelope.

SUGIMOTO: Is this a theory of speculation ?

EGGLETON: My feeling is that, based on the observations, particularly those observations of this set of systems, you would have a hard job explaining most of these by what I understand to be the common envelope scenario as put forward I think basically by Paczynski and as elaborated by several other people subsequently.

THE EVOLUTIONARY STATUS OF THE SECONDARIES OF CATACLYSMIC BINARIES

H. Ritter
Max-Planck-Institut für Physik und Astrophysik
Karl-Schwarzschild-Straße 1, D-8046 Garching
Fed. Rep. Germany

ABSTRACT

It is shown that the secondary components of cataclysmic binaries with orbital periods of less than ~10 hours are indistinguishable from ordinary low-mass main-sequence stars and that, therefore, they are essentially unevolved. On the other hand, it is shown that, depending on the mass ratio of the progenitor system, the secondary of a cataclysmic binary could be significantly evolved. The fact that nevertheless most of the observed secondaries are essentially unevolved can be accounted for by assuming that the probability distribution for the initial mass ratio is not strongly peaked towards unity mass ratio.

I. INTRODUCTION

In a number of recent observational studies of cataclysmic binaries (hereafter CB's), in particular of BV Cen (Vogt and Breysacher, 1980; Gilliland, 1982), AE Aqr (Patterson, 1979; Chincarini and Walker, 1981), DQ Her (Smak, 1980; Young and Schneider, 1980, 1981) and U Gem (Wade, 1979, 1981) it has been found that the corresponding secondary is oversized for its mass when compared with the theoretical main-sequence mass radius relation. Based on this observation it has been concluded that the secondary is evolved. On the other hand, at least in the case of DQ Her and U Gem the mass of the secondary is so small that normal nuclear evolution cannot result in any significant evolutionary effects over the age of the universe. Therefore, if these stars are in fact evolved, they must have undergone an unusual evolution. Thus, with regard to the evolutionary history of a CB, it is essential to know whether its secondary is evolved or not. It is the purpose of this paper to discuss briefly the observational and theoretical evidence for evolved secondaries.

II. DISCUSSION OF THE OBSERVATIONAL DATA

Using published observational data, a mass radius and a mass lumino-

M. Livio and G. Shaviv (eds.), Cataclysmic Variables and Related Objects, 257–262.

sity diagram (hereafter MR and ML diagram respectively) for the secondaries of CB's have been constructed. In order to avoid a circular argument regarding the evolutionary state of these stars, only systems for which the mass of the secondary can be inferred without assuming it to be on the main sequence have been used. After a critical review of the available observational data, the following systems have been taken into account for the MR diagram: BV Cen, AE Aqr, RU Peg, EM Cyg, Z Cam, SS Cyg, RW Tri, DQ Her, U Gem, LX Ser, AM Her, HT Cas and Z Cha. The ML diagram is derived from the MR diagram by computing the corresponding luminosity L from the radius R and the effective temperature T_{eff} of the secondary: T_{eff} in turn has been determined from the known spectral type using an observationally calibrated relation between the spectral type and T_{eff} given by Popper (1980). Acceptable spectral types for the secondary are available for all but two, namely LX Ser and HT Cas, of the above-mentioned systems. The resulting MR and ML diagrams have recently been published (Ritter, 1982 a). Furthermore, a preliminary version of the MR diagram has been discussed earlier (Ritter, 1980). A more detailed discussion of the observational data and of the selection criteria as well as the justification of the values for the parameters which have been adopted for a particular system will be published elsewhere (Ritter, 1982 b).

Conclusions on the evolutionary status of the secondaries of CB's are derived from comparing the MR and ML diagrams of CB's with the corresponding diagrams of the theoretical low-mass main sequence (Copeland, Jensen and Jørgensen, 1970; Grossman, Hays and Graboske, 1974) and of observed low-mass main-sequence stars (Popper, 1980). The main results of such a comparison (Ritter, 1980, 1982 a) can be summarized as follows:

a) The MR diagram

When comparing the MR diagram of the secondaries of CB's with theoretical MR relations, it turns out that, in fact, a number of secondaries have a radius which is larger by about $\Delta \log R \approx 0.1$ than the radius of the theoretical models with the same mass. However, when compared with the MR diagram of observed low-mass main-sequence stars (Popper, 1980), the two sets of data match perfectly within the observational errors. Thus, in the MR diagram, the secondaries of CB's with orbital periods of less than about 10 hours are indistinguishable from normal low-mass main-sequence stars. The discrepancy between the observed and the theoretical MR relations must be attributed to inadequacies of the theoretical models.

b) The ML diagram

The ML diagram is a much more sensitive probe of the evolutionary status of a star than the MR diagram. Therefore, evoluionary effects which might escape detection in the MR diagram might at the same time become visible in the ML diagram. However, when the ML diagram of the secondaries of CB's is compared with the ML diagram of the observed low-mass main-sequence stars, one finds again that the two sets of data match perfectly within the errors. The agreement is even better than one could have expected regarding the way the luminosities of the secondaries have been determined. The agreement between the theoretical and the observed data is not perfect but much better than in the case of the MR diagrams.

The conclusion to be drawn from these comparisons is that the secondaries of CB's with orbital periods of less than about 10 hours are indistinguishable from ordinary low-mass main-sequence stars. This, in turn, means that these stars are essentially unevolved. On the other hand, this does not imply that they are zero-age main-sequence stars.

III. THEORETICAL CONSIDERATIONS

Despite the fact that some of the secondaries of CB's have a very low mass, the observation that these stars are essentially unevolved nevertheless calls for a theoretical explanation. The reason for this is that the standard theory for the formation of CB's allows for rather strongly evolved secondaries even of low mass (Ritter, 1982 a). This conclusion is based on the following theoretical considerations:

a) The Mass Loss of the Secondaries

If the standard theory for the formation of CB's is correct, then most of the secondaries must have lost a significant fraction of their mass during the formation of the CB's. This is seen from the following argument: let $M_{1,i}$ and $M_{2,i}$ be the initial masses of a CB progenitor on the main sequence. As a result of the formation of the CB, a fraction f of the mass of the primary ends as a white dwarf of mass $M_{WD} = M_{1,f} = f \cdot M_{1,i}$. Because CB's are secularly stable against mass transfer, the mass, $M_{2,f}$ of the seconcary has to be smaller than that of the white dwarf primary, i.e. $M_{2,f} \lesssim M_{1,f} = M_{WD}$. This in turn implies that the secondary, in the course of its evolution, must have lost the mass $\Delta M_2 = M_{2,i} - M_{2,f} = M_{2,i}\,(1-f.q_i/q_f)$, where $q_i = M_{1,i}/M_{2,i}$ and $q_f = M_{1,f}/M_{2,f}$. Since for reasons of secular stability q_f 1, the secondary can keep its original mass only if $q_i \gtrsim f^{-1}$. Depending on $M_{1,i}$, f is in the range 0.2 to 0.5. Thus mass loss from the secondary can only be avoided for rather large values of the initial mass ratio q_i.

b) The Possibility of Evolved Secondaries

If the initial mass ratio q_i of the progenitor of a CB is not too far from unity, the secondary has spent a significant fraction of its main-sequence life-time when the binary enters the common-envelope phase. From the mass luminosity relation for main-sequence stars, i.e. $L = \text{const.}\ M^{\alpha}$, the fraction of the main-sequence life-time which the secondary can spend before the CB is formed, is estimated to be $\varepsilon \approx q_i^{1-\alpha}$. Since the systems which produce the most evolved secondaries, i.e. those with $q_i \approx 1$, are also those where the secondary suffers the largest mass loss, there is the possibility that some of the secondaries of CB's now essentially consist of the former hydrogen burning central region and even may expose nuclear processed matter at their surface. In fact, the secondaries of CB's are like the remnants of a Case A mass transfer (see e.g. Horn, Kriz and Plavec,

1970). Because $\alpha \approx 3.5$, no significantly evolved secondaries result if $q_i \gtrsim 2$.

c) The Probability of Evolved Secondaries

Since rather highly evolved secondaries are theoretically possible, the fact that the observed secondaries are essentially unevolved calls for an explanation. As has been shown above, the secondary is the more evolved the closer the initial mass ratio is to unity. Thus the absence of significantly evolved secondaries could be accounted for by assuming that systems with $q_i \approx 1$ are very rare. In fact it can be shown that it is already sufficient if the probability distribution of q_i is not strongly peaked towards unity mass ratio (Ritter, 1982 a).

ACKNOWLEDGEMENTS

The author is grateful to Dr. J. Kirk for improving the language.

REFERENCES

Chinarini, G., Walker, M.F.: 1981, Astron. Astrophys. 102, pp. 24 - 32.
Copeland, H., Jensen, J.O., Jørgensen, H.E.: 1970, Astron. Astrophys. 5, pp. 12 -34.
Gilliland, R.L.: 1982, preprint.
Grossman, A.S., Hays, D., Graboske, Jr., H.C.: 1974, Astron. Astrophys. 30, pp. 95 -103.
Horn, J., Kriz, S., Plavec, M.: 1970, Bull. Astron. Inst. Csl. 21, pp. 45 - 54.
Patterson, J.: 1979, Astrophys. J. 234, pp. 978 - 992.
Popper, D.M.: 1980, Ann. Rev. Astron. Astrophys. 18, pp. 115 - 164.
Ritter, H.: 1980, The Messenger 21, pp. 16 - 18.
Ritter, H.: 1982 a, preprint MPA 18, to appear in the proceedings of the workshop on "High Energy Astrophysics", Nanking 1982, Springer Verlag.
Ritter, H.: 1982 b, in preparation.
Smak, J.: 1980, Acta Astron. 30, pp. 267 - 283.
Vogt, N., Breysacher, J.: 1980, Astrophys. J. 235, pp. 945 - 954.
Wade, R.A.: 1979, Astron. J. 84, pp. 562 - 566.
Wade, R.A.: 1981, Astrophys. J. 246, pp. 215 - 222.
Young, P., Schneider, D.P.: 1980, Astrophys. J. 238, pp. 955 - 963.
Young, P., Schneider, D.P.: 1981, Astrophys. J. 247, pp. 960 - 968.

DISCUSSION FOLLOWING H. RITTER'S TALK

SUGIMOTO: What is the mechanism to strip off the mass from the scondary?

RITTER: I can't answer this question because our knowledge about common envelope is probably nearer to speculation than to actual knowledge, so this is only a boundary condition for any theory which wants to explain these objects, it has also to account at least in some cases, for a significant loss in the mass of the secondary star. What the mechanism is, I don't know.

SHAVIV: What is the average progenitor mass for the secondary you have?

RITTER: I have not estimated the typical progenitor mass, the argument I have made is essentially independent of that.

FINZI: Doesn't the space distribution of Cataclysmic Variables in the galaxy lead to significant upper bounds for the masses?

RITTER: Rather long ago it has been claimed by Kraft that the space distribution is similar to that of W Ursa Majoris stars, so I don't know whether one can really draw reliable conclusions from this space distributions because we really don't look very far when we observe Cataclysmic Variables. The typical distance is a few hundred parsecs and with the few novae we look at a kiloparsec or so, so I would hesitate to draw conclusions from statistical arguments.

EGGLETON: The diagram that you showed of the observed radii and masses, superimposed on the theoretical radii and masses, I thought you were going to say how good the theory was to give such a good agreement, but you drew a line between them and seemed to show that the agreement was not good. Were those 1σ error bars or 2σ error bars?

RITTER: The error bars for the mass-luminosity diagram of the visual binary components are 2σ error bars and Popper is very conservative. My error bars are rather more boundaries and probability ranges and not proper error bars.

EGGLETON: But one sees two observed stars there which disagree more with the trend than I think even the theoretical models do and this may well be evidence for instrinsic variability such as different compositions and so on. A variation in those parameters, such as the composition could presumably move things around, I would see the theoretical things as also having a consideraboy greater scatter and really I would have thought, that was a good agreement.

SHAVIV: I would like to mention that most of this separated low mass main sequence stars are peculiar in terms of magnetic activity and so on, so it is not clear a priori that they are quite main sequence stars of the type our programs calculate. Recently I have carried out a calculation with Art Cox, trying to fit a single point and we found that it was necessary to reduce the mixing length to scale height ratio to something like 10^{-3} to get anywhere near a good agreement between the mass, the luminosity and the radius of that star, assuming no evolutionary effects.

EGGLETON: Mixing length is less important for these low mass stars, but I am suggesting that metallicity may still be fairly important.

RITTER: I can only answer that it is because these low mass stars are probably not the models we compute, that it is much better probably to compare observed stars with other observed values, rather than theory with observations.

ORIGIN OF VERY-SHORT ORBITAL-PERIOD BINARY SYSTEMS

Shigeki Miyaji
Department of Natural History, College of Arts and Sciences
Chiba University, Chiba 260, Japan

1. INTRODUCTION

Recent observations of four close binaries have established that there is a group of very-short orbital-period (VSOP) binaries whose orbital periods are less than 60 minutes. The VSOP binaries consist of both x-ray close binaries (4U1626-67; Middleditch et al. 1981 and 4U1916 -0.5; White and Swank 1982) and cataclysmic variables (AM CVn; Faulkner et al. 1972 and G61-29; Nather et al. 1981). Their orbital periods are too short to have a main-sequence companion. However, four binaries, none of them belongs to any globular cluster, are too abundant to be explained by capturing mechanism of a white dwarf. Therefore it seemed to be worth to present an evolutionary scenario from an original binary system which can be applied for all of VSOP binaries.

Evolutional scenarios of binary systems have been discussed by many authors (e.g. van den Heuvel 1981 and Tutukov 1981). However VSOP binaries can not be explained by their scenario. One reason is that scenarios so far proposed have treated only extream cases, e.g. mass conservative case and spiral-in case in the huge common envelope. So it is hard to explain highly compact but not coalesced system. The other reason is that the VSOP binaries consist of two evolved stars and their separations are less than one solar radius. (The companion of AM CVn should be a helium or carbon-oxygen white dwarf, and the case of G61 -29 may be a helium white dwarf; Nather et al. 1981.) Such system can not be formed by the first phase of mass transfer only, because 3 x 10^7 years (life of a neutron star progenitor; 8 $M_\odot$ star) is so short to make a slight change in abundances of companion star even if it is as heavy as a $0.7M_\odot$ star initially. We discuss detailed process of mass and angular momentum losses in section 2 and present a scenario of on the origin of VSOP binaries in section 3.

2. MASS AND ANGULAR MOMENTUM LOSS

Stable mass transfer including angular momentum loss by gravitational wave radiation was well studied by Paczyński and Sienkiewicz

M. Livio and G. Shaviv (eds.), Cataclysmic Variables and Related Objects, 263–267.

(1981) and Rappaport et al. (1981). They mimicked evolutions of cataclysmic variables (the case of main-sequence secondary) and showed that there is a minimum orbital period at (60-80) minutes. This minimum period is the results from the facts that mass transfer is driven by angular momentum loss by gravitational wave radiation and that structure of mass losing star becomes isentrope, i.e., Kelvin-Helmholtz timescale of mass losing star is longer than gravitational timescale. Stable mass loss from the system does not affect this situation because its timescale is Kelvin-Helmholtz timescale of mass losing star. In fact, Rappaport et al. (1982) computed models with slow mass loss and showed that there is not any significant difference from other results without mass loss. Moreover, as the companion star loses its mass, resultant degenerate star keeps its initial composition ratio. Therefore VSOP binaries should not be the products of stable mass transfer.

On the other hand, if mass transfer is rapid, well accepted spiral-in scenario poses a serious difficulty, i.e., the system should merge into a single star. However, when two components have similar mass, spiral-in episode is not correct because the mass in common envelope is also the same. Once angular momentum was transmitted to the envelope from the secondary, drag force does not work any more.

When the envelope overflows from the second Lagrangian point (L_2 point), angular momentum is lost by the acceleration of particles by the dipole gravitational potential of the binary. Following the trajectory of zero verocity particles, Nariai and Sugimoto (1977) showed that the particles takes away about 1.7 times of mean angular momentum if mass ratio $q \equiv M_A/(M_A+M_B) \gtrsim 0.05$. Therefore if common envelope overflows from L_2 point, the system begins to shrink unstably.

This acceleration mechanism, of course, works even for stable mass loss. However, stellar wind particle has, at least, escape velocity, so angular momentum loss is much smaller than the case of zero velocity particle. For the case of wind from a thin common envelope (which does not fill L_2 point), angular momentum loss is moderate because it starts from much shallower gravitational potential. We can not give exact timescale of shrinkage here but it is much longer than 10^{3-4} years (spiral-in timescale given by van den Heuvel 1981).

3. A SCENARIO OF VERY-SHORT ORBITAL-PERIOD BINARIES

Our scenario is illustrated in figure 1. As VSOP binaries contain x-ray close binaries, we choose $8M_\odot$ main-sequence star as initial primary (star A) and $1M_\odot$ secondary (star B). Initial separation of components is about $200R_\odot$.

After 3×10^7 years, star A begins to expand and fills up its Roche robe. Since OB star has lost large amount of its mass by stellar wind, the system shrinks a tenth of its initial separation. At 3.2×10^7 years, $1.3M_\odot$ O-Ne-Mg white dwarf and $1M_\odot$ main-sequence star are left and their separation is about $15R_\odot$.

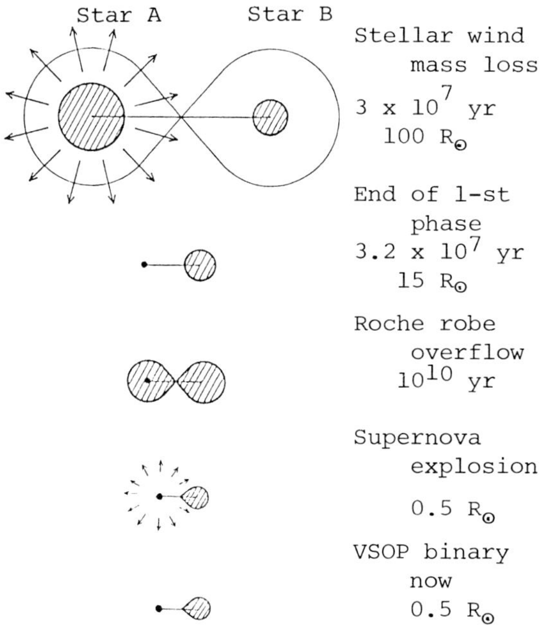

Fig. 1 A Scenario of VSOP Binaries

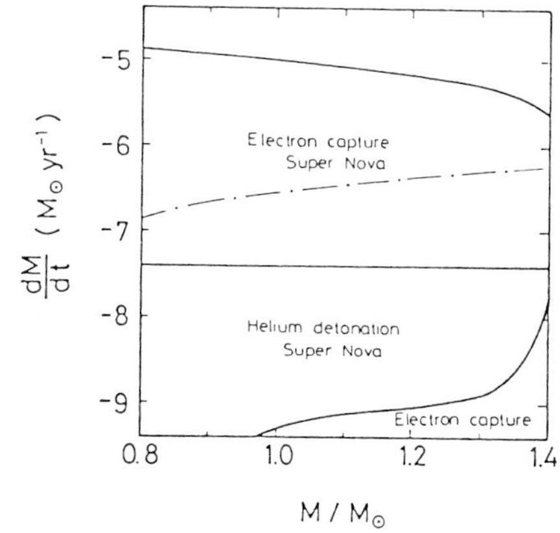

Fig. 2. Final Fate of Helium Accreting O-Ne-Mg White Dwarf.

After 10^{10} years, star B begins to expand and fills up its Roche robe. As $1M_\odot$ star needs 10^8 years to expands up to $100R_\odot$, mass transfer rate from star B to star A is the order of 10^{-7} $M_\odot$/yr. If mass recieving star is a main-sequence star of $0.75M_\odot$ and mass accretion rate is $\dot{M}=2\times10^{-7}$ $M_\odot$/yr, the radius of star B becomes only three times of its initial radius at $1.0M_\odot$ (Neo et al. 1979). However, in this case, mass recieving star is a white dwarf so that it cannot accept all the amount of accreted mass. Because, on white dwarf, hydrogen burning releases about 10 times energy than that of gravitational energy release. From the figure 4 of Fujimoto (1982), $\dot{M}=3\times10^{-7}$ $M_\odot$/yr is enough large rate to form a stable hydrogen burning layer. If mass accretion rate is less than $2\times10^{-7}M_\odot$/yr, the white dwarf recurs weak shell-flashes. These shell-flashes do not grow into nova outburst but the envelope of white dwarf expands up to red giant size (Yasutomi et al. 1982). This red giant phase continues about a tenth of recurrence time of shell-flashes, e.g., about 10 years. Therefore, it is easy to fill up the Roche robe of white dwarf even at moderate Case B mass transfer. The rest of amount should overflow into a common envelope but its mass is smaller than that of white dwarf. Since angular momentum is lost by the wind from the common envelope, the system shrinks by the evolutional timescale of mass losing star; about 10^7 years. This timescale is long enough to accumurated some amount of mass ($\sim 1M_\odot$) onto a white dwarf.

Final fate of such white dwarf is studied by the model of helium accreting white dwarf (Nomoto et al. 1982), because mass accreting white dwarf grows in its mass by stable hydrogen burning or weak hydrogen shell-flashes. For the case of stable hydrogen burning, succeeding helium shell-flashes are weak and finally the white dwarf makes electron capture supernova explosion (figure 2). Then the common envelope is blown off so that a neutron star and a helium core are left finally. For the case of weak hydrogen shell-flashes ($\dot{M} \gtrsim 5 \times 10^{-8} M_{\odot}/yr$), helium shell-flash grows to a detonative burning and the white dwarf makes a supernova explosion as the case of helium accreting carbon-oxygen white dwarf (Nomoto 1980). In this case a white dwarf and a helium core are left behind.

This is a scenario on the origin of VSOP binaries. However, we here notice that this scenario is not the unique one. Because star A can be 4-8 $M_{\odot}$ main-sequence star and supernova explosion can be a carbon deflagration supernova. Although, in this case, x-ray binaries can not be explained because carbon deflagration supernova does not leave a neutron star.

The author is grateful to Dr. J. Swank for stimulating his interest on this problem and to Dr. D. Sugimoto for discussions.

REFERENCES

Faulkner, J., Flannery, B. P., and Warner, B., 1972, Astrophys. J. Letters, 175, L79.
Fujimoto, M. Y., 1982, Astrophys. J., 257, 767.
Middleditch, J., Mason, K. O., Nelson, J. E., and White, N. E., 1981, Astrophys. J., 244, 1001.
Nariai, K. and Sugimoto, D., 1977, Publ. Astron. Soc. Japan, 28, 593.
Nather, R. E., Robinson, E. L., and Stover, R. J., 1981, Astrophys. J., 244, 269.
Neo, S., Miyaji, S., Nomoto, K., and Sugimoto, D., 1979, Publ. Astron. Soc. Japan, 29, 249.
Nomoto, K. 1980, Space Sci. Rev., 27, 563.
Nomoto, K., Miyaji, S., and Sugimoto, D., 1982, private communication.
Paczynski, B. and Sienkiewicz, R., 1981, Astrophys. J. Letters, 248, 127.
Rappaport, S., Joss, P. C., and Webbink, R. F., 1982, Astrophys. J., 254, 616.
Tutukov, A. V., 1981, in 'Fundamental Problems in Stellar Evolution', eds.: D. Sugimoto et al., Reidel Publ. Co., Dordrecht, p. 137.
van den Heuvel, E. P. J., 1981, in 'Fundamental Problems in Stellar Evolution', eds.: D. Sugimoto et al., Reidel Publ. Co., Dordrecht, p. 165.
White, N. E. and Swank, J. H., 1982, Astrophys. J. Letters, 253, L61.
Yasutomi, M., Miyaji, S., and Sugimoto, D., 1982, in preparation.

DISCUSSION FOLLOWING S. MIYAJI'S TALK

LAMB: I did not understand how the common envelope stage is reached after the white dwarf is formed. Could you explain that again?

MIYAJI: If the mass accretion is larger than a certain value, hydrogen is burning stably so that it forms something like a red giant's envelope, so that transferred mass cannot accumulate anymore, because it has say, 1 R_o or 10 R_o. The Roche lobe is filled by such a red-giant-like envelope and then the rest of the transferred mass should form a common envelope.

MEYER: The common envelope, which you would like to blow away by the supernova explosion, is that a very low density envelope or an ordinary red giant type envelope?

MIYAJI: I think it is a red giant gype envelope.

MEYER: The reason I ask is that the computations that we did indicate that the spiralling-in times are very short. The reason for that is, that basically you create from this friction of the spiralling-in, a luminosity of the order of the Eddington luminosity and that gives you timescales (for orbital periods of a few hours or fractions of an hour), of the order of only 1000 years, and I doubt whether a supernova explosion would occur on such a short timescale just at the right moment. So might this create a problem for that part of your scenario?

MIYAJI: Yes. But we have to form a common envelope, so it needs a mass accretion rate larger than 10^{-7} M_o/yr and the main sequence star also has to evolve to become large, so that its timescale is of the order of 10^8 years, this gives you 10 M_o or so. Of course, the spiralling-in timescale is very short, but after the spiralling the white dwarf becomes very close to 1.4 M_o or so, so there may be a possibility.

FINZI: I wanted to ask why do you need to create a common envelope before the supernova explosion, isn't it enough that you have an inflow of mass onto a white dwarf that is close to the Chandrasekhar limit.

MIYAJI: Because we need to have some rapid mass and angular momentum loss process. However, once such a rapid angular momentum loss works, the main problem is how to halt the coalescence.

STAR-PLANET SYSTEMS AND THEIR RELATION TO LOW MASS CATACLYSMIC VARIABLES

Mario Livio
Department of Physics
Technion-Israel Institute of Technology
Haifa 32000, Israel

ABSTRACT. We investigate the possibility of forming a low mass cataclysmic binary from a star-planet system. Using certain simplifying assumptions it is shown that an analytic solution for the evolution of the separation can be obtained. The fate of the system is determined by the competition between mass loss and accretion drag. While it is found that some fine tuning is required for the formation of a close binary, such a binary can still be formed, for parameters that are not too exotic.

1. INTRODUCTION

Following an idea of Eggleton (1978) we have carried out an investigation of a possible scenario for the formation of a low mass cataclysmic binary. We describe here the results of a preliminary study (see Livio 1982).

The general idea is to start with a star-planet system. In the course of its evolution the star becomes a red giant, thus allowing the planet to start to accrete, first from a stellar wind. The increase in the planet's mass leads to an even more efficient accretion. In the more advanced stages the planet is actually embedded in the giant's envelope, accreting while spiralling in. By this process the planet is transformed into a low mass star which, as the giant loses its envelope, forms a close binary companion to the giant's core (which later becomes a white dwarf).

2. PHYSICAL ASSUMPTIONS AND EQUATIONS

We shall assume the following typical initial parameters for the star-planet system: Mass of the star $M_1 \sim M_\odot$, mass of the planet $M_2 \sim 0.01\ M_\odot$, a separation of $a \sim 10^{13}$ cm. We shall denote the mass ratio by $q \equiv M_1/M_2$ and the velocity of the wind from the giant (assumed radial) by V_W. We assume that the planet accretes a fraction β of the mass lost by the giant, namely

M. Livio and G. Shaviv (eds.), Cataclysmic Variables and Related Objects, 269–273.

$$\dot{M}_2 = -\beta \dot{M}_1 \tag{1}$$

The equation for relative motion can be written as (Alexander et al. 1976)

$$\ddot{\vec{r}} = -\frac{G(M_1+M_2)}{r^3}\vec{r} + \frac{1}{M_2}\vec{F} \tag{2}$$

where $\vec{F}$ is the force due to the accretion drag. Equation (1) can be transformed into an equation for the separation (Alexander et al. 1976, Choi and Vila 1981, Livio 1982)

$$\frac{\dot{a}}{a} = -\frac{(\dot{M}_1+\dot{M}_2)}{M_1+M_2} - \frac{2\dot{M}_2}{M_2}\left(1 + \ln\frac{R_{max}}{R_{acc}}\right) \tag{3}$$

where R_{max} can be taken as the radius of the planets Roche lobe (Kopal 1959)

$$R_{max} \sim 10^{12}\left(\frac{q+1}{100}\right)^{-1/3}\left(\frac{a}{10^{13}\,\mathrm{cm}}\right)\ \mathrm{cm} \tag{4}$$

and R_{acc} is the accretion radius (Bondi and Hoyle 1944)

$$R_{acc} \sim 2\times10^{11}\left(\frac{a}{10^{13}\,\mathrm{cm}}\right)\left(\frac{q+1}{q}\right)\left(\frac{q}{100}\right)^{-1}(1+\delta)^{-1}\ \mathrm{cm} \tag{5}$$

where δ is a small correction (Livio 1982). It turns out that

$$\varepsilon(q) \equiv 1+\ln\frac{R_{max}}{R_{acc}} \simeq 2.6+\ln\left[\left(\frac{q}{q+1}\right)^{4/3}\left(\frac{q}{100}\right)^{2/3}(1+\delta)^{-1}\right] \tag{6}$$

is an extremely slowly varying function of q, so that we make the simplifying assumption $\varepsilon(q)$ = const. = 2.587. We note that if we further assume for the moment that β=const. (which is clearly not accurate and we shall discuss this point later), then equation (3) admits an analytic solution

$$\frac{a}{a_o} = \left(\frac{1+\beta q}{1+\beta q_o}\right)^{2\varepsilon+1}\left(\frac{q_o+1}{q+1}\right) \tag{7}$$

where a_o and q_o are the initial values. The separation as a function of q is presented in Fig. 1. It can be easily checked that the behaviour of the separation is determined by the following condition

$$\beta q \gtrsim \frac{1}{2\varepsilon} \qquad \text{reduction in separation} \tag{8}$$

This condition has a very obvious physical interpretation in terms of the relevant timescales describing the problem

$$\tau_{mass\ loss} = \frac{M_1}{\dot{M}_1} ; \quad \tau_{acc} = \frac{M_2}{\dot{M}_2}; \quad \tau_{drag} \simeq \frac{\tau_{acc}}{2\varepsilon} \tag{9}$$

Using these definitions condition (8) can be expressed as a simple statement of the relative importance of the two competing processes

$$\tau_{drag} \lesssim \tau_{mass\ loss} \tag{10}$$

It turns out that even if $\beta=\beta(q)$, a solution of (3) can be obtained in the limit $\beta(q)\cdot q>>1$, which coincides asymptotically with (7) (Livio 1982).

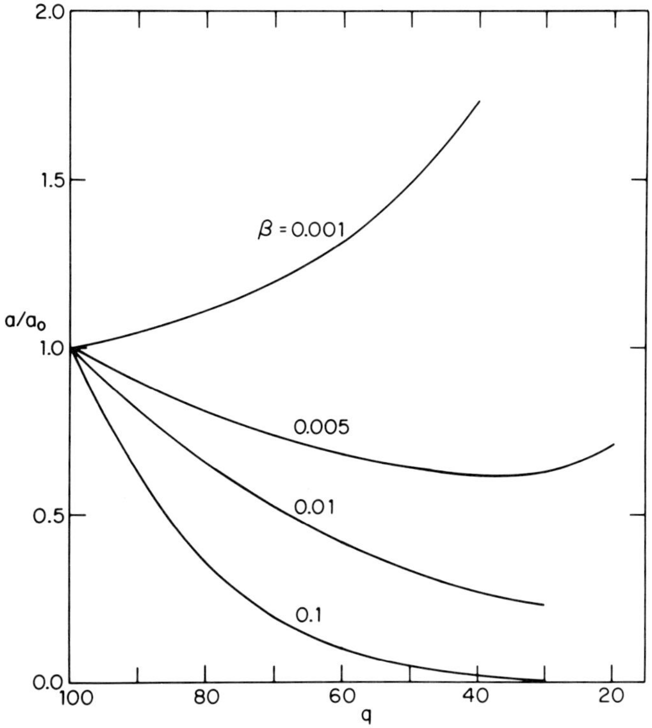

Fig. 1. The evolution of the separation as a function of the mass ratio, for several values of the fraction of accreted mass, β.

For accretion from a wind the various timescales can be estimated as (see also Kudritzki and Reimers 1978)

$$\tau_{mass\ loss} \sim 1.8\text{x}10^6 \left(\frac{M_1}{M_\odot}\right)^2 \left(\frac{L_1}{10^4 L_\odot}\right)^{-1} \left(\frac{R_1}{100 R_\odot}\right)^{-1} \text{yr} \tag{11}$$

$$\tau_{drag} \sim 1.4\times10^{7}\ \varepsilon^{-1}\left(\frac{M_2}{0.01M_\odot}\right)^{-1}\left(\frac{\dot{M}_1}{10^{-6}M_\odot/yr}\right)^{-1}\left(\frac{a}{10^{13}cm}\right)^{1/2}\left(\frac{V_W}{10^{6}cm\ s^{-1}}\right)\times$$

$$\times\left(\frac{M_1}{M_\odot}\right)^{3/2}\left(\frac{q}{q+1}\right)^{3/2}(1+\ )^{3/2}\ yr \tag{12}$$

and we find

$$\beta q \sim 0.037\ \left(\frac{q}{100}\right)^{-1}\left(\frac{q+1}{q}\right)^{3/2}\left(\frac{V_W}{10^{6}cm\ s^{-1}}\right)^{-1}\left(\frac{a}{10^{13}cm}\right)^{-1/2}\left(\frac{M_1}{M_\odot}\right)^{1/2}(1+\delta)^{-3/2} \tag{13}$$

which implies that under normal conditions accretion from a stellar wind <u>will not</u> result in a decrease of the separation and a close binary system <u>will not</u> be formed.

In contrast to this, the planet may start at such an initial separation that it will be moving inside the giant's envelope as that expands. This can lead to an accretion rate of

$$\dot{M}_2 \simeq 7.2\times10^{-7}\left(\frac{M_2}{0.01M_\odot}\right)^{2}\left(\frac{\rho}{10^{-10}gcm^{-3}}\right)\left(\frac{M_1}{M_\odot}\right)^{-3/2}\left(\frac{a}{10^{13}cm}\right)^{3/2}\left(\frac{q+1}{q}\right)^{3/2}\ M_\odot/yr \tag{14}$$

where ρ is the local atmospheric density. Under these conditions the planet will spiral into the giant on a drag timescale that can be as short as ~600 years and may be completely dissipated, thus, again not forming a close binary system.

3. CONCLUSIONS

On the basis of this simplified preliminary study the following conclusions can be drawn:

1) A certain amount of fine tuning is required for the formation of a low mass binary system. Specifically, for $a_o \gtrsim 2000\ R_\odot$ very little accretion will take place and no reduction in the separation will result. For $a_o \lesssim 500\ R_\odot$ a spiralling-in of the planet, on a timescale that is short compared to evolutionary is inevitable, the planet will be totally dissipated in the giant.

2) In spite of conclusion (1), for a quite reasonable initial separation, $a_o \sim 900\ R_\odot$, a low mass secondary can be formed in $\tau \sim 10^5 - 10^6$ years, which is of the order of the time spent by the star in the giant phase (Eggleton 1973).

3) The amount of fine tuning needed for the formation of the binary is not excessive. A change in the initial separation by 100 $R_\odot$ results in

a change in the accretion timescale by less than an order of magnitude.

4) The virial temperature of a 0.01 $M_\odot$ planet is of the order of (Zapolski and Salpeter 1969) $T_V \sim 1.6 \times 10^6$ °K so that the planet can plunge into the giant very deeply (to $r \sim 1.9\ R_\odot$, Harpaz 1981) before being evaporated.

5) The tidal evolution of the system should be explored (see Livio 1982 for a preliminary study).

6) Any star-planet system in which the initial separation is not greater than a few thousands solar radii, must inevitably undergo a situation of the type described here, the processes involved (in particular evaporation of the planet), therefore, deserve further study.

REFERENCES

Alexander, M.E., Chau, W.Y., Henriksen, R.N.: 1976, *Astrophys. J.* 204, 879.
Bondi, H. and Hoyle, F.: 1944, *Mon. Not. R. Astr. Soc.* 104, 273.
Choi, K.H. and Vila, S.C.: 1981, *Astrophys. Space Sci.* 77, 319.
Eggleton, P.P.: 1973, *Mon. Not. R. Astr. Soc.* 163, 279.
Eggleton, P.P.: 1978, *Science Today* 13, 22.
Harpaz, A.: 1981 (private communication).
Kopal, Z.: 1959, *Close Binary Systems*, Chapman and Hall, London.
Kudritzki, R.P. and Reimers, D.: 1978, *Astron. Astrophys.* 70, 227.
Livio, M.: 1982, *Astron. Astrophys.* 112, 190.
Zapolski, H.S. and Salpeter, E.E.: 1969, *Astrophys. J.* 158, 809.

DISCUSSION FOLLOWING M. LIVIO'S TALK

LAMB: Did you restrict your accretion rate onto the planet to be less than the Eddington limit at all times?

LIVIO: Yes. In fact, I never went to more than one fifth or so, of the Eddington limit.

LANGER: You seem to need to take this planet down inside the envelope of the red giant, can you cool the accreted mass fast enough to sort of keep settling it down onto the planet and then use the evaporation estimate, based on the radius that the planet has, for a given mass?

LIVIO: I don't know, I have not looked into this yet. However, you must realize also that the temperatures that you get from accretion onto a planet are not that high, you get to about 10^5 °K and this may agree quite well with the temperature you already have, inside your medium. So this whole thing may even not have observational effects from the outside, because all the luminosity you get may be absorbed.

CATACLYSMIC BINARIES CONTAINING A BLACK-DWARF SECONDARY

H. Ritter
Max-Planck-Institut für Physik und Astrophysik
Karl-Schwarzschild-Str. 1, D-8046 Garching
Fed. Rep. Germany

ABSTRACT

Cataclysmic binaries with a black-dwarf secondary can in principle be distinguished from those containing a low-mass main-sequence secondary because of the differences of the corresponding eclipse light curves. However, due to the very low intrinsic luminosity of the systems containing a black dwarf, the probability of detection turns out to be smaller than the corresponding probability for an ordinary cataclysmic binary by about 1.5 to 3 orders of magnitude. Therefore, despite the fact that the predicted space density of these objects is quite high, systems containing a black-dwarf secondary must remain essentially undetectable.

I. INTRODUCTION

Recent theoretical investigations of the consequences of angular momentum loss in ultra-short-period cataclysmic binaries (hereafter USPCB's) due to gravitational radiation predict that the corresponding decay of the orbit forces the Roche lobe-filling secondary to mass loss which eventually brings it out of thermal equilibrium and finally transforms it into a hydrogen-rich black dwarf (Paczyński and Sienkiewicz, 1981; Rappaport, Joss and Webbink, 1982; hereafter referred to as PS and RJW respectively). During the transition of the secondary from a low-mass main-sequence star to a black dwarf, the binary period first decreases, goes through a minimum and finally increases when the secondary has become a black dwarf. Therefore, for every orbital period longer than the minimum period there are two possible configurations for an USPCB, one with a main-sequence secondary (hereafter called ordinary systems) and one with a black-dwarf secondary (hereafter called black-dwarf systems). According to RWJ, as many as 20% of all the USPCB's could be black-dwarf systems. The question then arises as to why, up to now, no black-dwarf system has been identified. Is it because the two types of USPCB's cannot be distinguished with present observational techniques or because of observational selection effects? It is the

M. Livio and G. Shaviv (eds.), Cataclysmic Variables and Related Objects, 275–278.

purpose of this paper to discuss briefly these questions.

II. SOME PROPERTIES OF BLACK-DWARF SYSTEMS

One of the predictions of PS and RJW is that the change of the orbital period has opposite sign for the two types of systems: it is decreasing for ordinary systems, i.e. as long as the secondary is at or near to the main sequence, and increasing for black-dwarf systems. Based on this property one could therefore identify black-dwarf systems by determining the sign of the period change. In fact, recent observations of the USPCB Z Cha by Cook and Warner (1981) show that the orbital period of this system is increasing. Based on this fact Faulkner and Ritter (1982) tried to work out the physical parameters of this system assuming that the secondary is a black dwarf. However, this assumption leads to inconsistencies with other observational properties. Therefore it was concluded that Z Cha is not a black-dwarf system, despite its increasing period. This, however, demonstrates that the nature of the secondary cannot be reliably inferred from the sign of the period change.

One of the reasons why, in the case of Z Cha, the black-dwarf hypothesis has to be rejected is that it results in contradictions with observed eclipse light curves (Warner, 1974; Bailey, 1979). Turning this argument around, this implies that *black-dwarf systems which undergo a double eclipse produce light curves which are different from those produced by ordinary systems*. In order to study these differences systematically and to investigate whether they can be used as a means of distinguishing between the two types of USPCB's, synthetic eclipse light curves for both types have been computed. The synthetic light curves and details of how they have been computed will be published elsewhere. The main properties of these light curves can be summarized as follows: In ordinary systems the phase lag between the white-dwarf eclipse and the hot-spot eclipse is small and the two eclipses interfere strongly. Many examples of such light curves for different objects are known, e.g. for Z Cha (Warner, 1974; Bailey, 1979), OY Car (Vogt et al. 1981), HT Cas (Patterson, 1981). If, on the other hand, the secondary is a black dwarf, the phase lag between the two eclipses is much larger and in the extreme case the eclipse of the hot spot begins only when the eclipse of the white dwarf has already ended. The difference in the light curves, i.e. in the phase lag of the two eclipses, can be traced back to the difference in mass ratio of the two types of USPCB's. The less massive the secondary star is (keeping the mass of the primary constant), the larger is the phase lag between the two eclipses. This in turn is due to the change with mass ratio of the trajectory of a single particle leaving the inner Lagrangian point, along which the hot spot approximately lies. Since the difference in mass ratio between ordinary systems and black-dwarf systems increases with increasing orbital period, the difference in the light curves is strongest for systems with the longest orbital period and it vanishes for systems having the minimum period. Thus, in principle, the two possible configurations of USPCB's can be distinguished using the morphology of the eclipse light curves.

III. OBSERVABILITY

The above-mentioned distinction involves a double eclipse. Therefore this method applies only to a rather small fraction among the USPCB's. Since in cataclysmic binaries the secondary fills its critical Roche lobe, the probability for the occurence of a double eclipse depends only on the mass ratio of the system and decreases with increasing mass ratio. For a given orbital period and primary mass, the black-dwarf system always has a higher mass ratio than the corresponding ordinary system. Therefore, the probability for the occurence of a double eclipse in a black-dwarf system is always lower, by as much as a factor of 3, when compared with the corresponding value for the ordinary system.

In addition to this not very strong selection effect there is, however, another and much stronger one, favouring the detection of ordinary systems. This is due to the fact that the luminosity of an USPCB is essentially accretion luminosity and that the accretion rate, i.e. the mass transfer rate, in a black-dwarf system is lower by one to two orders of magnitude when compared with the corresponding ordinary system (PS, RJW). As a consequence, in a magnitude limited sample of USPCB's the expected fraction of black-dwarf systems is between ~ 0.03 and ~ 0.001. Considering the still rather small number of known USPCB's (about 25) it is unlikely that a black-dwarf system is among them.

There is yet another effect which could act against the identification of black-dwarf systems. It is again a consequence of the extremely low mass transfer rate in black-dwarf systems. For the predicted values of $\sim 10^{-11}$ to $\sim 10^{-12}$ $M_{\odot}$/yr (PS,RJW), the outer parts of the accretion disk, where the hot spot is formed, are optically thin (see e.g. Meyer and Meyer-Hofmeister, 1981). It could therefore well be that the hot spot itself is also optically thin, so that there is no bright spot which can be eclipsed. In this case at most a single rather than a double eclipse occurs. This, however, does not allow a distinction to be made between the two configurations of USPCB's.

In conclusion, taking all these selection effects into account, black dwarf systems must remain essentially unobservable or at least unidentifyable, despite the fact that, according to RJW, their space density might be quite high.

ACKNOWLEDGEMENTS

I am grateful to Dr. F. Meyer for stimulating discussions and to Dr. J. Kirk for improving the English.

REFERENCES

Bailey, J.: 1979, Monthly Notices Roy. Astron. Soc. 187, pp. 645-653.

Cook, M.C., Warner, B.: 1981, Monthly Notices Roy. Astron. Soc. 196, pp. 55P-57P.

Faulkner, J., Ritter, H.: 1982, in "Binary Stars and Multiple Stars as Tracers of Stellar Evolution", proceedings of the IAU Coll. No. 69, eds. Z. Kopal and J. Rahe, D. Reidel Publishing Company, pp. 483-488.
Meyer, F., Meyer-Hofmeister, E.: 1981, Astron. Astrophys. 106, pp. 34-42.
Paczyński, B., Sienkiewicz, R.: 1981, Astrophys. J. Letters 248, pp. L27-L30.
Patterson, J.: 1981, Astrophys. J. Suppl. 45, pp. 517-539.
Rappaport, S., Joss, P.C., Webbink, R.F.: 1982, Astrophys. J. 254, pp. 616-640.
Vogt, N., Schoembs, R., Krzeminski, W., Pedersen, H.: 1981, Astron. Astrophys. 94, pp. L29-L32.
Warner, B.: 1974, Monthly Notices Roy. Astron. Soc. 168, pp. 235-247.

DISCUSSION FOLLOWING H. RITTER'S TALK

EGGLETON: Doesn't a lot of this analysis in fact depend on the assumption that only gravitational radiation is operating to remove angular momentum? If these systems were evolving faster than that it would not, for instance, be surprising if the period increased, even if you are on the non degenerate portion of that curve.

RITTER: If we have objects on the degenerate branch evolving ten times faster, they increase the mass accretion rate by a factor ten, then we get a factor of 30 less of antiselection. So, there might be a small chance to find one or the other of these objects. Also, if the accretion rates are higher, of course, conditions become better with respect to having an observable hot spot which undergoes an eclipse.

MIYAJI: I agree with Eggleton that your definition of degenerate and non degenerate is somewhat misleading. The meaning of these two branches is, that one part is an isentrope and the other part is an n=3 polytrope or something like that. So that if the envelopes are removed from normal stars very rapidly, then the stars will become isentrope even near the main sequence branch, the period can be increased.

RITTER: I do not completely agree because if you remove a lot of matter in a short time from the secondary star, the size of the Roche lobe does not depend on the radius of the star, but the radius of the star has to care about the size of the Roche lobe. These systems are not free in choosing the mass transfer rate, even if you go to another mass, angular momentum loss mechanism. I think you can only increase the orbital period on the, what I call nondegenerate branch, if you lose the mass out of the system with less than the mean angular momentum.

FAST OSCILLATIONS IN VARIABLE X-RAY SOURCES AND X-RAY BURSTERS

Mario Livio
Department of Physics
Technion-Israel Institute of Technology
Haifa 32000, Israel

ABSTRACT. The properties of coherent oscillations with periods of tens of milliseconds that have been observed in seven variable X-ray sources and bursters are reviewed. Theoretical models for the generation of the oscillations are discussed. Difficulties with a magnetic rotator model are indicated. It is suggested that at least in several cases, g-mode oscillations of the neutron star provide a satisfactory explanation of the observations.

1. OBSERVATIONS

Recent observations have revealed the presence of periodicities of the order of tens to hundreds of milliseconds in several X-ray sources that are either variables or that have been identified as burst sources. Table I summarizes some of the properties of these sources.

Theoretical models for the observed periodicities are discussed in section 2 and a comparison of these models is carried out in section 3. A comparison of the properties of the oscillations with the coherent oscillations of dwarf novae is conducted in section 4.

2. THEORETICAL MODELS

We will now attempt to study possible models for the generation of the observed periodicities, following the work of Livio and Bath (1982, see also Livio, 1982).

Three major classes of models will be discussed;

a) Neutron star magnetic rotator model.
b) Luminous blobs in the accretion disk.
c) Neutron star oscillations

M. Livio and G. Shaviv (eds.), Cataclysmic Variables and Related Objects, 279–291.

TABLE I

Observed Coherent Oscillations

Source	Type	Period (ms)	$\dot{P}$	Duration of Pulsations	Pulsed Fraction	Ref.
MXB 1728-34	X-ray Burster	12.25	-10^{-6}	~40 sec	3%	1
4U1907+09	Variable X-ray source, binary	15.2	$+10^{-5}$	~10 sec (several times)	12%	2
A0538-66	Recurrent X-ray transient, binary	69.2	$+5\mathrm{x}10^{-10}$	~3900 sec	26%	3,4
4U1254-69	Optical burst observed	27.47	No data	~40 sec	5%	5
2103+09	No known source. Increase in X-ray intensity.	67.55	Not observed	~5000 sec	1.5%	6
2100-79	Coincides with weak HEAO source	15.85 (evidence)	$\lesssim 10^{-16}$	Observed for 8 days	~70%	7
MXB1730-335	Rapid burster	503,508	No data	~240 sec	No data	8

REFERENCES: 1. Sadeh et al. (1982). 2. Sadeh and Livio (1982a). 3. Skinner et al. (1982a) 4. Skinner et al. (1982b). 5. Mason et al. (1980). 6. Sadeh and Livio (1982b) 7. Sadeh and Livio (1982c). 8. Tanaka (1981).

2.1 Magnetic Rotator

Rotation is probably the easiest way to obtain a singly periodic signal. The possibility that the periodicities are generated by a rotating magnetic neutron star should, therefore, be investigated. Three points that have to be discussed in the context of such a model are: (i) The generation of a modulated signal, (ii) Period changes, (iii) The duration of the periodicities.

2.1.1. Modulation of signal. Magnetic fields are believed to be the agent via which the accreted matter is funnelled onto a small cap on the neutron star's surface, thus generating a "spot" which in turn, as the neutron star rotates, produces the modulated emission. The area of the caps decreases with increasing magnetic field strength.

This same magnetic field, however, can enforce plasma corotation up to the Alfven radius and thus may inhibit accretion altogether, if the Alfven radius is larger than the corotation radius, because of the matter's inability to overcome the centrifugal barrier.

In order to illustrate the competition between these two conflicting effects we shall use for the area of the magnetic caps the expression obtained by Arons and Lea (1980), although their formula is really inapplicable in the case of fast rotators of the type we are discussing. Assuming typical values for the X-ray luminosity-L_X, temperature-T_X and for the neutron star radius R_X and taking a representative period of P=0.02 sec, the two requirements that: (a) the caps do not cover the whole neutron star surface and (b) the Alfven radius is smaller than the corotation radius, lead to:

$A_{mag.caps} \lesssim A_{n.s.}$	$R_{Alf.} \lesssim R_{cor.}$
$L_x \sim 10^{37}$ ergs s^{-1}	$L_x \sim 10^{37}$ ergs s^{-1}
$T_x \sim 7$ keV	$R_x \sim 10^6$ cm
$R_x \sim 10^6$ cm	$P \sim 0.02$ sec
$\mu_{30}^{1.652} + 0.34\,\mu_{30}^{2.566} \gtrsim 0.1$	$R_A \sim 3\times10^8 \mu_{30}^{4/7} L_{x37}^{-2/7} R_{x6}^{-2/7} M_{x\odot}^{1/7}$ cm
	$R_C \sim 1.1\times10^7 M_{x\odot}^{1/3} P_{0.02}^{3/3}$ cm
	$R_A \lesssim R_C$
$\mu_{30} \gtrsim 0.25$	$\mu_{30} \lesssim 0.003$

Clearly the two requirements are inconsistent. This is probably not an insurmountable difficulty because of the inapplicability of the formula used for the cap area. In this respect it is enough to note that simple, kinematical estimates give much smaller values (e.g. Davidson and Ostriker 1973) for the same parameters

$$\frac{A_{mag.caps}}{A_{n.s.}} \sim 8\times10^{-4}\, \mu_{30}^{-0.57} \qquad (1)$$

2.1.2. Funnelling and bursts. A second problem that may be associated with strong funnelling relates to the possible inhibition of X-ray bursts. On one hand, as a result of strong funnelling onto a small fraction of the neutron star surface area, the effective accretion rate is increased to

$$\dot{M}_{EFF} \sim 7.5\times10^{18} L_{x37} R_{x6} M_{x\odot}^{-1} \left(\frac{A_{n.s.}/A_{mag\ caps}}{10}\right) g\ s^{-1} \quad (2)$$

On the other hand, an increase in the accretion rate results in strong compressional heating, relatively early helium ignition and therefore, weak electron degeneracy. Consequently, as it has been found in the detailed study of Ayasli and Joss (1982), the thermonuclear flashes are essentially suppressed at accretion rates above 3×10^{18}g s^{-1} thus, leading to a non bursting situation. Now, among the sources in which the periodicities have been detected, at least two, MXB 1728-34 and MXB 1730-335 and possibly three (an optical burst has been observed from 4U1254-69) are bursters. The magnetic rotator scenario may have, therefore, additional difficulties in this respect.

2.1.3. Changes in period. When a rotator model is considered an important point that has to be addressed is that of the observed period changes. At least three different mechanisms can be invoked to produce changes in the period: (i) A changing Doppler shift due to orbital motion. In this case the rate of change in the period is given by

$$\dot{P} \sim 9\times10^{-8} \left(\frac{d}{10^{11} cm}\right)^{-2} P_{0.02} \left(\frac{M_{tot}}{10\ M_{\odot}}\right) \quad (3)$$

where d is the binary separation and M_{tot} is the total mass of the system. Expression (3) gives very good agreement with the observations of A0538-66 (Skinner et al. 1982b, Charles et al. 1982), however, it definitely cannot explain the $\dot{P}$'s of 4U1907+09 (which is known to have an orbital period of 8.38 days, Marshall and Ricketts 1980) and of MXB 1728-34. (ii) Braking by magnetorotational radiation in the initial spin-down phase. In this case the $\dot{P}$ is given by (e.g. Lipunov 1982, Illarionov and Sunyaev 1975).

$$\dot{P} \sim 5\times10^{-14} P_{0.02}^{-1} \mu_{30}^{2} I_{45}^{-1} \quad (4)$$

where I_{45} is the moment of inertia (in units of 10^{45}g cm^2), so that besides being too low for at least two of the observed $\dot{P}$'s, it clearly cannot account for the spin-up observed in MXB 1728-34. (iii) Spin-up in an accreting X-ray pulsar. The maximal spin-up rate, for a neutron star accreting from a disk is given by (Ghosh and Lamb 1979)

$$-\dot{P} \sim 6.3\times10^{-16} \mu_{30}^{2/7} \alpha(\omega_s) R_{x6}^{6/7} M_{x\odot}^{-3/7} I_{45}^{-1} P_{0.02}^{2} L_{x37}^{6/7} \quad (5)$$

where

$$\omega_s \sim 67.5\ \mu_{30}^{6/7} R_{x6}^{-3/7} M_{x\odot}^{-2/7} P_{0.02}^{-1} L_{x37}^{-3/7} \quad (6)$$

and the function $\alpha(\omega_s)$ was computed by Ghosh and Lamb (1979). This spin-up rate cannot, again, explain some of the observed $\dot{P}$'s. Furthermore, it does, in fact, apply only to slow rotators. For the fast rotators discussed here, the near cancellation of magnetic and material stresses will, in fact, result in much lower rates.

Finally, we would like to point out that even from energy considerations alone, the difficulty associated with period changes can be illustrated by noting that

$$\dot{P} \sim 2\text{x}10^{-13} \left(\frac{\dot{E}}{10^{39} \text{ergs s}^{-1}}\right) I_{45}^{-1} P_{0.02}^{3} \qquad (7)$$

and recalling that the energy involved in X-ray bursts is of the order of 10^{39} ergs, making high $\dot{P}$'s very difficult to explain.

2.1.4. Duration of periodicities. In any magnetic rotator scenario it is expected that the flux will be pulsed at all times, including at quiescence (if bursters are involved). Difficulties in the detectability of the periodic signal may weaken the last statement to the requirement that the periodicities should be detectable in all bursts and increased luminosity states. This requirement is essentially model independent, as far as details of the rotator are concerned. The observational situation for some of the sources is summarized in Table II. It clearly demonstrates that severe difficulties are posed to the magnetic rotator model (for these sources).

All the problems associated with rotation that have been discussed motivate the search for other models for the origin of the oscillations. We will discuss here two additional possible scenarios.

2.2. Luminous Blobs in Disks

The Keplerian motion of hot, luminous blobs in the accretion disk has been originally suggested by Bath (1973) as the source of the coherent oscillations observed in dwarf novae. It is possible therefore, in principle, that similar blobs orbiting the neutron star produce the periodicities under discussion. The Keplerian period of such a blob is given by

$$P_{Kep} \sim 17\, M_{x\odot}^{-1/2} \left(\frac{R_{orb}}{10^{7} \text{cm}}\right)^{3/2} \text{ ms} \qquad (8)$$

where R_{orb} is the orbital radius of the blob. Since the transition between the gas pressure dominated and radiation pressure dominated regions in standard steady disk models occurs at (Novikov and Thorne 1973)

$$R_{mid\text{-}in} \sim 1.2\text{x}10^{7}\, \alpha^{2/21} M_{x\odot}^{1/3} \dot{M}_{17}^{16/21} \text{ cm} \qquad (9)$$

where α is the usual viscosity parameter (Shakura and Sunyaev 1973), Keplerian periods in the middle or inner parts of the disk agree well

TABLE II

Observational situation concerning the duration of periodicities

Source	Observations
MXB 1728-34	1) No 12 ms period in DC flux 2) Periodicity observed only in one out of four bursts. 3) Most of the periodic signal coincides with bursts.
A0538-66	Pulsations detected at 40 MPC counts/sec, not detected at 10 counts/sec (although should have been detected at the same pulsed fraction).
4U1907+09	Periodicity observed only in 3 (possibly 4) out of 20 passes. In two of them the source was in a high state.
4U1254-69	No periodic modulation was found in entire data set. The period was found in data only from the immediate vicinity of the burst.
MXB 1730-335	1) Oscillations discovered only in two trapezoidal bursts. 2) Do not appear as due to neutron star rotation.

with the observed ones.

In view of the apparent phase stability of the oscillations (in several cases), however, the lifetime of such blobs against smearing by differential rotation should be considered. This lifetime is roughly given by (Bath, Evans and Papaloizou 1974)

$$\tau \lesssim \frac{2}{3}\,\frac{PR_{orb}}{D} \tag{10}$$

where D is the radial dimension of the blob. If we assume that

$$D \sim h(R_{orb}) \tag{11}$$

where h is the local disk thickness, we obtain for steady disk models a lifetime of

$$\tau \sim \begin{cases} 1.1\ \alpha^{1/10} P_{0.02}^{29/30} M_{x\odot}^{1/3} \dot{M}_{17}^{-1/5}\ \text{sec} & \text{in middle region} \\ 1.5\ P_{0.02}^{5/3} M_{x\odot}^{1/3} \dot{M}_{17}\ \text{sec} & \text{in inner region} \end{cases} \tag{12}$$

which seems too short to account for the observed duration of the periodicities, even when considering the uncertainties involved in (12).

2.3 Non Radial Oscillations of Neutron Star

The properties of several non radial modes of neutron stars are summarized in Table III (see Van Horn 1980). The shortness of the periods and the fact that gravitational radiation causes the damping of p-modes (for $\ell \gtrsim 2$) on an extremely short timescale $\tau_{Grav.} \lesssim 2$ sec, probably excludes them as a possible origin of the oscillations.

TABLE III

Properties of neutron star oscillation modes (low order)

	p-modes	g-modes	r-modes	Torsional Modes
Period	0.1 - 1 ms	10 - 1000 ms	P_{rot}	~20 ms
Damping	τ_{Grav}	τ_{Therm}	τ_{Rot}	τ_{Crust}
Excitation	Thermonuclear events	Thermonuclear events	Thermonuclear events	Thermonuclear events
	Mass transfer events	Mass transfer events	Mass transfer events	Mass transfer events
		Shear	Shear	Shear

We will now examine more closely the properties of g-mode oscillations. Analytic estimates of the periods give (Van Horn 1980, Livio and Bath 1982)

$$P_g \sim \frac{49}{[\ell(\ell+1)]^{1/2}} R_{x6}\, \Pi_{22}^{-1/2}\, \rho_6^{1/2} \quad \text{ms} \tag{13}$$

where Π_{22} is the pressure in units of 10^{22} dynes cm^{-2} and ρ_6 is the density in units of 10^6 g/cc.

Numerical calculations (Van Horn et al. 1982) using neutron star models give for g_1-g_3 periods of $P_g \sim 66$-81 ms for a neutron star mass of 1.326 $M_\odot$ and $P_g \sim 371$-692 ms for a mass of 0.503 $M_\odot$ (which may however be too small to be formed in a collapse followed by a supernova explosion).

An additional feature that was demonstrated by the numerical calculations was the confinement of the oscillations to the outermost layers of the neutron star. This fact may have two important consequences: (i) It makes the excitation of the modes by surface phenomena (such as thermonuclear or accretion events) easier and (ii) It may explain the observed changes in the period.

With respect to the first point here, we note that Van Horn et al. (1982) have concluded that the energy released in bursts is sufficient to excite the low order g-modes. Furthermore, the simulation of an accretion event by Starrfield et al. (1982) has resulted in a "ringing" of the neutron star's envelope with a period of ~200 ms. As for the second point (period changes), this represents a change in the structure of the region of mode concentration. If, based on the observed timescale for changes in several sources, one demands $\tau_{Therm} \sim 10$ sec, then, one obtains for the mass involved in the changes

$$\Delta M \sim 10^{21} \ L_{x37}\tau_{10}\Pi_{22}^{-1}\rho_6 \ \text{g} \qquad (14)$$

which is a typical mass accreted prior to the burst (Ayasli and Joss 1982).

We thus conclude that g-modes have most of the observed properties for several sources. An important observational test of the g-mode hypothesis will be to determine whether several periods are ever excited simultaneously. So far, only the strongest signal in the power spectra has been studied, in most of the observations.

3. AGREEMENT OF THEORETICAL MODELS WITH OBSERVATIONS

In Tables IVa and IVb we have attempted to summarize some of the difficulties and agreements met by what we feel are the representatives of two major classes of models, with observations. With a total number of 7 sources so far, in which periodicities have been detected, an over division into subgroups does not seem wise, however, there does seem to emerge a suggestive distinction between two subclasses of objects. One contains the sources with short lived lower Q oscillations, for which the g-mode hypothesis seems to fit the observations better than rotation (Table IVa). Rotation seems to work quite well for the second group (Table IVb). One source, A0538-66, could perhaps be classified in both groups (the fact that the oscillations were not observed at turn-on are a strong point against rotation, which works very well otherwise). Clearly, many more observations will be required to decide among the various possible models.

It is very tempting to compare the observed periodicities in variable X-ray sources and bursters with the familiar coherent oscillations observed in cataclysmic variables. Such a comparison is presented in Table V.

TABLE IVa

Agreement of Theoretical Models with Observations

Source	Agreement and Difficulties with Rotation	Agreement with Difficulties with g-modes
MXB1728-34	Difficulties: 1) $\dot{P}$ too large 2) Duration of periodicities 3) Bursts not favoured by funnelling	Difficulties: 1) Only one period observed Agreement: 1) Excitation by bursts 2) $\dot{P}$ agrees with mass accreted between bursts
4U1907+09	Difficulties: 1) $\dot{P}$ too large 2) Duration of periodicities	Difficulties: 1) Only one period observed 2) Excitation not clear (once seen in low state) Agreement: 1) Excitation by accretion events?
MXB1730-335	Difficulties: 1) Period change 2) Duration of periodicities	Agreement: 1) Excitation by bursts
4U1254-69	Difficulties: 1) Duration of periodicities	Agreement: 1) Excitation by burst

TABLE IVb

Agreement of Theoretical Models with Observations

Source	Agreement and difficulties with rotation	Agreement and difficulties with g-modes
A0538-66	Difficulties: 1) Not seen at turn on Agreement: 1) $\dot{P}$ agrees well with orbital Doppler shift 2) Duration quite long	Agreement: 1) Could be excited by accretion event.
2103+09	Agreement: 1) No observed $\dot{P}$ 2) Duration long	
2100-79	Agreement: 1) $\dot{P}$ very low 2) Duration long 3) Large pulsed fraction	

4. Comparison with "classical" cataclysmic variables

One cannot escape noticing a few quite striking similarities between the two types of oscillations. A question that therefore naturally arises (although caution is always advisable in making such analogies) is whether we are not observing the same phenomenon in the two classes of objects, the only difference being the nature of the compact object, a neutron star (in the "new" CV's) instead of a white dwarf (in the "classical" CV's).

TABLE V

Comparison with "Classical" Cataclysmic Variables

	"Classical" CV's	"New" CV's
Periods	8-39 sec	12-508 ms
Lifetime/Period	$\sim 10^3$	$\sim 10^3$-10^6
P/P_{radial}	~ 10	~ 10
$P/P_{g\text{-}mode}$	~ 1	~ 1
$P/\dot{P}$	10^5-10^7 sec	10^3-10^{14} sec
Relation to Activity	Seen only during outburst	In several sources seen at outburst or sudden increase of luminosity

REFERENCES

Arons, J. and Lea, S.M.: 1980, *Astrophys. J.* 235, 1016.
Ayasli, S. and Joss, P.C.: 1982, *Astrophys. J.* 256, 637.
Bath, G.T.: 1973, *Nature Phys. Sci.* 264, 84.
Bath, G.T., Evans, W.D. and Papaloizou, J.: 1974, *Mon. Not. R. Astr. Soc.* 167, 7.
Charles, P.A., Booth, L., Denshaw, R.H., Bath, G.T., Thorstensen, J.R., Willis, A.J., Howarth, I.D., Skinner, G.K. and Olszewski, E.: 1982, preprint.
Davidson, K. and Ostriker, J.P.: 1973, *Astrophys. J.* 179, 585.
Ghosh, P. and Lamb, F.K.: 1979, *Astrophys. J.* 234, 296.
Illarionov, A.F. and Sunyaev, R.A.: 1975, *Astron. Astrophys.* 39, 185.
Lipunov, V.M.: 1982, *Astrophys. Space Sci.* 82, 343.
Livio, M.: 1982, *Astrophys. Space Sci.* 82, 167.
Livio, M. and Bath, G.T.: 1982, *Astron. Astrophys.* in press.
Marshall, N. and Ricketts, M.J.: 1980, *Mon. Not. R. Astr. Soc.* 193, 7p.
Mason, K.O., Middleditch, J., Nelson, J.E. and White, N.E.: 1980, *Nature* 287, 516.
Novikov, I.D. and Thorne, K.S.: 1973, in De Witt and De Witt (eds.), *Black Holes*, Les Houches, Gordon and Breach, New York.

Sadeh, D., Byram, E.T., Chubb, T.A., Friedman, H., Hedler, R.L., Meekins, J.F., Wood, K.S. and Yentis, D.J.: 1982, *Astrophys. J.* 257, 214.
Sadeh, D. and Livio, M.: 1982a, *Astrophys. J.* 258, 770.
Sadeh, D. and Livio, M.: 1982b, *Astrophys. J.*, in press.
Sadeh, D. and Livio, M.: 1982c, *Astrophys. J.*, submitted.
Shakura, N.I. and Sunyaev, R.A.: 1973, *Astron. Astrophys.* 24, 337.
Skinner, G.K., Leahy, D., Elsner, R.F. and Weisskopf, M.C.: 1982a, IAUC No. 3671.
Skinner, G.K., Bedford, D.K., Elsner, R.F., Leahy, D., Weisskopf, M.C. and Grindlay, J.: 1982b, *Nature* 297, 568.
Starrfield, S., Kenyon, S., Sparks, W.M. and Truran, J.W.: 1982, *Astrophys. J.* 258, 683.
Tanaka, Y.: 1981, in *X-Ray Astronomy in the 1980's*, NASA Technical Memorandum 83848, p. 219.
Van Horn, H.M.: 1980, *Astrophys. J.* 236, 899.
Van Horn, H.M., McDermott, P.N. and Carroll, B.W.: 1982, paper presented at Meeting on Pulsation in Classical and Cataclysmic Variable Stars at JILA, Boulder Colorado, June 1-4, 1982.

DISCUSSION FOLLOWING M. LIVIO'S TALK

ROBINSON: I don't think you should be particularly worried about there just being one oscillation there, what you really have is one detected oscillation and limits on the amplitudes of any other oscillations and my impression in looking at your power spectra is that those limits are very poor, you could easily have oscillations in there, with amplitudes up to 50% as large as the main one. In the case of the single white dwarfs in the field, that we know are pulsating, like ZZ Ceti stars, every one is multi-periodic, but many of them have power spectra that are dominated by just one oscillation, and if you would put a lot of noise on top of their light curves you would detect just that one oscillation. Then there is a question, why do you restrict yourself to g-modes? There are lots of different things.

LIVIO: I picked g-modes just to illustrate my point. In fact, the torsional modes may fit just as well. I also agree of course, with your former point, it strengthens the case for oscillations.

WARNER: I agree with Rob that perhaps the comparison with the isolated white dwarfs is better rather than the very impressive correlation that you have between the properties of the cataclysmic variables and your neutron stars. The interpretation of the oscillations in cataclysmics is certainly not agreed on and I doubt if very many people would say that they are g-mode oscillations. If you accept that the cataclysmics are more likely blobs in disks, can you make any comment on the possibility of that model applying to the neutron stars?

LIVIO: All that I can say is that like in cataclysmic variables, the periods are always longer than the Keplerian period right near the surface of the neutron star. There is a problem with the lifetime. The disks in cataclysmics are different than those around neutron stars in that the latter have also "middle" and "inner" zones in the standard disk models and this causes the short lifetimes. Another possibility is that this is not a blob, perhaps, but rather an oscillation mode of the disk, because there, always the periods are of the order of the Keplerian one. The problem is that very little is known about the excitation of those modes and their decay times.

CHANMUGAM: I just want to make a comment that while it is true that the periods of radial oscillations of neutron stars are much shorter than what you have here, you can have low mass neutron stars, say of mass 0.25 $M_{\odot}$, which have radial oscillation periods of the right order of magnitude. I don't know if such stars can be formed.

MAZEH: Do you expect such a high percentage of the modulation in the g-modes?

LIVIO: I don't know, it is not that high in all of them, there is one with 1.5% and one with 3%. For the two that have a very high pulsed fraction I have indicated that rotation probably works well.

LAMB: I have a couple of comments. One is that the group of objects that you have listed seems to be fairly heterogeneous and I am a little concerned about trying to find a common explanation for what may be a real menagerie of phenomena. There are possibilities that are not even explored here. For example, if you have accretion columns formed by magnetic flux tubes with plasma in them, the flux tubes can oscillate. There are also possible instabilities in the flow and so on.

LIVIO: I agree with you. I have tried to indicate the fact that different phenomena may be involved in my division of the objects into two classes and there may be more. It looks to me that the sources that are bursters or alike, look better with an oscillation explanation and the others look perhaps better with the rotation or something else. Regarding your second point, I have looked also into the type of shock-height oscillations in column accretion, which Steve Langer is going to talk about. It is not clear whether that phenomenon will work at all for neutron stars, where cyclotron cooling is important. The periods one obtains from their results, using, as an exercise, neutron star parameters, are a little bit shorter than the observed ones.

LAMB: I was really mentioning another possibility, that the plasma-filled flux tube could oscillate by itself.

BATH: I just want to point out that A0538-66 is an incredibly interesting object, in itself, it is a transient X-ray source with a strictly periodic outbursting behaviour. So one could program one's observations for the first time with transient sources.

MASS LOSS ASSOCIATED WITH X-RAY BURSTS OF NEUTRON STARS*

D. Sugimoto, T. Ebisuzaki, T. Hanawa, and M. Kato
Department of Earth Science and Astronomy,
College of General Education, University of Tokyo

1. INTRODUCTION

Basic features of observed Type I X-ray bursts are thought to be interpreted in terms of helium shell-flash near the surface of the accreted neutron stars (Lewin and Joss 1981). Numerical models of the helium shell-flash show that the luminosity grows very close to the Eddingotn luminosity just after the peak of energy generation.

The Eddington luminosity is given by

$$L_{Ed} = 4\pi c G M_r/\kappa \quad . \tag{1}$$

Here M_r is the mass contained within the sphere of the radius $\underline{r}$, which is well approximated by the total mass of the neutron star $\underline{M}$. The opacity κ is described by the Compton scattering opacity

$$\kappa = 0.20/(1+\alpha T) ; \qquad \alpha = 2.2 \times 10^{-9} \ K^{-1} , \tag{2}$$

where the chemical compositions devoid of hydrogen is assumed, and $\underline{T}$ is the temperature. From equations (1) and (2) we see that the Eddington luminosity is a local quantity which is proportional to $(1+\alpha T)$; it decreases outward as the temperature decreases.

What happens when a radiation flux comes which exceeds the Eddington luminosity at the surface of the star $L_{Ed}^{(0)}$? It must push the outer shells to lead the mass loss from the star. It is the aim of the present paper to solve for the stellar structure with such steady mass outflow.

Many years ago Zytkow (1972) considered a similar problem and calculated stellar envelopes with mass outflow. However, her solutions could not be fitted consistently to any star or core of main-sequence- or

---

*Details of this work are submitted for publication by Ebisuzaki et al. (1982) and by Kato (1982).

M. Livio and G. Shaviv (eds.), Cataclysmic Variables and Related Objects, 293–298.

of white-dwarf-type. Here, we will show that such envelopes can be fitted to the accreted neutron stars in the stages near the peak of the helium shell-flash.

We shall solve for the structure of the inner shells with temperatures as high as $10^8 - 10^9$ K through the outermost shells of lower temperatures, the latter of which is outflowing as Parker-type stellar wind. Ruggles and Bath (1979) computed such Parker-type stellar wind solutions which simulated the mass outflow at nova explosion of accreted white dwarfs. They were interested in the flow patterns etc. of the stellar wind. Therefore, they assumed that a given value of mass flux came from the interior, and they solved only for the part of the Parker-type stellar wind. In this sense their computations did not give any insight into the driving mechanism of the mass loss. On the contrary, the main aim of our present study is to clarify the driving mechanism of the mass loss operating rather deep interior where αT of equation (2) is not negligible as compared with unity ($T \simeq 10^8 - 10^9$ K). Therefore, our outer boundary conditions of the Parker-type solution is much less important than the solutions in the deep interior.

2. BASIC EQUATIONS AND ASSUMPTIONS

Stellar structure with a steady mass outflow with velocity v is described by

$$v\frac{dr}{dr} + \frac{GM}{r^2} + \frac{1}{\rho}\frac{dP}{dr} = 0 \quad , \tag{3}$$

$$\frac{d}{dr}(4\pi r^2 \rho v) = 0 \quad , \tag{4}$$

$$\kappa L_r^{(dif)} = -\frac{16\pi a c r^2 T^3}{3\rho}\frac{dT}{dr} \quad , \tag{5}$$

$$vT\frac{ds}{dr} + \frac{\rho}{4\pi r^2}\frac{dL_r}{dr} = 0 \quad , \tag{6}$$

where P, ρ, s and $L_r^{(dif)}$ are, respectively, the pressure, the density, the specific entropy and the diffusive energy flux flowing through a shell at r. The energy is transported not only as the diffusive luminosity but also as the advection luminosity $L_r^{(adv)}$. Since we shall treat the shells exterior to the helium convective zone, equation of state is well described by those of the ideal gas plus radiation.

As the inner boundary conditions, the solution is fitted to the core of the neutron star at a shell where the ratio of the radiation pressure to the gas pressure y is equal to unity; the temperature should be equal to T_b, and the radius should be equal to the radius of the inert nuetron star. (We have found that the choice of the boundary value of y affects but very slightly to our results.) As the outer boundary conditions, the solution is fitted to the Parker-type stellar wind at the sonic point. Thus, our problem is a boundary value problem which contains a parameter T_b at our disposal. The mass flux $\dot{\Phi}$ is obtained as an eigenvalue. Then the velocity profile v , the diffusive flux

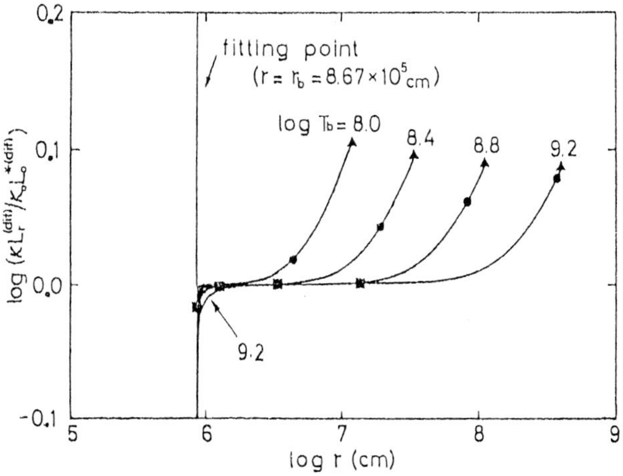

Figure 1. The ratio of the computed $\kappa L_r^{(dif)}$ to the assumed value of $\kappa_0 L_0^{*(dif)}$ is plotted against r for solutions with different T_b. Filled square, circle, and triangle denote the color photosphere, scattering photosphere (see below), and the sonic point, respectively.

$L_r^{(dif)}$ etc. are determined.

In general, equations (3)-(6) can be solved only numerically. However, we can solve them analytically if we approximate

$$\kappa L_r^{(dif)} = \kappa_0 L_0^{*(dif)} = \text{const.} \qquad (7)$$

In addition to this equation we obtain analytical expressions of three integrals which describe the constancy of the mass flux, the Bernoulli's theorem, and a barotropic relation. Differentiating temperature distribution of such analytical solution, we can compute $\kappa L_r^{(dif)}$ by means of equation (5). Then, we can check the consistency of our procedure, as is done in figure 1. We have found that equation (7) holds very precisely except for very outer region and for the region close to the inner boundary.

In order to check the validity of such approximation Kato (1982) has solved equations (3)-(6) numerically for similar problem. Her results show that $\kappa L_r^{(dif)}$ is constant in the region above the shell where the local Eddington luminosity becomes almost equal to the incoming diffusive luminosity. This constancy holds upto the scattering photosphere which is defined below. Here we refer to such Kato's (1982) results in order only to demonstrate the validity of our approximation. Now we shall return to discuss the results of our analytical treatment, since it will give a better insight into physical processes involved in our problem.

3. RESULTS

For the inert neutron star we used the mass of $0.476\ M_\odot$ and the radius of $r_b = 8.67$ km. Numerical results are summarized in table 1, where the subscript <u>cp</u> and <u>sp</u> denote the color- and the scattering-photosphere, respectively. The color photosphere is defined with

$$\int_{r_{cp}}^{\infty} (3\kappa_{ff}\kappa_E)^{1/2} \rho\, dr = 1 , \qquad (8)$$

Table 1. Models for steady-state mass loss from neutron star.

	model	1	2	3	4
	$\log T_b$ (K)	9.20	8.80	8.40	8.00
	$\log \Phi$ (g s^{-1})	18.76	18.35	17.94	17.51
	$\log r_s$(cm) (sonic point)	8.60	8.04	7.54	7.08
color pht-sph	$\log T_{cp}$ (K)	7.29	7.64	7.91	7.98
	$\log \tau_{cp}$	2.30	2.51	2.68	2.61
	$\log r_{cp}$ (cm)	7.14	6.54	6.11	5.94
	$\log R_{bb}$ (cm)	6.03	5.33	4.79	4.65
scatt. photosph.	$\log T_{sp}$ (K)	5.89	6.22	6.54	6.84
	$\log r_{sp}$ (cm)	8.54	7.88	7.26	6.67
	$\log L_{sp}^{*(dif)}$ (erg s^{-1})	38.08	38.08	38.08	38.08
	$\log L_{sp}^{(adv)}$ (erg s^{-1})	36.19	36.41	36.59	36.69
	$\log v_{sp}$ (cm s^{-1})	8.43	8.65	8.81	8.87

where κ_{ff} and κ_E are the free-free and the electron scattering opacity, respectively. The scattering photosphere is defined by $\tau(r_{sp}) = 3/2$ where τ is the optical depth computed with κ_E.

Model 2 with $\log T_b(K) = 8.80$ is a typical solution for the star just after the peak of the helium shell-flash. We see that the diffusive luminosity is very close to the Eddington luminosity at the surface [$\log L_{Ed}^{(0)}$ (erg sec^{-1}) = 38.08]. The mass flux amounts to 10^{18} erg sec^{-1}, but the velocity is rather low as compared with the light velocity. This is the reasons why the diffusive luminosity can not exceed the Eddington luminosity appreciably, and why the advection luminosity is relatively low at the scattering photosphere. We notice also that the surplus flux, i.e., $L_r^{(dif)}(T_b) - L_{Ed}^{(0)} = \alpha T_b L_{Ed}^{(0)}$, [$\simeq 1.4 L_{Ed}^{(0)}$ for model 2], is almost equal to the rate of potential energy fed into the escaping gas, i.e., $(GM/r_b)\Phi$. This high conversion efficiency does not yield to any appreciable super-Eddington luminosity.

The observed color temperature will be rather close to T_{cp} than T_{sp}. The diffusive luminosity is expressed as $(16\pi/3)\, r_{cp}^2\, T_{cp}^4/\tau_{cp}$. If we interpret the luminosity in terms of the usual black-body law with the effective temperature equal to T_{cp}, the black-body radius is calculated as $R_{bb} = (4/3\ \tau_{cp})^{1/2} r_{cp}$. Since we have $\tau_{cp} \simeq 300$, the black-body radius is regarded as only about $r_{cp}/15$, and we can not recognize the large radius of the extended envelope directly from observations. Such effect has to be paid attention to when we analyze observed reslults.

REFERENCES

Ebisuzaki, T., Hanawa, T., and Sugimoto, D. 1982, submitted for publication to Publ. Astron. Soc. Japan.

Kato, M. 1982, submitted for publication to Publ. Astron. Soc. Japan.

Lewin, W.H.G., and Joss, P.C. 1981, Space Sci. Rev., 28, 3.

Ruggles, C.L.N., and Bath, G.T. 1979, Astron. Astrophys., 80, 97.

Zytkow, A. 1972, Acta Astron., 22, 103.

DISCUSSION FOLLOWING D. SUGIMOTO'S TALK

BATH: I did some work with a student of mine, Ruggles, in which we computed steady outflowing optically thick wind solutions from white dwarfs. With no approximations at all, the general Parker type solutions, but with the interaction of the radiation with matter included. Have you compared your solutions with your approximations with those solutions?

SUGIMOTO: We have not compared with your solution. We have compared with an old calculation by Zytkow.

BATH: The Zytkow work is not right. I would be interested to know how your work compares with ours. Because, when I was doing that work I was struck by the fact that the solutions were extremely sensitive to the eigenvalue nature of the problem and that unless you have your boundary conditions satisfied at the critical point and at the photosphere simultaneously, precisely, then the solutions were very disturbed.

SUGIMOTO: Concerning this problem, one of my colleagues, computed the same problem, without assuming that κL is constant. The result is essentially the same as ours. Concerning your discussion of the sensitivity of the solution to boundary conditions, we have to think of the physical situation, for example, if we change the boundary condition at the sonic point, it will make changes in the whole structure of the solution, but physically speaking, the layer is pushed up by surplus radiation, the pressure of the surface region is very low, so if we state, physically, that surface boundary condition has little to do with determining the mass flux, in such a case I think the assumption of steady mass flow is bad instead of that the boundary condition is important.

BATH: That may be true, but steady solutions are the only ones we can solve in a realistic way at the moment.

LAMB: I have a comment and a question. The comment refers to low mass neutron stars. While it is true that the problem of supernovae has not yet been solved, in all the cases that have been calculated it is impossible to form neutron stars with masses less than 0.7-0.9 $M_\odot$. This is because when collapse occurs, part of the core falls in homologously and the reverse shock has to begin further out. In all calculations, the homologous core is in the range 0.7-0.9 $M_\odot$. As for my question, I am trying to understand why the approximation $\kappa L_{diffusion}$= Const. is roughly satisfied. I can imagine an electron scattering atmoswhere where κ = Const. and where the total luminosity might be constant in the envelope. But I would have thought that as soon as the advective part of the luminosity begins to dominate, the approximation must fail. Can you give us a physical picture of why the approximation is valid?

SUGIMOTO: The part of the advection energy and the energy which is necessary to push up the matter against the gravitational field, are all taken into account in the equation. People say that κL_r = const. is a polytrope of n=3, but the polytrope of n=3 is something like a singular point when the ratio of radiation pressure to gas pressure goes to infinity, so our solution is very different from the so-called κL = const, n=3. My understanding of the physical picture is that if an excessive energy comes, it cannot be transported by diffusion, the layer will be pushed up in order to keep the diffusive luminosity just equal or lower than the Eddington luminosity, it implies κL = const.

LAMB: The electron scattering optical depth from the photosphere is quite high in your solutions. Compton scattering is important if one is dealing with photon energies greater than mc^2/τ^2 . Even for $\tau \simeq 20$, Compton degradation is going to be significant for the temperatures you are dealing with at the color photosphere. Therefore, I would think that degradation of the spectrum would be rather severe in the outer parts of the outflowing envelope and would affect the results significantly.

SUGIMOTO: Yes, that is true, it is only an order of magnitude estimate we are concerned with now, not the detailed quantitative solution, but the physical interpretation.

LAMB: It would seem that in the picture you are talking about one might expect strong magnetic fields and therefore opacities that could be significantly less than the usual Thomson scattering opacity. This would make possible super-Eddington fluxes.

SUGIMOTO: Yes, I agree that we should proceed to think about such situations.

MEYER: I think it is claimed that the situation with respect to a black body interpretation gives you a cooling solution, where you keep the surface area constant and you decrease the temperature. Would your models also fit to these observations within the range of uncertainties?

SUGIMOTO: Our models are concerned with only the very peak of the X-ray burst so we need very detailed observations near the peak of the burst, the black body radiation at constant radius observation concerns the decaying phase of the X-ray burst.

X-RAY EMISSION FROM CATACLYSMIC VARIABLES

D. Q. Lamb
Harvard-Smithsonian Center for Astrophysics

ABSTRACT

We review X-ray observations of cataclysmic variables and theoretical work on X-ray emission from them, emphasizing recent developments.

1. INTRODUCTION

The first degenerate dwarf X-ray source was discovered in 1974 when Rappaport et al. (1974) detected an unexpected soft X-ray source during a brief rocket flight and deduced that the source was SS Cyg in outburst.

Two years later Berg and Duthie (1976) suggested that the cataclysmic variable AM Her was the optical counterpart of the hard X-ray source 4U1814+50. This identification was confirmed by Hearn et al. (1976), who detected AM Her in soft X rays. Soon after the optical identification, Szkody and Brownlee (1977) and Cowley and Crampton (1977) found that AM Her has a binary period of 3.1 hours. More remarkably, Tapia (1977a) discovered that the optical light from AM Her is nearly 10% circularly and linearly polarized and that the degenerate dwarf is strongly magnetic. The periods of the circular and linear polarization curves, the radial velocity curves, and the optical, soft and hard X-ray light curves are identical. Thus the rotation period of the accreting magnetic degenerate dwarf is phase-locked to the binary orbital period of 3.1 hours. Within less than a year, two other similar sources, AN UMa and VV Pup, were identified from their optical emission line spectrum and confirmed by detection of linear and circular polarization (Krzeminski and Serkowski 1977, Tapia 1977b).

Soft and hard X-rays were soon also detected from the well-known cataclysmic variables U Gem (Mason et al. 1978; Swank et al. 1978), EX Hya (Watson, Sherrington, and Jameson 1978; Cordova and Riegler 1979), and GK Per (King, Ricketts, and Warwick 1979), in addition to SS Cyg (Heise et al. 1978; Mason, Cordova, and Swank 1979).

Subsequent examination of nearby cataclysmic variables turned up many more X-ray sources. A few of these were discovered by Ariel 5 and HEAO-1, but most detections required the high sensitivity of the focusing instrument on Einstein. Other cataclysmic variables have been found by looking for the optical counterparts of faint galactic X-ray sources, and this method promises to become increasingly important in the future. More

M. Livio and G. Shaviv (eds.), Cataclysmic Variables and Related Objects, 299–322.

than 53 of these accreting degenerate dwarf X-ray sources are now known (Cordova and Mason 1982).

The study of degenerate dwarf X-ray sources can yield many returns. For example, these sources afford a laboratory in which to explore the physics of hot, dense plasmas, often in strong magnetic fields (the parameter regime is, in fact, similar to that of interest in plasma fusion reactors). We can also learn from them a great deal about the masses, internal structure, and magnetic fields of degenerate dwarfs. Potentially, the pulsing sources can provide as much information as has been obtained from the pulsing neutron star X-ray sources. Noise measurements can be used to probe the accretion process, reflection and reprocessing effects provide clues to the geometry of the disk and the binary system, and time delay curves yield the parameters of the binary and thereby lend insight into its formation and evolution.

However, because most degenerate dwarf X-ray sources were found only recently, we know very little about their X-ray properties. Only three (AM Her, SS Cyg, and U Gem) have been studied in any detail. The situation is similar in this respect to that of the stellar X-ray sources also found by Einstein (Linsky 1981). Exploration of the X-ray emission from both has only begun, and future X-ray astronomy missions must provide the data with which to understand it.

In this review, we concentrate on the soft and hard X-ray emission produced by accreting degenerate dwarfs. For reviews of the optical properties of cataclysmic variables, see Robinson (1976) and Warner (1976); for reviews of the X-ray observations, see Garmire (1979) and Cordova and Mason (1982). Lamb (1979a) and Kylafis et al. (1980) contain earlier reviews of theoretical work.

2. OBSERVATIONAL PROPERTIES

2.1. Luminosities and Space Densities

Probably all cataclysmic variables are X-ray sources. The ones detected so far have X-ray luminosities $L \sim 10^{31} - 10^{33}$ ergs s^{-1}. None of the bright ($L \sim 10^{36}$-10^{38} ergs s^{-1}) galactic X-ray sources have been identified with degenerate dwarfs. Thus the known accreting degenerate dwarf X-ray sources are $\sim 10^5$ times fainter than, e.g., the pulsing neutron stars (Lamb 1982) but $\sim 10^3$ times brighter than ordinary stars (Linsky 1981).

The nearest cataclysmic variable X-ray sources lie at distances d of only 75 - 100 pc (Cordova and Mason 1982). This implies a space density $n \sim 3 \times 10^{-7}$ (d/100 pc)$^{-3}$ pc^{-3}. Assuming a uniform distribution of sources throughout the galaxy and a galactic volume $V \sim 1 \times 10^{12}$ pc^3, the above space density implies that the total number of sources in the galaxy is $N \sim 3 \times 10^5$ (d/ 100 pc)$^{-3}$. Thus the total number of degenerate dwarf X-ray sources in the galaxy may exceed a million. This compares with a total number of bright ($L \sim 10^{36}$-10^{38} ergs s^{-1}) neutron star sources of ~ 100.

2.2. X-Ray Spectra and Temporal Behavior

Among accreting degenerate dwarf X-ray sources, there are two recognized classes involving magnetic degenerate dwarfs: the AM Her stars and the DQ Her stars (Lamb 1979a, Patterson and Price 1981). The remaining systems do not manifest a magnetic field. However, some of these sources will no doubt be reclassified as AM Her or DQ Her stars on the basis of future observations. Below we discuss the X-ray spectra and temporal behavior of the AM Her stars, the DQ Her stars, and the remaining cataclysmic variables in turn.

2.2.1. AM Her stars. Eleven AM Her stars are now known (EF Eri, E1114+182, VV Pup, E1405-451, E1013-447, H0139-68, PG1550+191, CW1103+254, AN UMa, AM Her, and E2003+225, in order of increasing period). These stars show strong (> 10%) circular and linear polarization of their infrared and visible light, and are believed to be accreting magnetic degenerate dwarfs (Chanmugam and Wagner 1977, 1978; Stockman et al. 1977). The polarization (Tapia 1977a) of the visible light from AM Her, the prototype of this class, is shown in Figure 1. The X-ray spectra of these stars typically have two distinct components: an apparent blackbody component with T_{bb} < 100 ev and a bremsstrahlung component with T_{br} > 10 kev. The measured soft X-ray flux is greater than the hard X-ray flux, often by a factor of 10 or more (cf. Tuohy et al. 1978, 1981; Szkody et al. 1981; Patterson et al. 1982). Figure 2 shows the soft and hard X-ray spectrum of AM Her recently constructed from HEAO-1 observations by Rothschild et al. (1981). The bremsstrahlung spectra of these sources also show strong iron line emission at ~ 7 keV, as is evident in Figure 2. In these systems, the periods of the polarized light, the optical and X-ray light, and the orbital velocity curves are all the same. Thus the rotation period of the degenerate dwarf is synchronized with the orbital period of the binary system, due to interaction of the magnetic field of the degenerate dwarf with the companion star (Joss, Katz, and Rappaport 1979; Lamb et al. 1983). Figure 3 shows the resulting 3.1 hour "pulse profile" of AM Her in soft X-rays (Tuohy et al. 1978).

2.2.2. DQ Her stars. We have classified nine systems as DQ Her stars (AE Aqr, V533 Her, DQ Her, V1223 Sqr, H2252-035, 3A0729+103, H2215-086, EX Hya, and TV Col in order of increasing period). DQ Her, the prototype of this class, has long been believed to be an accreting magnetic degenerate dwarf (Bath, Evans, and Pringle 1974; Lamb 1974). However, it shows little, if any, polarization of its infrared and visible light (Swedlund, Kemp, and Wolstencroft 1974). This system underwent a nova outburst in 1934 and shows coherent small amplitude optical pulsations at 71 seconds, which are believed to represent the rotation period of the degenerate dwarf (Patterson, Robinson, and Nather 1978, and references therein). Two other members of this class are V533 Her, which underwent a nova outburst in 1963 and shows coherent small amplitude optical pulsations at 63 seconds (Patterson 1979a), and AE Aqr, which shows similar pulsations at 33 seconds (Patterson 1979b). Embarrassingly, neither DQ Her nor V533 Her have been detected in X rays. In the case of DQ Her, it has been suggested that the X rays are blocked by the disk because we are nearly in the orbital plane of the system, while in the case of V533 it can be argued that the system is too far away, and therefore too faint, to have been detected. Thankfully (for the theorists), pulsed X rays have been detected from a third member of the class, AE Aqr, at the 33 second optical period (Patterson et al. 1980).

Several faint galactic X-ray sources have now been identified with systems that are optically similar to cataclysmic variables. They exhibit large amplitude optical and X-ray pulsations with periods $\gtrsim$ 1000 seconds that are believed to represent the rotation period of the accreting magnetic star The source H2252-035 was the first of these systems to be optically identified (Griffiths et al. 1980) and have its character recognized (Patterson and Price 1981; Warner, O'Donoghue, and Fairall 1981; White and Marshall 1981, Hassall et al. 1981). Figure 4 shows its optical light curve (Patterson and Price 1981). Clearly visible are the optical pulsations with a period of 859 seconds, which are thought to be produced by reprossessing of the 805 second X-ray pulse. Figures 5 and 6 show the pulse profile and the spectrum of the hard X-rays (White and Marshall 1981). The hard X-ray spectrum

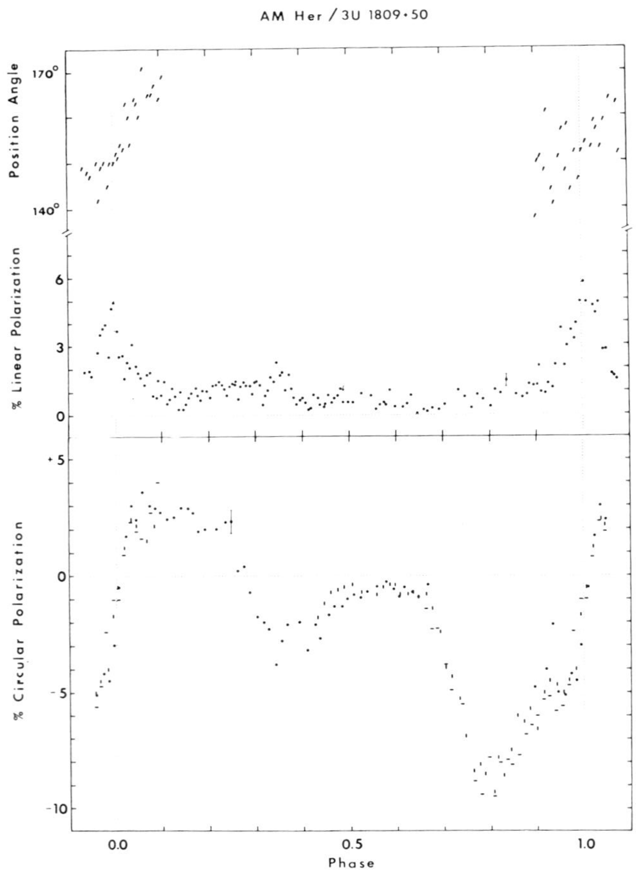

Fig. 1--Circular and linear polarization of the optical light from AM Her as a function of the phase of the 3.1 hour rotational period of the degenerate dwarf (from Tapia 1977a).

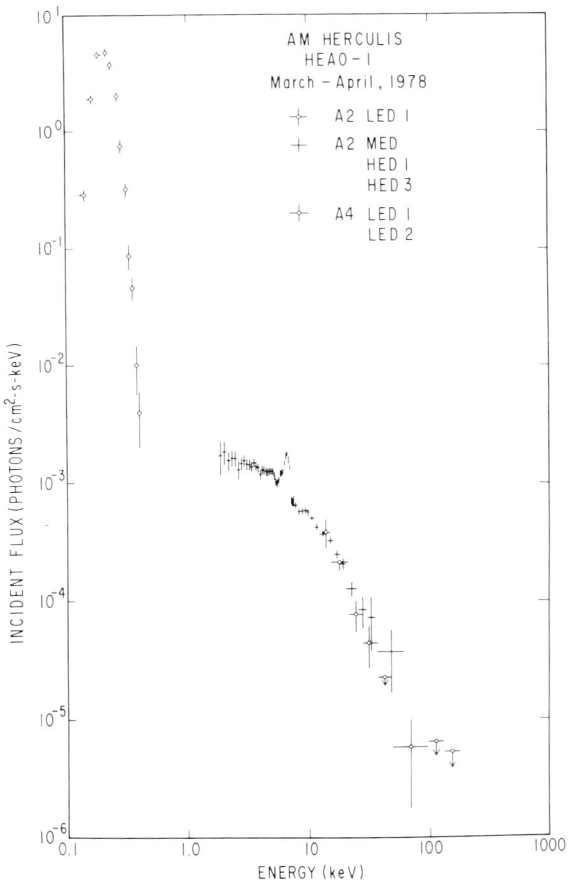

Fig. 2--Hard and soft X-ray spectrum of AM Her (from Rothschild et al. 1981). The two distinct components with T ≃ 30 keV and T < 40 eV are clearly visible. Note also the iron emission line at ≃ 7 keV.

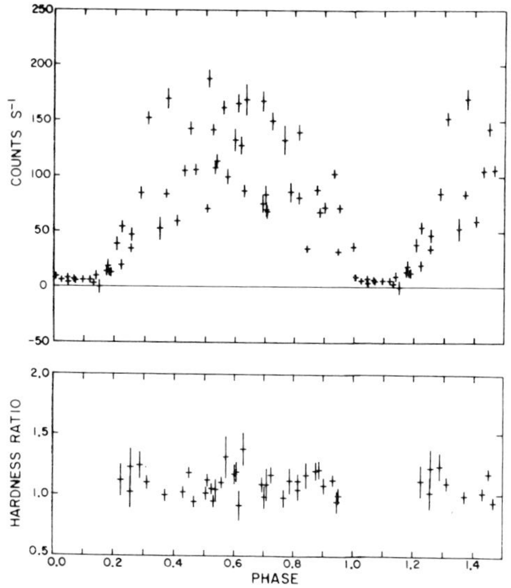

Figure 1

Fig. 3--Soft X-ray pulse profile and hardness ratio of AM Her as a function of the phase of the 3.1 hour rotational period of the degenerate dwarf (from Tuohy et al. 1978).

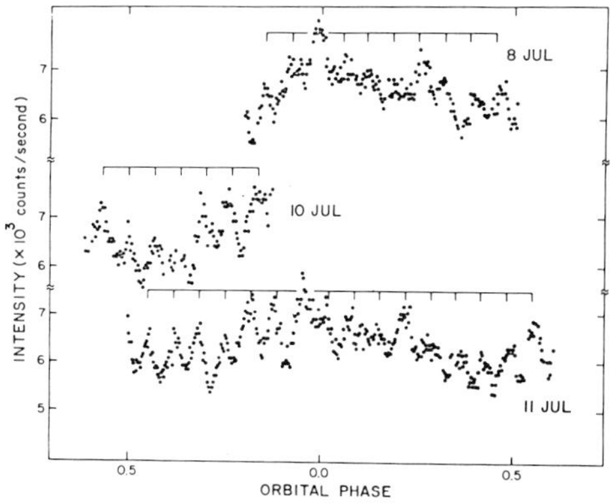

Fig. 4--Fast photometry of the optical light from H2252-035 (from Patterson and Price 1981). The 859 second pulsations, corresponding to reprocessed light from a point stationary in the rotating frame of the binary system, are clearly visible.

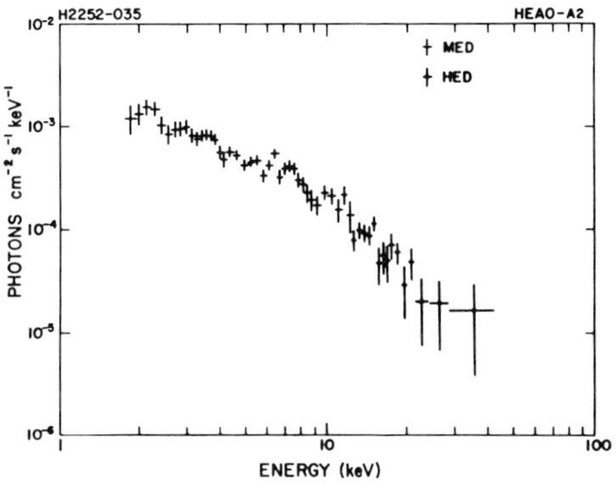

Fig. 5--Hard X-ray spectrum of H2252-035 (from White and Marshall 1981). The iron emission line at $\simeq$ 7 keV is again clearly visible.

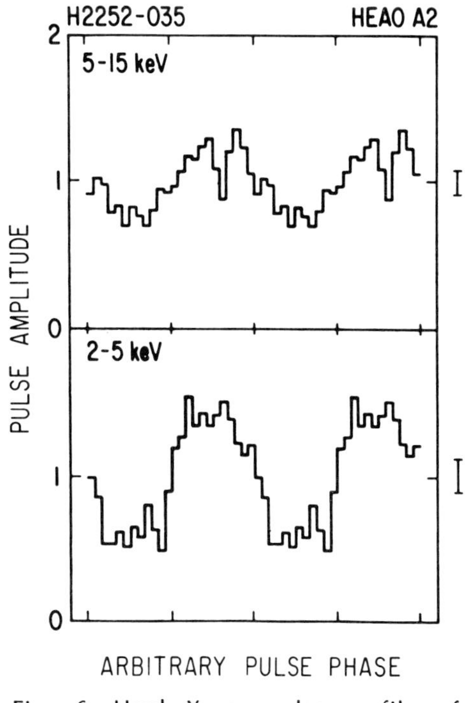

Fig. 6--Hard X-ray pulse profile of H2252-035 through the 805 second rotation period of the degenerate dwarf (from White and Marshall 1981).

exhibits iron line emission at ~ 7 keV. Another star, now known as TV Col, had already been optically identified (Charles et al. 1979). It was recognized as possibly a similar system after photometry (Motch 1981) and spectroscopy (Hutchings et al. 1981) showed that its orbital period of 5^h11^m differed from the principal photometric period of 5^h29^m. Interestingly, a 67 minute period has also recently been found in the optical light curve of the well-studied cataclysmic variable X-ray source EX Hya (Vogt, Krzeminski, and Sterken 1980; Gilliland 1982). The period is present in soft X-rays but not in hard (Swank and White 1981). The coherence of the period over many years suggests that it is also due to rotation of a magnetic star.

2.2.3. Other cataclysmic variables. Forty-four other cataclysmic variable X-ray sources are currently known (Cordova and Mason 1982). Among these are the prototypical dwarf novae, SS Cyg and U Gem, which undergo outbursts every ~ 100 days. During quiescence, both exhibit a hard X-ray spectrum with T_{br} ~ 10 - 20 keV (Mason, Cordova, and Swank 1979; Swank 1979). During outburst, the hard X-ray luminosity first increases and then decreases, the spectral temperature of the hard X-rays decreases, and an intense blackbody component with temperature $T_{bb} < 100$ eV appears in soft X-rays (cf. Mason, Cordova, and Swank 1979). The lack of a pronounced low or high energy cutoff in the spectrum indicates that this behavior can not be due to a larger absorption or scattering optical depth during the outburst (Kylafis and Lamb 1982a), as had been earlier proposed (Ricketts, King, and Raine 1979). The origin of this behavior is therefore not yet understood.

Most of the remaining cataclysmic variables show only a hard X-ray component. It is not known whether the failure to detect a soft component during quiescence, or even during outburst in some sources, is due to its absence or due to the fact that it may have so low a spectral temperature that it is unobservable in soft X-rays. Essentially all of them exhibit small amplitude quasi-periodic or coherent optical pulsations, usually during the onset of an outburst (Robinson 1976). Of special interest are the ~ 8 - 10 second quasi-periodic pulsations in SS Cyg. They are strongly present in soft X-rays during outburst, yet their coherence persists for only 3-5 pulse periods (Cordova et al. 1980, 1981).

2.3. X-Ray Emitting Cataclysmic Variables in Globular Clusters

Very recently, Hertz and Grindlay (1983) have detected faint X-ray sources in a number of globular clusters, as shown in Figure 7. In fact, several of the clusters contain multiple sources. They argue convincingly that these detections represent the "tip of the iceberg" of a large population (~10^4 per cluster) of cataclysmic variable X-ray sources. Such a population offers exciting new prospects for studying the formation and evolution of cataclysmic variables as well as the properties of X-ray emission from degenerate dwarfs.

3. THEORY

3.1. Qualitative Picture of X-Ray Emission

The basic idea of X-ray emission from accreting degenerate dwarfs is that matter falling into the deep gravitational potential well of the star supplies a large amount of energy. The resulting rate of energy production, or luminosity, is

$$L_{acc} = \frac{GM\dot{M}}{R} = 8 \times 10^{32}\ (M/M_\odot)\ (R/10^9\ \mathrm{cm})^{-1}\ (\dot{M}/10^{-10}\ M_\odot\ \mathrm{yr}^{-1})\ \mathrm{ergs\ s}^{-1}, \quad (1)$$

where M and R are the mass and radius of the star, and $\dot{M}$ is the mass accretion rate. The maximum possible shock temperature of the matter, and hence of the resulting radiation, is

$$T_s = \frac{3}{8}\,T_{ff} = 2 \times 10^8\ (M/M_\odot)\ (R/10^9\ \mathrm{cm})^{-1}\ \mathrm{K}, \quad (2)$$

where T_{ff} is the freefall temperature. On the other hand, if the X-ray emission region radiates as a blackbody, the temperature of the resulting radiation is

$$T_{bb} = (L/4\pi R^2\sigma)^{1/4} = 3 \times 10^4\ (M/M_\odot)^{1/4}\ (R/10^9\ \mathrm{cm})^{-3/4}\ (\dot{M}/10^{-10}\ M_\odot\ \mathrm{yr}^{-1})^{1/4}\ \mathrm{K}. \quad (3)$$

Therefore, whether an accreting degenerate dwarf is an X-ray source and, if it is, whether it emits primarily soft or hard X rays depends critically on the manner in which the kinetic energy of infall is converted into radiation.

3.1.1. Disk accretion. Many cataclysmic variables show clear optical and UV evidence of accretion disks. If the disk extends down to the stellar surface, viscous dissipation in it releases approximately half of the available gravitational energy. This energy appears as blackbody radiation from the disk surfaces and produces a characteristic power law spectrum $\propto\nu^{1/3}$ in the optical and UV (cf. Shakura and Sunyaev 1973). The other half of the available gravitational energy is released in a boundary layer at the inner edge of the disk where it encounters the surface of the star, unless the star is rotating near breakup. This luminosity is $L_{bdry} \sim L_{acc}/2$. If the boundary layer is optically thick, the temperature of the emitted radiation is given by eq.(3), but with an appropriately smaller area. At moderate or high accretion rates, such a boundary layer is capable of producing soft X-rays by blackbody emission (Pringle 1977).

More recently, Pringle and Savonije (1979) proposed that the boundary layer might be optically thin and produce hard X-ray emission by bremsstrahlung if shocks occurred in it, thereby accounting for the hard X-rays observed from many cataclysmic variables. Since the maximum possible shock temperature is that given by eq.(2), the shocks must be strong. However, it is not obvious, as in the case of radial accretion (see below), that shocks play a dominant role in the dissipation of kinetic energy. The disk can, in principle, join onto the star without any shocks occurring. Supersonic flow is characteristic of the disk as a whole, yet in spite of that the transport of angular momentum and dissipation of energy in the disk are generally assumed to be due to subsonic turbulence rather than shocks. Of course the rate of shearing in the boundary layer is much stronger than that in the disk. However, even if shocks were to form, the geometry of the flow would tend to favor production of a large number of weak, oblique shocks. To achieve the required strong shocks, Pringle and Savonije (1979) postulate a two-stage process in which gas that is initially mildly shocked in the boundary layer expands into the path of, and collides with, gas still circulating in the inner disk.

Tylenda (1981b), however, argues that turbulent viscosity will be a more efficient mechanism than shocks for dissipating energy in the boundary layer and suggests that this can account for the observed high temperatures without resorting to complicated flows. Unfortunately, the temperature of the boundary layer in this picture is extremely sensitive to the magnitude of the assumed turbulent viscosity, about which we know little. The temperature might approach the maximum value given by eq.(3) but could be much smaller. The situation is further clouded by the fact that, even if the Reynolds number is large, it is far from clear that the flow is turbulently unstable.

Knowledge of whether the boundary layer can produce hard X-rays and, if so, how, is important for understanding cataclymsic variable X-ray sources. But the ideas proposed have not as yet been worked out in any detail. For excellent recent discussions of disks, see the review by Pringle (1981) and the paper by Tylenda (1981).

3.1.2. Radial accretion. If the degenerate dwarf has a magnetic field,

$$B \gtrsim 2 \times 10^3 (10^{-10}\ M_\odot\ yr^{-1})^{1/2}\ (R/10^9\ cm)^{-5/4}\ (M/M_\odot)^{1/4}\ \text{gauss}, \tag{4}$$

the field will disrupt the disk and lead to approximately radial inflow near the star. This picture certainly applies to the AM Her and DQ Her stars, and may apply to other cataclysmic variables if magnetic fields are present in them. Radial inflow may also occur if mass transfer takes place via a stellar wind rather than via Roche lobe overflow. Most theoretical work has assumed radial inflow because it is far more tractable; in the remainder of this review, we will concentrate on such inflow.

As accreting matter flows radially toward a star, a strong standoff shock forms far enough above the star for the hot, post-shock matter to cool and come to rest at the stellar surface (Hoshi 1973; Aizu 1973; Fabian, Pringle, and Rees 1976). The total luminosity is given by eq.(1). The standoff distance is

$$d \equiv r_s - R \simeq 1/4\ v_{ff}(r_s)\ t_{cool}(r_s), \tag{5}$$

where r_s is the shock radius, R is the stellar radius, v_{ff} is the free-fall velocity, and t_{cool} is the time scale for cooling, due to bremsstrahlung and, if a magnetic field is present, cyclotron emission. The temperature of the bremsstrahlung radiation is approximately that given by eq.(2). Roughly half of it is emitted outward and forms a hard X-ray component. Roughly half of the cyclotron flux is also emitted outward and forms a blackbody-limited component in the UV. The other halves of the bremsstrahlung and cyclotron fluxes are emitted inward and are reflected or absorbed by the stellar surface. The luminosity of the resulting blackbody radiation is $L_{bb} \simeq L_{cyc} + L_{br}$, where L_{cyc}, and L_{br} are the luminosities in the cyclotron and bremsstrahlung components. The temperature of the blackbody radiation is approximately that given by eq.(3).

If we allow for the possible presence of a magnetic field, the accreting matter may be channeled onto the magnetic poles and accretion may occur over only a fraction f of the stellar surface. The effective accretion rate of the accreting sector is $\dot{M}/f$, and the corresponding luminosity is L/f. X and UV radiation from magnetic degenerate dwarfs is thus a function of stellar mass M, magnetic field strength B, and effective luminosity L/f. The dependence on stellar mass is significant but is less than on the other two variables. If we specify the mass of the star, the parameter regimes encountered are conveniently displayed on a (B,L/f)-plane, as shown in Figure 8. The upper left of the plane corresponds to low magnetic field strengths and high effective luminosities (and thus high densities in the emission region). In this portion of the plane, bremsstrahlung cooling dominates cyclotron cooling in the hot, post-shock emission region, and the character of the X-ray emission is essentially the same as that of a nonmagnetic degenerate dwarf. As one increases B or lowers L/f, moving toward the lower right in Figure 8, cyclotron cooling becomes more important until eventually it dominates (Masters et al. 1977). The solid line

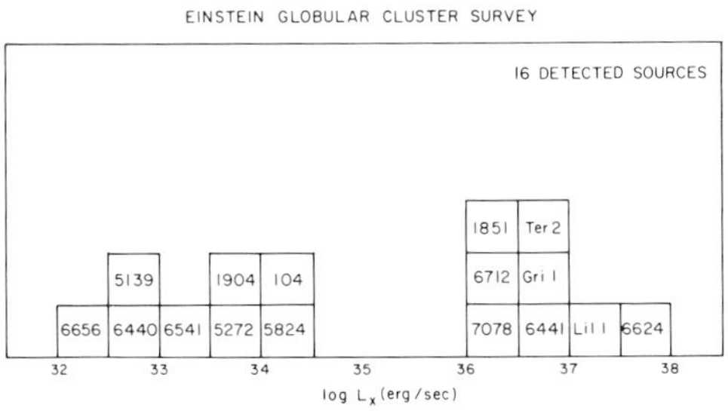

Fig. 7--Luminosity histogram of the 16 X-ray emitting globular clusters detected with Einstein (from Hertz and Grindlay 1983). Each source is identified by the NGC number or common name of the cluster. The clusters NGC 5139 and 6656 contain multiple sources, but only the source located within the core of each has been plotted for purposes of clarity.

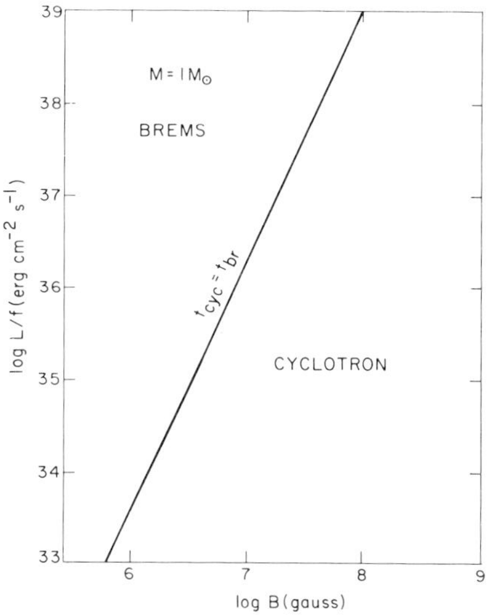

Fig. 8--Bremsstrahlung and cyclotron emission regimes in the (L/f,B)-plane for a 1 $M_\odot$ star (after Lamb and Masters 1979).

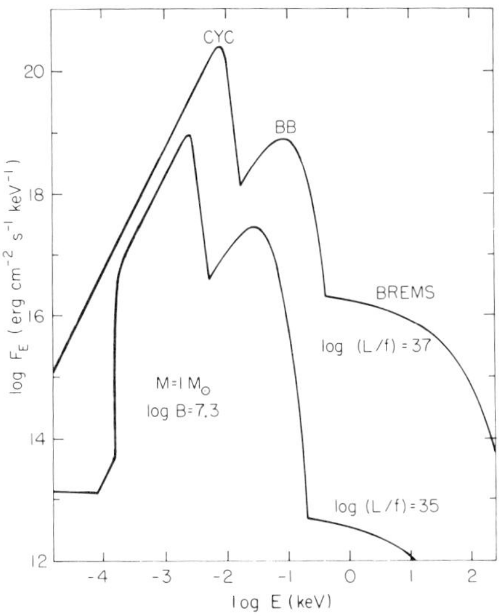

Fig. 9--X-ray and UV spectra produced by accretion at two different accretion rates onto a 1 $M_\odot$ with a magnetic field of 2 x 10^7 gauss. The spectrum with L/f = 10^{37} ergs s^{-1} is in the bremsstrahlung dominated regime, while the spectrum with L/f = 10^{35} ergs s^{-1} is in the cyclotron dominated regime (from Lamb and Masters 1979).

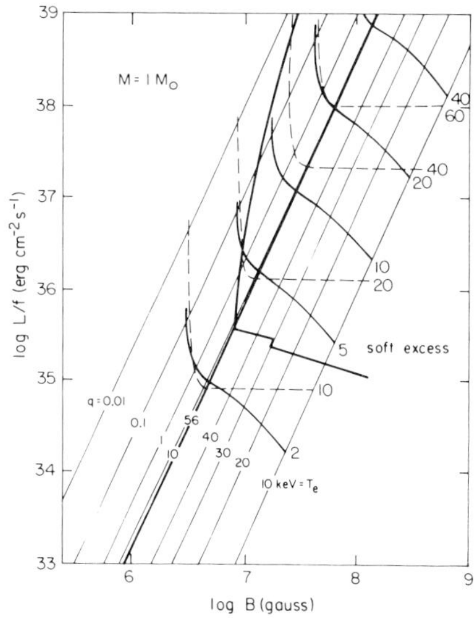

Fig. 10--Contours in the (L/f,B)-plane for a 1 $M_\odot$ star. For an explanation of the various lines, see the text.

shows the location at which this occurs, as determined from detailed numerical calculations equating t_{cyc} and t_{br}, the cyclotron and bremsstrahlung cooling time scales. This line is approximately given by

$$B = 6 \times 10^6 \, (Lf^{-1}/10^{36} \text{ ergs s}^{-1})^{2/5} \text{ gauss.} \tag{6}$$

To the lower right of this solid line, the magnetic field qualitatively alters the character of the X-ray emission.

3.2. Magnetic Stars

Fabian, Pringle, and Rees (1976), Masters et al. (1977), and King and Lasota (1979) have discussed the qualitative features of X-ray emission by magnetic degenerate dwarfs. Lamb and Masters (1979; see also Masters 1978) carried out detailed numerical calculations of high harmonic cyclotron emission from a hot plasma, and from them developed a self-consistent, quantitative model of the X-ray and UV emission. Wada et al. (1981) carried out a few calculations for the regime in which bremsstrahlung, not cyclotron emission, dominates (see Figure 7).

3.2.1. Spectra. The X and UV spectrum produced by accretion onto magnetic degenerate dwarfs generally has four components: 1) a blackbody-limited UV cyclotron component produced by the hot emission region, 2) a hard X-ray bremsstrahlung component also produced by the hot emission region, 3) a hard UV or soft X-ray blackbody component produced by cyclotron and bremsstrahlung photons that are absorbed by the stellar surface and re-emitted, and 4) secondary radiation from infalling matter above the shock or, possibly, from the stellar surface around the emission region. The first three components are clearly visible in Figure 9, which shows spectra produced by the hot, post-shock emission region alone. Since the secondary radiation is not included, the spectra do not accurately represent the observed spectrum below ~5 eV.

The spectra in Figure 9 illustrate two important features. First, strongly magnetic degenerate dwarfs should be intense UV sources with only a few percent of the total accretion luminosity ordinarily appearing as optical or soft and hard X-rays, and therefore easily accessible. Second, the position and relative strength of the spectral components depend sensitively on the accretion rate. For example, the variation in accretion rate shown in Figure 9 moves the blackbody component from the UV into the soft X-ray region. Further, the luminosity of the bremsstrahlung hard X-ray component increases by nearly 4 orders of magnitude, although the total accretion luminosity increases by only 2.

3.2.2. Correlation between spectral temperature and luminosity. Variations in the shape and the strength of the spectral components are a function of both mass accretion rate and magnetic field strength. They can be conveniently displayed by plotting contours on a (B,L/f)-plane, as shown in Figure 10. Bremsstrahlung and cyclotron emission dominate in the same regions as in Figure 8. The thin solid lines in the bremsstrahlung-dominated region show contours of constant $q \equiv L_{cyc}/L_{br}$, while those in the cyclotron-dominated region show contours of constant T_e, the temperature of the bremsstrahlung hard X-ray component. Contours of constant E^*, the peak of the blackbody-limited cyclotron component, are shown as thick solid lines and contours of constant T_{bb}, the temperature of the blackbody component, are shown as dashed lines. To the upper right of the curve labelled "soft excess", the blackbody luminosity in soft X-rays exceeds the bremsstrahlung luminosity in hard X-rays. Figure 10 illustrates that the qualitative features of the observed X and UV spectrum determine fairly accurately the physical conditions in the emission region, including the value of the magnetic field.

Near and above $L/f = L_E = 1.4 \times 10^{38}$ erg s^{-1}, radiation pressure can be important and modify the results, but because photons can easily scatter out of the accretion column if f << 1, the Eddington luminosity does not represent the stringent upper limit to the luminosity that it does in the case of nonmagnetic degenerate dwarfs. If the geometry of the hot, post-shock emission region is such that most of the flux escapes through the face rather than through the edges of the emission region (i.e., $d << \sqrt{(2f)}R$), then Compton degradation of the bremsstrahlung hard X-ray component will occur if L/f exceeds $\sim 10^{37}$ erg s^{-1}, as in nonmagnetic stars (see below). Scattering and degradation affect not only the spectrum but also the pulse profile, and Imamura and Durisen (1983) have carried out a study of the resulting behavior of both.

3.3. Nonmagnetic Stars

Studies of X-ray emission from accreting nonmagnetic degenerate dwarfs include those by Hoshi (1973), Aizu (1973), Hayakawa (1973), DeGregoria (1974), Hayakawa and Hoshi (1976), Fabian, Pringle, and Rees (1976), Katz (1977), and Kylafis and Lamb (1979, 1982a,b). These calculations are applicable, even if a magnetic field is present, as long as the accretion flow is approximately radial and bremsstrahlung cooling dominates cyclotron cooling in the X-ray emission region (recall Figure 8). Thus they are relevant to the AM Her stars, such as AM Her itself, which has a magnetic field $B \sim 2 \times 10^7$ gauss (Lamb and Masters 1979; Schmidt, Stockman, and Margon 1981; Latham, Liebert, and Steiner 1981), and VV Pup, which has a magnetic field $B \sim 3 \times 10^7$ gauss (Visvanathan and Wickramasinghe 1979; Stockman, Liebert, and Bond 1979), as well as to the DQ Her stars.

3.3.1. Spectra. The X and UV spectrum produced by accretion onto nonmagnetic degenerate dwarfs generally has three components: 1) a hard X-ray bremsstrahlung component produced by the hot, post-shock emission region, 2) a soft X-ray blackbody component produced by bremsstrahlung photons that are absorbed by the stellar surface and re-emitted, and 3) secondary radiation produced by Compton heating of infalling matter above the shock.

These components are clearly visible in Figure 11, which shows six spectra that span the entire range of accretion rates. Figure 12 shows for comparison three similar spectra when nuclear burning occurs at the accretion rate (see below). At low accretion rates, $\tau_{es} < 1$ and the observed hard X-ray spectrum is essentially the same as that produced in the emission region. As the accretion rate is increased, τ_{es} exceeds unity and Compton scattering begins to degrade the spectrum (Illarionov and Sunyaev 1972). The blackbody component then contains a contribution from bremsstrahlung photons which are backscattered by the accreting matter and absorbed by the stellar surface. The secondary radiation, which arises from accreting matter heated by the Compton scattering of the bremsstrahlung photons, is important only when degradation of the bremsstrahlung is substantial. As the accretion rate is increased further, this degradation becomes more severe. Finally, due to the combined effects of degradation and weakening of the shock by radiation pressure, the bremsstrahlung component disappears altogether. The star then ceases to be a hard (i.e., $T_{obs} > 2$ keV) X-ray source.

Figure 11 illustrates two important features of X-ray emission from nonmagnetic degenerate dwarfs. First, an intense blackbody soft X-ray component is always present. Second, at high accretion rates Compton degradation leads to low spectral temperatures even for high mass stars.

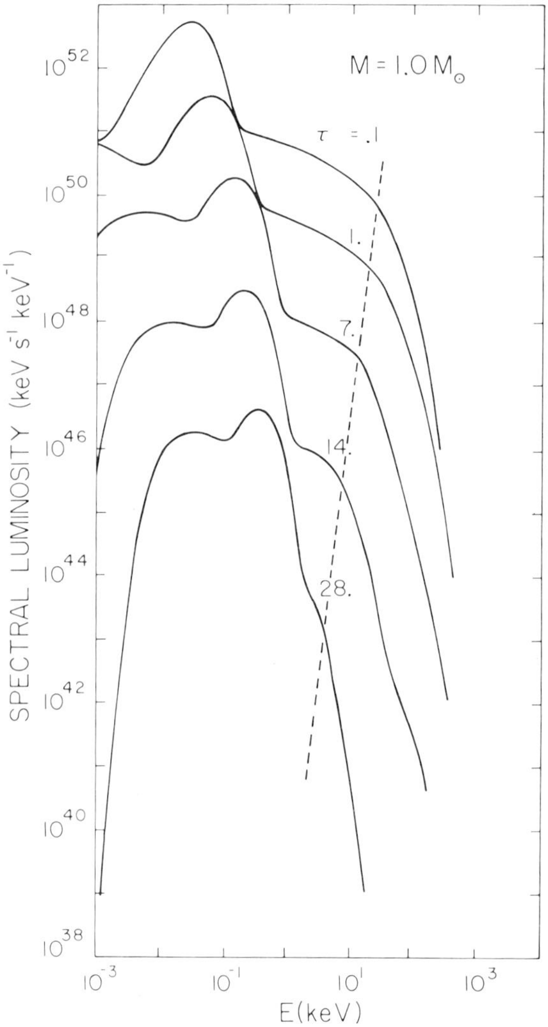

Fig. 11--X and UV spectra produced by accretion onto a 1 $M_\odot$ star for six different accretion rates (from Kylafis and Lamb 1982a). The curves are labeled by the value of the electron optical depth through the accreting matter. The dashed line shows the changing cutoff due to Compton degradation.

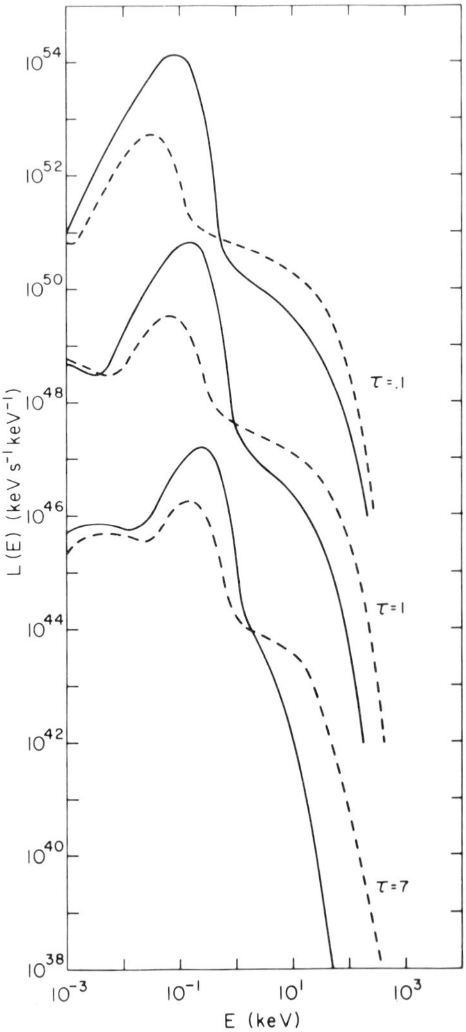

Fig. 12--Comparison of X and UV spectra produced by accretion onto a 1 $M_\odot$ star with nuclear burning at the accretion rate (solid curves) and without nuclear burning (dashed curves) (from Weast et al. 1982). The curves are again labeled by the value of the electron optical depth through the accreting matter.

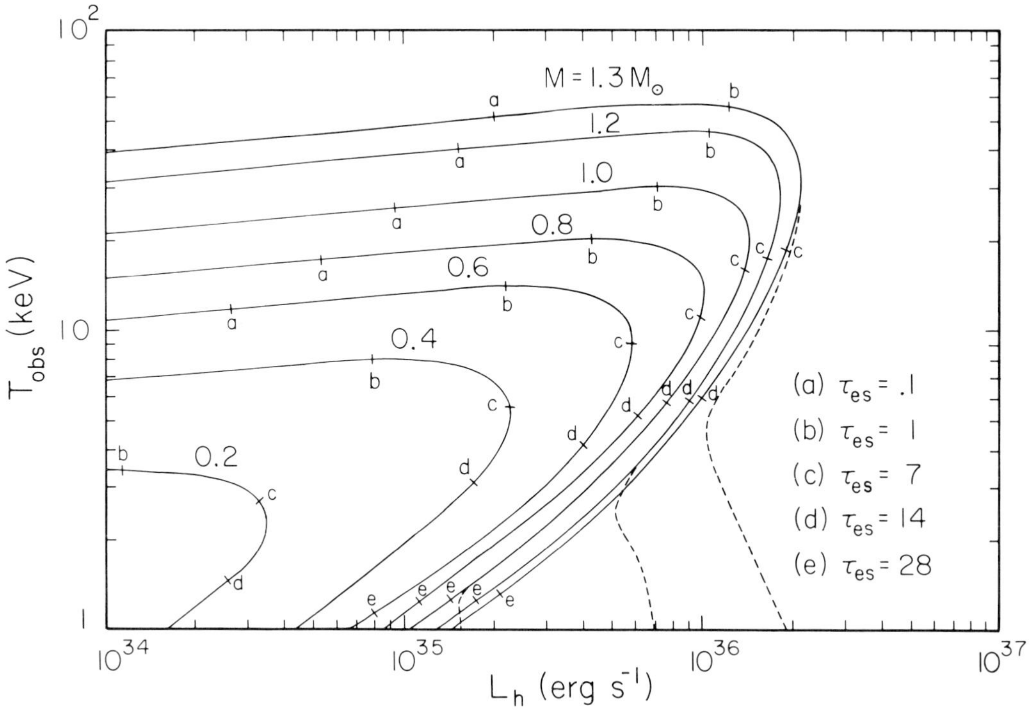

Fig. 13--Correlation between T_{obs} and L_h for stars with masses 0.2 - 1.2 $M_\odot$ (from Kylafis and Lamb 1982a). The dashed lines give the same correlation when the contribution of the blackbody component is included in L_h.

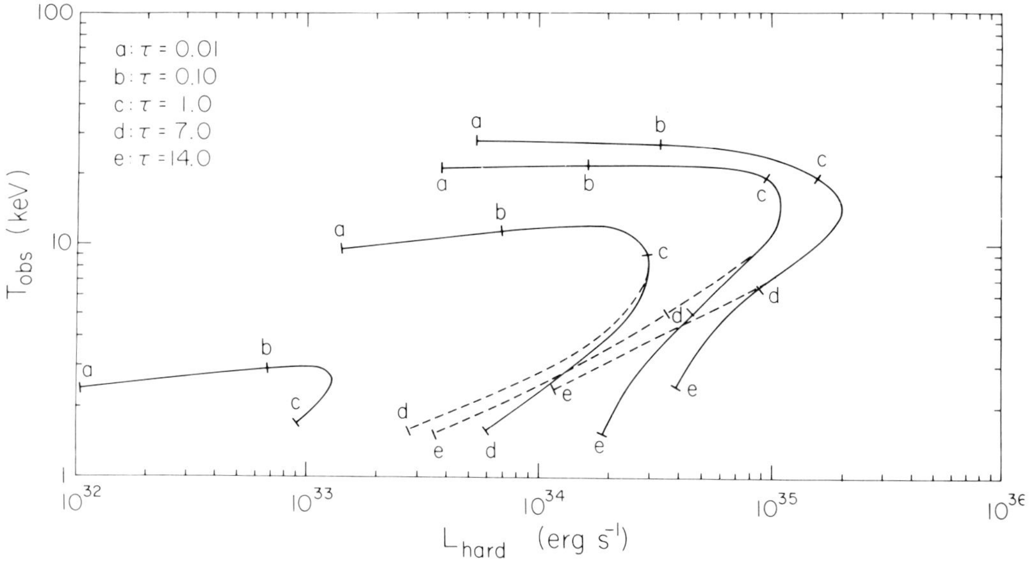

Fig. 14--Correlation between T_{obs} and L_h for stars with masses 0.2, 0.6, 1.0 and 1.2 $M_\odot$ and nuclear burning at the accretion rate (from Weast et al. 1982). The dashed lines show the correlation when the contribution of the blackbody component is omitted.

3.3.2. Correlation between spectral temperature and luminosity. Figure 13 shows the resulting dramatic variation in the spectral temperature at high accretion rates and the pronounced correlation between X-ray spectral temperature T_{obs} and luminosity L_h. Figure 14 shows for comparison this correlation when nuclear burning occurs at the accretion rate (see below). Note that the accretion rate increases as one moves from upper left to lower right along the curves. For sources found in the lower right of the figure, an increase in T_{obs} and L_h therefore corresponds to a decrease in the accretion rate: T_{obs} and L_h increase since the smaller accretion rate lessens Compton degradation of the hard X-ray spectrum.

3.4. Ionization Structure and Line Features

3.4.1. Ionization structure. The circumstellar ionization structure of degenerate dwarf X-ray sources has been calculated analytically by Hayakawa (1973) and more recently by Kylafis and Lamb (1982b). These calculations assume spherical symmetry, and that the optical depth to absorption is small. The calculations by Kylafis and Lamb (1982b), which take self-consistent values for the hard and soft X-ray luminosities and spectral shapes based on their earlier detailed calculations, show that the blackbody soft X-ray flux ionizes H and He out to distances large compared with a typical binary separation. For high mass stars and low accretion rates, the bremsstrahlung hard X-ray flux ionizes heavy elements out to considerable distances.

3.4.2. Emission lines. The temperatures in the X-ray emission regions of degenerate dwarf X-ray sources are high enough (> 10 keV) to produce thermal emission lines, including those of Fe at ~ 7 keV, with significant equivalent widths. Emission lines can also be produced by fluorescence in the accreting matter above the X-ray emission region. Fluorescent emission lines may also be produced by X-rays striking the stellar surface surrounding the emission region, the disk, and even the companion star. The emission lines may be broadened by thermal Doppler broadening, Compton scattering, and Doppler broadening due to bulk streaming velocities.

3.4.3. Absorption lines. According to the analytical calculations of Kylafis and Lamb (1982b), the absorption optical depth at the ionization edges of heavy elements increases rapidly when the accretion rate exceeds about 3×10^{-3} $\dot{M}_E$. These calculations also show that Compton scattering and the resulting degradation of the hard X-ray spectrum occur close to the star, while most of the absorption occurs relatively far from the star. Thus the amount of degradation is less sensitive, while the amount of absorption is more sensitive, to the distribution of accreting matter around the star.

3.4.4. Spectra. Ross and Fabian (1980) have carried out detailed numerical calculations of the emergent spectrum from a 1.0 $M_\odot$ star for three different accretion rates. These calculations treat the atomic physics carefully and are valid even for large absorption optical depths. Their results for τ_{es} = 6 and 10 are shown in Figures 15 and 16. Note both the absorption K-edges due to O VIII (0.87 keV), Si XIV (2.7 keV), and Fe XXI-XXVI (8.2 - 9.3 keV), and the emission lines, broadened by Compton scattering, due to the Kα lines of O VIII (0.65 keV), Si XIV (2.0 keV), and Fe XXV (6.7 keV).

3.5. Nuclear Burning

3.5.1. Nonmagnetic stars. The effects of nuclear burning on X-ray emission by nonmagnetic degenerate dwarfs were first discussed by Katz (1977) and have been

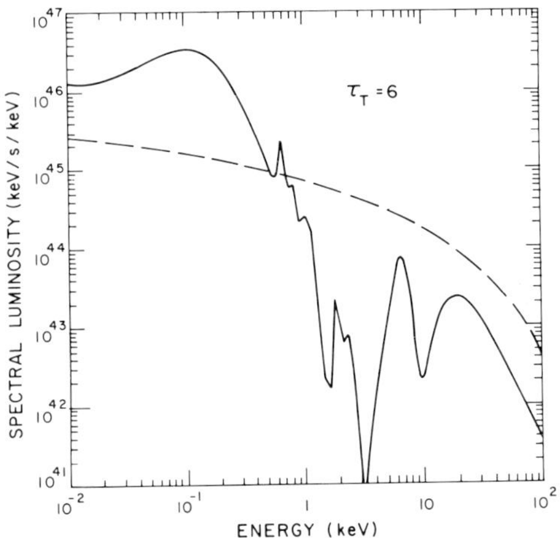

Fig. 15--X and UV spectrum produced by a 1 $M_{\odot}$ star at an accretion rate 0.090 $\dot{M}_E$ (τ = 6) taking into account the effects of photoabsorption (from Ross and Fabian 1980).

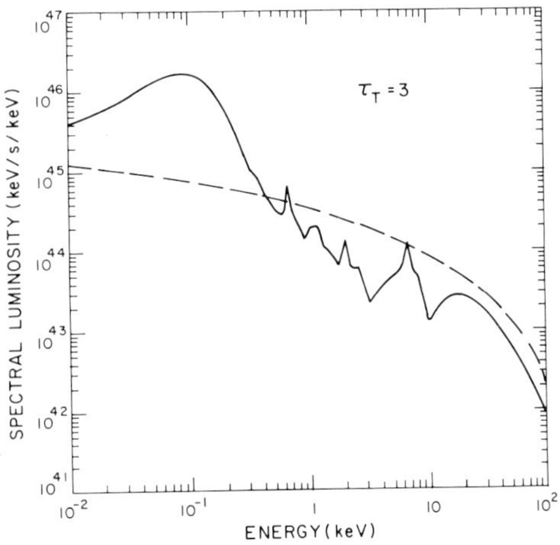

Fig. 16--Same as Figure 15 for an accretion rate 0.15 $\dot{M}_E$ (τ = 10) (from Ross and Fabian 1980).

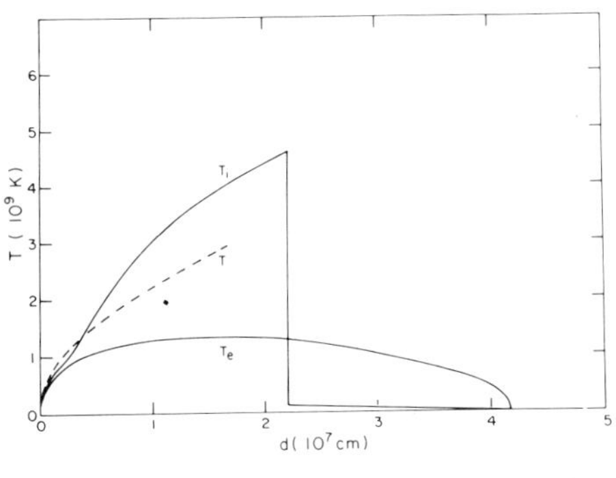

Fig. 17--Temperature structure of the emission region for a 1.4 $M_{\odot}$ star for an accretion rate 0.0253 $\dot{M}_E$ (τ =1) without nuclear burning (from Imamura et al. 1982). Here d is the distance above the surface of the star, and T_e and T_i are the electron and ion temperatures.

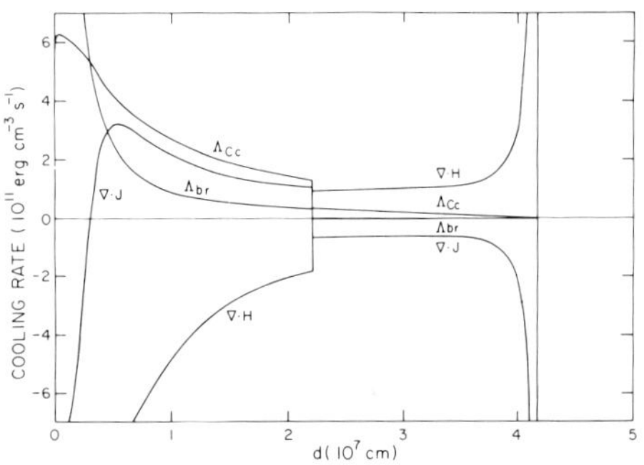

Fig. 18.--Cooling rates throughout the emission region of Figure 17. The curve labeled $\nabla\cdot H$ is the divergence of the enthalpy of the matter and corresponds to cooling (or heating) due to the flow of matter. The curves labeled Λ_{br} and Λ_{Cc} show the cooling due to bremsstrahlung and Compton cooling. The conduction curve ($\nabla\cdot J$) shows that conduction cools the region just behind the shock and heats the region near the stellar surface.

investigated in detail by Imamura et al. (1979, 1982) and Weast et al. (1979, 1982). The accreting matter does not burn in the hot X-ray emission region, but may do so deeper in the envelope of the star. The energy thus liberated is transported to the stellar surface and enhances the blackbody flux in soft X-rays. This flux of soft X-ray photons cools the X-ray emission region by inverse Compton scattering. As a result, the hard X-ray luminosity is often an order of magnitude less than it would be in the absence of nuclear burning, the hard X-ray spectrum is softer, and the soft X-ray luminosity can be 100 times the hard X-ray luminosity. Figure 12 compares the X-ray spectra of a 1 $M_{\odot}$ star when the accreting matter burns at the accretion rate and the spectra when it does not. The three spectra shown span the entire range of accretion rates. Figure 14 shows the correlation between T_{obs} and L_h when nuclear burning occurs at the accretion rate. These curves should be compared with those in Figure 13, which assumes no nuclear burning.

3.5.2. Magnetic stars. The effects of nuclear burning on X-ray emission by magnetic degenerate dwarfs are not expected to be as dramatic. As long as cooling by cyclotron emission dominates cooling by inverse Compton scattering of the blackbody photons, the cyclotron UV and bremsstrahlung hard X-ray luminosities will be little changed. The spectral temperatures of these components will also be little affected. The blackbody soft X-ray luminosity will, however, be much larger.

3.6. Electron Conduction

Energy transport from the X-ray emission region into the star by electron conduction has been proposed (King and Lasota 1980) as a solution to the "soft X-ray puzzle" in the AM Her stars (see § 4.3 below). According to this suggestion, which is based on a one fluid (i.e. $T_e \equiv T_i$) picture of the X-ray emission region, most of the energy liberated by accretion is conducted into the star, leading to intense blackbody emission from the stellar surface and only weak bremsstrahlung hard X-ray emission from the hot, post-shock region above the surface.

The calculations of X-ray emission with nuclear burning described above entailed two-fluid (i.e. $T_e \neq T_i$) calculations incorporating electron conduction in a fully self-consistent way (Imamura et al. 1979, 1982; Weast et al. 1979; 1982). Figure 17 compares the temperature profile of the X-ray emission region in the one- and two-fluid treatments for a 1.4 $M_{\odot}$ star accreting at a rate corresponding to $\tau_{es} = 1$. Figure 18 shows the rates of heating and cooling due to mass flow (enthalpy), bremsstrahlung, Compton cooling, and electron conduction. The results show that in some cases conduction transports energy from one part of the emission region to another, but does not transport energy into the star.

Recently, Frank, King, and Lasota (1982) have confirmed this conclusion, within the approximation of a one-fluid treatment, and have gone on to propose a new mechanism of energy trasport into the star by nonthermal electrons. However, such a process is not viable in the X-ray emission regions of degenerate dwarfs, in which the mean free path of ions to Coulomb scattering is small and necessitates an ion shock and in which the diffusion approximation remains valid.

4. ISSUES

Among the important issues concerning X-ray emission from cataclysmic variables are the following.

4.1. Magnetic Fields

We have long held that magnetic fields are crucial to understanding cataclysmic variables, especially their X-ray emission (cf. Lamb 1974, 1979a,b, 1981). Recent observations have increasingly borne out our view. In AM Her stars, the strong magnetic field locks the degenerate dwarf in synchronous rotation with the binary, channels the flow of accreting matter to emission regions at the magnetic poles, and produces highly polarized light at optical wavelengths. In DQ Her stars, the magnetic field disrupts the disk, leads to more rapid spin-up of the degenerate dwarf, again channels the flow to emission regions at the magnetic poles, and is predicted (see below) to produce polarized light at infrared wavelengths. Although there is as yet no clear need to invoke magnetic fields in the other cataclysmic variables, we conjecture that smaller fields may play an important role in many, perhaps all, of them.

4.2. Origin of Hard X Rays

A crucial role in the emission of hard X rays by the AM Her and DQ Her stars is played by the magnetic field, which leads to radial inflow near the star and the formation of a strong shock. What is the origin of the hard X rays emitted by the other cataclysmic variables? Are the X rays produced by optically thin emission in the boundary layer between the disk and the star? If so, are the required high temperatures achieved by strong shocks, turbulence, or some other mechanism? Alternatively, are there weak magnetic fields present in these sources sufficient to disrupt the disk very near the star, producing quasi-radial inflow and a strong shock?

4.3. Soft X-Ray Puzzle in AM Her Stars

Much of the soft X-ray emission from the AM Her stars is inaccessible to observation because of interstellar absorption (cf. Tuohy et al. 1978, 1981; Rothschild et al. 1981; Patterson et al. 1983). UV observations of AM Her several years ago unexpectedly showed a turn-up at short wavelengths (Raymond et al. 1979). The most elegant explanation of the data attributed both the UV turn-up and the soft X-rays to an intense blackbody component with a temperature T_{bb} = 27 eV. However, this implied $L_{bb} >> L_{br}$ whereas theory predicted $L_{bb} \simeq L_{br}$ (Kylafis and Lamb 1979, 1982a; Lamb and Masters 1979; King and Lasota 1979). A similar discrepancy was inferred in AN UMA (Szkody et al. 1981) and possibly for the other AM Her stars. This conflict has become known as the "soft X-ray puzzle." Ideas ranging from the transport of energy from the X-ray emission region into the star by electron conduction (King and Lasota 1980) or nonthermal electrons (Frank, King, and Lasota 1982) to nuclear burning (Fabbiano et al. 1981) have been put forward as possible explanations. Energy transport into the star by electron conduction or nonthermal electrons is not viable (recall the discussion in Section 3.6), while nuclear burning also faces difficulties (see below).

Recently, the results of nearly simultaneous observations of AM Her with the Einstein objective grating spectrometer and with IUE have become available (Kahn 1982; see also Heise and Brinkman 1982). They show that the blackbody temperature of the soft X-ray emission from AM Her is $\simeq$ 45 eV, and that the UV and soft X-ray fluxes cannot be connnected by a blackbody. The results are consistent with $L_{bb} \simeq L_{br}$, and indicate that the turn-up in the UV flux has a separate origin. Recent Einstein soft X-ray observations of VV Pup are also consistent with $L_{bb} \simeq L_{br}$ (Patterson et al. 1982), as are earlier observations of EF Eri (Patterson et al. 1981). These results suggest that the famous "soft

X-ray puzzle" may have gone away.

4.4. Origin of Soft X Rays

Although a soft X-ray component has not been detected in most of the other cataclysmic variables, SS Cyg and U Gem exhibit intense soft X-ray emission at outburst. Is its origin similar to the soft X-ray emission seen in the AM Her stars (which may, after all, be just reprocessed hard X-rays)? Is it emission from the hot degenerate dwarf itself? Or is it something more exotic, like nuclear burning (see below)?

4.4. Nuclear Burning

The energy liberated by nuclear burning of matter accreting onto degenerate dwarfs can be more than an order of magnitude greater than that available from the release of gravitational energy. If burning occurs quiescently, the resulting energy is transported to the stellar surface and produces an intense blackbody soft X-ray flux. Steady nuclear burning has therefore recently received a great deal of attention as a possible explanation of intense blackbody soft X-ray components inferred in the AM Her stars (Raymond et al. 1979) and in the cataclysmic variables SS Cyg and U Gem during outburst (Fabbiano et al. 1981).

Unfortunately, the conditions under which steady nuclear burning can occur are poorly understood. Detailed spherically symmetric calculations by Paczynski and Zytkow (1978), Sion, Acierno, and Turnshek (1978), and Sion, Acierno, and Tomczyk (1979) show that if the degenerate dwarf is initially cold and the accretion rate is not too high, the accreting matter becomes highly degenerate before it ignites. Electron conduction then rapidly transports energy away into the core, and it must be heated before ignition can occur. If the degenerate dwarf is hot, or if the accretion rate is high, the hydrogen in the accreting matter soon ignites due to compressional heating. In either case, eventually a violent nuclear outburst ensues. Such outbursts are believed to account for novae (cf. Starrfield, Sparks, and Truran 1974).

The outbursts are separated by quiescent periods, in which nuclear burning occurs steadily at only a small fraction of the accretion rate. The quiescent periods are shorter for higher accretion rates and can last from ~ 20 years or less (Sion et al. 1979) to $> 10^7$ years (Paczynski and Zytkow 1978). For a narrow range of higher accretion rates, steady nuclear burning is possible at the rate of accretion (e. g. $1.0 - 2.7 \times 10^{-7}\ M_\odot\ yr^{-1}$ for a $0.8\ M_\odot$ star; Paczynski and Zytkow 1978). Still higher accretion rates lead to envelope expansion and the formation of a red giant with a degenerate core.

Depletion of CNO nuclei in the accreting matter and the burning region by diffusion can lead to burning via the p-p chain rather than via the more temperature sensitive CNO cycle (Starrfield, Truran, and Sparks 1981), and stabilize the burning at higher accretion rates. However, theoretical investigations show that such rapid depletion is unlikely (Fujimoto and Truran 1982; Papaloizou, Pringle, and MacDonald 1982). Effects due to non-spherical geometries also warrant investigation. For example, in the AM Her and DQ Her stars a strong magnetic field channels the accreting matter onto the magnetic poles. If the matter is confined and burns over only a small fraction of the stellar surface, the burning might be stabilized by the rapid transport of energy horizontally.

4.5. Long-Period Pulsing Sources

Recently, several faint galactic X-ray sources have been found to pulse in both X-rays and optical light with periods $\gtrsim$ 1000 seconds (see section 2.2.2). There has been controversy about whether these objects are degenerate dwarfs or neutron stars (cf. Patterson and Price 1981, White and Marshall 1981). Their optical appearance and their X-ray to optical luminosity ratio of order unity suggest that they are degenerate dwarfs (Patterson and Price 1981). We have carried out a detailed comparison of their spin-up rate and that expected from theory (Lamb and Patterson **1983**). This analysis confirms that they are indeed degenerate dwarfs. It also shows that they have magnetic fields $B \sim 3 \times 10^6$ G, an order of magnitude smaller than those of AM Her stars, and that the three previously known DQ Her stars with rotation periods < 100 seconds have still smaller fields. On this basis, we predict that the long-period sources will show significant polarization in the infrared while the short-period ones will show polarization only at longer wavelengths.

The newly discovered stars have been called "interlopers" (Patterson and Price 1981) between the previously known DQ Her stars, with rotation periods of 33 - 71 seconds, and the AM Her stars, with rotation periods synchronous with their orbital periods of 1.3 - 3.7 hours. However, spin-up theory shows that they have rotation periods appropriate to their magnetic fields and luminosities. We therefore believe that they and the DQ Her stars should be regarded as members of a single class, as are the short and long period pulsing neutron star X-ray sources.

4.6. Cyclotron-Dominated Sources

Although the AM Her stars are strongly magnetic ($B \simeq 2 \times 10^7$ G), bremsstralung, not cyclotron emission, is the dominant emission mechanism. This underscores the fact that whether bremsstrahlung or cyclotron emission dominates depends not only on the magnetic field, but also on the effective accretion rate $\dot{M}/f$ or, equivalently, the effective luminosity L/f (recall Figure 8). Bremsstrahlung dominates in AM Her stars because the accretion flow is channeled onto such a small fraction ($f \sim 10^{-3}$-10^{-6}) of the stellar surface that the effective accretion rate is very high.

Sources in which cyclotron emission dominates are characterized by weak X-ray emission and intense polarized cyclotron emission in the infrared, optical, or ultraviolet, depending on the strength of the magnetic field. We know of no reason why they cannot exist. Isolated degenerate dwarfs have magnetic fields ranging up to $\simeq 3 \times 10^8$G. If degenerate dwarfs with fields this large occur in cataclysmic variables, cyclotron emission would dominate even if the emitting area were as small as it is in the AM Her stars. Alternatively, cyclotron emission would dominate in sources with fields far smaller, provided that the area of the emission region is larger ($f \sim 10^{-1}$-10^{-2}). This could occur for magnetic degenerate dwarfs in cataclysmic variables with disks, such as the DQ Her stars, where the field lines accessible to accreting matter cover a substantial fraction of the stellar surface. So far no sources like these have been discovered, despite some searches (cf. Bond and Chanmugam 1982). Where are they? Will they be found by future UV or extreme UV surveys? Or is there some as yet unrecognized reason why they cannot occur?

4.7. High Luminosity Sources

Degenerate dwarfs with X-ray luminosities greater than $\simeq 10^{36}$ ergs s^{-1} cannot occur, due to Compton degradation and photoelectric absorption by the accreting matter. However, we know of no reason why sources with small X-ray luminosities but with total luminosities $L_{acc} \sim 10^{36}$ - 10^{38} ergs s^{-1} cannot occur. Such sources, whether magnetic or not, would be characterized by an intense soft X-ray component and a weak hard X-ray component, severely degraded at high energies and absorbed at low energies (cf. Figure 15 and Ross and Fabian 1980). No such X-ray source has been unequivocally identified with a degenerate dwarf. Cyg X-2 exhibited the X-ray behavior expected due to degradation (Branduardi et al. 1980) but not to absorption (Ross and Fabian 1980). And the distance, and hence luminosity, inferred from optical spectroscopy (Cowley et al. 1979) rules out a degenerate dwarf in the system. Do we not yet know how to recognize degenerate dwarf X-ray sources with high total (but not X-ray) luminosities? Or are there no such sources? And if not, why not?

4.8. Stability of Accretion Flow

Quasi-periodicities have been observed in the X-ray emission from several cataclysmic variables (cf. Patterson et al. 1981). Langer, Chanmugam, and Shaviv (1981, 1982) have considered the time dependence of accretion onto magnetized degenerate dwarfs. Their calculations show that the flow is unstable to periodic oscillations at the fundamental, and suggest that these may account for the observed quasi-periodicities. However, Chevalier and Imamura (1982) using a linear stability analysis find the fundamental to be stable, and only the overtones to be unstable. Further, such oscillations may be damped by Compton cooling and cyclotron emission, processes which have not yet been included in the calculations. Thus whether oscillations in the flow of accreting matter onto magnetic degenerate dwarfs account for the observed quasi-periodicities is not yet clear.

From the above remarks, it should be evident that we have only begun to explore the nature of degenerate X-ray sources. We must rely on future X-ray astronomy missions to provide the data and theorists to create the ideas needed to understand them.

This research was supported in part by NASA grant NAGW 246. I also gratefully acknowledge support from the Fluid Research Fund of the Smithsonian Institution.

REFERENCES

Aizu, K.: 1973, Prog. Theoret. Phys., 49, 1184.

Bath, G. T., Evans, W. D., and Pringle, J. E.: 1974, Mon. Not. R. Astron. Soc., 166, 113.

Berg, R., and Duthie, J.: 1977, Astrophys. J., 211, 859.

Bond, H. E., and Chanmugam, G.: 1982, in Advances in Ultraviolet Astronomy: Four Years of IUE Research, 530.

Branduardi, G., Kylafis, N. D., Lamb, D. Q., and Mason, K. O.: 1980, Astrophys. J. Letters, 235, L153.

Chanmugam, G., and Wagner, R. L.: 1977, Astrophys. J. Letters, 213, L13.

Chanmugam, G., and Wagner, R. L.: 1978, Astrophys. J., 222, 641.

Charles, P. A., Thorstensen, J., Bowyer, S., and Middleditch, J.: 1979, Astrophys. J. Letters, 231, L131.

Chevalier, R. A., and Imamura, J. N.: 1982, submitted to Astrophys. J.

Cordova, F. A., Chester, T. J., Mason, K. O., Kahn, S. M., Garmire, G. P., and Middleditch, J.: 1981, submitted to Astrophys. J.

Cordova, F. A., Chester, T. J., Tuohy, I. R., and Garmire, G. P.: 1980, Astrophys. J., 235, 163.

Cordova, F. A., and Mason, K. O.: 1982, in 'Accretion Driven Stellar X-Ray Sources', ed. W. H. G. Lewin and E. P. J. van den Heuvel (Cambridge, England: Cambridge

University Press), in press.
Cordova, F. A., Mason, K. O., and Nelson, J. E.: 1981, Astrophys. J., 245, 609.
Cordova, F. A., and Riegler, G. R.: 1979, Mon. Not. R. Astron. Soc., 188, 103.
Cowley, A., and Crampton, D.: 1977, Astrophys. J. Letters, 212, L121.
Cowley, A. P., Crampton, D., and Hutchings, J. B.: 1979, Astrophys. J., 231, 539.
DeGregoria, A. J.: 1974, Astrophys. J., 189, 555.
Fabbiano, G., Hartmann, L., Raymond, J., Steiner, J., Branduardi-Raymont, G., and Matilsky, T.: 1981, Astrophys. J., 243, 911.
Fabian, A. C., Pringle, J. E., and Rees, M. J.: 1976, Mon. Not. R. Astron. Soc., 173, 43.
Frank, J., King, A. R., and Lasota, J. P.: 1982, submitted to Mon. Not. R. Astron. Soc.
Fujimoto, M. Y., and Truran, J. W.: 1982, Astrophys. J., 257, 303.
Garmire, G.: 1979, in 'Compact Galactic X-Ray Sources', ed. F. K. Lamb and D. Pines (Urbana, Illinois: University of Illinois), p.62.
Gilliland, R. L.: 1982, Astrophys. J., in press.
Griffiths, R., Lamb, D. Q., Ward, M. M., Wilson, A., Charles, P. A., Thorstensen, J., McHardy, I. M., and Lawrence, A.: 1980, Mon. Not. R. Astron. Soc., 193, 25P.
Hassall, B. J. M., Pringle, J. E., Ward, M. J., Whelan, J. A. J., Mayo, S. K., Echerarria, J., Jones, D. H. P., Wallis, R. E., Allen, D. A., and Hyland, A. R.: 1981, Mon. Not. R.Astron. Soc., 197, 275.
Hayakawa, S.: 1973, Prog. Theoret. Phys., 50, 459.
Hayakawa, S., and Hoshi, R.: 1976, Prog. Theoret. Phys., 55, 1320.
Hearn, D. R., Richardson, J. A., and Clark, G. W.: 1976, Astrophys. J. Letters, 210, L23.
Heise, J., and Brinkman, A. C.: 1982, in Galactic X-Ray Sources, ed. P. W. Sanford, P. Laskarides, and J. Salton (Chichester: Wiley), p. 393.
Heise, J., Mewe, R., Brinkman, A. C., Groenschild, E. H. B. M., den Boggende, A. J. F., Schrijver, J., and Grindlay, J. E.: 1978, Astron. Astrophys., 63, L1.
Hertz, P., and Grindlay, J. E.: 1983, submitted to Astrophys. J. Letters.
Hoshi, R.: 1973, Prog. Theoret. Phys., 49, 776.
Hutchings, J. B., Crampton, D., Cowley, A. P., Thorstensen, J. R., and Charles, P. A.: 1981, Astrophys. J., 249, 680.
Illarionov, A. F., and Sunyaev, R. A.: 1972, Astrophys. Zh., 49, 58 (English transl. in Soviet Astr.-AJ, 16, 45, 1972).
Imamura, J. N., and Durisen, R. H.: 1983, submitted to Astrophys. J.
Imamura, J. N., Durisen, R. H., Lamb, D. Q., and Weast, G. J.: 1979, in IAU Colloquium 53, 'White Dwarfs and Variable Degenerate Stars', ed. H. M. Van Horn and V. Weidemann (Rochester, N.Y.: University of Rochester), p. 406.
Imamura, J. N., Durisen, R. H., Lamb, D. Q., and Weast, G. J.: 1982, submitted to Astrophys. J.
Joss, P. C., Katz, J. I., and Rappaport, S.: 1979, Astrophys. J., 230, 176.
Kahn, S.: 1982, private communication.
Katz, J. I.: 1977, Astrophys. J., 215, 265.
King, A. R., Ricketts, M. J., and Warwick, R. S.: 1979, Mon. Not. R. Astron. Soc., 187, 77P.
King, A. R., and Lasota, J. P.: 1979, Mon. Not. R. Astron. Soc., 188, 653.
King, A. R., and Lasota, J. P.: 1980, Mon. Not. R. Astron. Soc., 191, 721.
Krzeminski, W., and Serkowski, K.: 1977, Astrophys. J. Letters, 216, L45.
Kylafis, N. D., and Lamb, D. Q.: 1979, Astrophys. J. Letters, 228, L105.
Kylafis, N. D., and Lamb, D. Q.: 1982a, Astrophys. J. (Supp.), 48,239.
Kylafis, N. D., and Lamb, D. Q.: 1982b, Astrophys. J., in press.

Kylafis, N. D., Lamb, D. Q., Masters, A. R., and Weast, G. J.: 1980, 'Proc. Ninth Texas Symposium on Relativistic Astrophysics', Ann. N.Y. Acad. Sci., 336, 520.
Lamb, D. Q.: 1974, Astrophys. J. Letters, 192, L129.
Lamb, D. Q.: 1979a, in 'Compact Galactic X-Ray Sources', ed. F. K. Lamb and D. Pines (Urbana, Illinois: University of Illinois), p. 27.
Lamb, D. Q.: 1979b, in IAU Colloquium 53, 'White Dwarfs and Variable Degenerate Stars', ed. H. M. Van Horn and V. Weidemann (Rochester, N.Y.: University of Rochester), p. 508.
Lamb, D. Q.: 1981, in 'X-Ray Astronomy in the 1980's', ed. S. S. Holt (NASA TM83848), p. 37.
Lamb, D. Q., and Masters, A. R.: 1979, Astrophys. J. Letters, 234, L117.
Lamb, D. Q., and Patterson, J.: **1983**, these proceedings, **229**.
Lamb, F. K.: 1981, in 'X-Ray Astronomy in the 1980's', ed. S. S. Holt (NASA TM83848), p. 77.
Lamb, F. K., Lamb, D. Q., Aly, J.-J., and Cook, M.: 1983, to be submitted to Astrophys. J.
Langer, S. H., Chanmugam, G., and Shaviv, G.: 1981, Astrophys. J. Letters, 245, L23.
Langer, S. H., Chanmugam, G., and Shaviv, G.: 1982, submitted to Astrophys. J.
Latham, D. W., Liebert, J., and Steiner, J.: 1981, Astrophys. J., in press.
Linsky, J.: 1981, in 'X-Ray Astronomy in the 1980's', ed. S. S. Holt (NASA TM83848), p. 13.
Mason, K., Cordova, F., and Swank, J.: 1979, in '(COSPAR) X-ray Astronomy', ed. W. A. Baity and L. E. Peterson (Oxford and New York: Pergamon Press), p. 121.
Mason, K. O., Lampton, M., Charles, P., and Bowyer, S.: 1978, Astrophys. J. Letters, 226, L129.
Masters, A. R.: 1978, Ph.D. Thesis, University of Illinois, unpublished.
Masters, A. R., Fabian, A. C., Pringle, J. E., and Rees, M. J.: 1977, Mon. Not. R. Astron. Soc., 178, 501.
Motch, C.: 1981, Astron. Astrophys., 100, 277.
Paczynski, B., and Zytkow, A. N.: 1978, Astrophys. J., 222, 604.
Papaloizou, J. C. B., Pringle, J. E., and MacDonald, J.: 1982, Mon. Not. R. Astron. Soc., 198, 215.
Patterson, J.: 1979a, Astrophys. J. Letters, 233, L13.
Patterson, J.: 1979b, Astrophys. J., 234, 978.
Patterson, J., Branch, D., Chincarini, G., and Robinson, E. L.: 1980, Astrophys. J. Letters, 240, L133.
Patterson, J., Fabbiano, G., Lamb, D. Q., Raymond, J., Horne, K., White, N., and Swank, J.: 1982, submitted to Astrophys. J.
Patterson, J., and Price, C. M.: 1981, Astrophys. J. Letters, 243, L83.
Patterson, J., Robinson, E. L., and Nather, R. E.: 1978, Astrophys. J., 224, 570.
Patterson, J., Williams, G., and Hiltner, W. A. 1981, Astrophys. J., 245, 618.
Pringle, J. E.: 1977, Mon. Not. R. Astron. Soc., 178, 195.
Pringle, J. E.: 1981, Ann. Rev. Astron. Astrophys., 19, 137.
Pringle, J. E., and Savonije, G. J.: 1979, Mon. Not. R. Astron. Soc., 187, 777.
Rappaport, S., Cash, W., Doxsey, R., McClintock, J., and Moore, G.: 1974, Astrophys. J. Letters, 187, L5.
Raymond, J. C., Black, J. H., Davis, R. J., Dupree, A. K., Gursky, H., Hartmann, L., and Matilsky, T. A.: 1979, Astrophys. J. Letters, 230, L95.
Ricketts, M. J., King, A. R., and Raine, D. J.: 1979, Mon. Not. R. Astron. Soc., 186, 233.
Robinson, E. L.: 1976, Ann. Rev. Astron. Astrophys., 14, 119.
Ross, R. R., and Fabian, A. C.: 1980, Mon. Not. R. Astron. Soc., 193, 1P.

Rothschild, R. E., et al.: 1981, Astrophys. J., 250, 723.
Shakura, N. I, and Sunyaev, R. A.: 1973, Astron. Astrophys., 24, 337.
Schmidt, G. D., Stockman, H. S., and Margon, B.: 1981, Astrophys. J. Letters, 243, L157.
Sion, E. M., Acierno, M. J., and Tomczyk, S.: 1979, Astrophys. J., 230, 832.
Sion, E. M., Acierno, M. J., and Turnshek, D. A.: 1978, Astrophys. J., 220, 636.
Starrfield, S., Sparks, W. M., and Truran, J. W.: 1974, Astrophys. J. Suppl., 28, 247.
Starrfield, S., Truran, J. W., and Sparks, W. M.: 1981, Astrophys. J. Letters, 243, L27.
Stockman, H. S., Liebert, J., and Bond, H. E.: 1979, in IAU Colloquium 53, 'White Dwarfs and Variable Degenerate Stars', ed. H. M. Van Horn and V. Weidemann (Rochester, N.Y.: University or Rochester), p. 334.
Stockman, H. S., Schmidt, G. D., Angel, J. R. P., Liebert, J., Tapia, S., and Beaver, E. A.: 1977, Astrophys. J., 217, 815.
Swank, J. H.: 1979, IAU Colloq. No. 53, 'White Dwarfs and Variable Degenerate Stars', ed. H. M. Van Horn and V. Weidemann (Rochester, NY: Univ. of Rochester Press), p. 135.
Swank, J. H., Boldt, E. A., Holt, S. S., Rothschild, R. E., and Serlemitsos, P. J.: 1978, Astrophys. J. Letters, 226, L133.
Swank, J. H. and White, N.: 1981, private communication.
Swedlund, J. B., Kemp, J. C., and Wolstencroft, R. D.: 1974, Astrophys. J. Letters, 193, L11.
Szkody, P., and Brownlee, D. E.: 1977, Astrophys. J. Letters, 212, L113.
Szkody, P., Schmidt, E., Crosa, L., and Schommer, R.: 1981, Astrophys. J., 246, 233.
Tapia, S.: 1977a, Astrophys. J. Letters, 212, L125.
Tapia, S.: 1977b, IAU Circ. No. 3054.
Tuohy, I. R., Lamb, F. K., Garmire, G. P., and Mason, K. O.: 1978, Astrophys. J. Letters, 226, L17.
Tuohy, I. R., Mason, K. O., Garmire, G. P., and Lamb, F. K.: 1981, Astrophys. J., 245, 183.
Tylenda, R.: 1981a, Acta Astron., 31, 127.
Tylenda, R.: 1981b, Acta AStron., 31, 267.
Visvanathan, N., and Wickramasinghe, D. T.: 1979, in IAU Colloquium 53, 'White Dwarfs and Variable Degenerate Stars', ed. H. M. Van Horn and V. Weidemann (Rochester, N.Y.: University of Rochester), p. 330.
Vogt, N., Krzeminski, W., and Sterkin, C.: 1980, Astron. Astrophys., 85, 106.
Wada, T., Shimizu, A., Suzuki, M., Kato, M., and Hoshi, R.: 1981, Prog. Theoret. Phys., 66, 1771.
Warner, B.: 1976, IAU Symposium 73, 'Structure and Evolution of Close Binary Systems', ed. P. Eggleton, S. Mitton, and J. Whelan (Dordrecht: Reidel), p. 85.
Warner, B., O'Donoghue, D., and Fairall, A. P.: 1981, Mon. Not. R. Astron. Soc., 196, 705.
Watson, M. G., Sherrington, M. R., and Jameson, R. F.: 1978, Mon. Not. R. Astron. Soc., 184, 79P.
Weast, G. J., Durisen, R. H., Imamura, J. N., Kylafis, N. D., and Lamb, D. Q.: 1979, in IAU Colloquium 53, 'White Dwarfs and Variable Degenerate Stars', ed. H. M. Van Horn and V. Weidemann (Rochester, N.Y.: University of Rocheter), p. 330.
Weast, G. J., Durisen, R. H., Imamura, J. N., Kylafis, N. D., and Lamb, D. Q.: 1982, submitted to Astrophys. J.
White, N. E., and Marshall, F. E.: 1981, Astrophys. J. Letters, 249, L25.

DISCUSSION FOLLOWING D. LAMB'S TALK

KING: Just a clarification. The calculations with the Compton degradation are spherically symmetric, the geometry is spherically symmetric.

LAMB: That is correct.

KING: Is it then not very different if the accretion is funnelled down a column, in particular the photons will not tend to see a very large scattering optical depth, but will tend to try and escape through the sides of the column.

LAMB: That is correct, but it depends sensitively on the actual geometry.

KING: But they can escape through the sides, once they get to the sides they don't scatter back.

LAMB: In fact, the luminosities of most of these sources are surprisingly low. Even if one has a picture based on the spherically symmetric results, AM Her is on the borderline of the regime where Compton degradation would be significant. When you take into account the reduction due to the nonspherical geometry in AM Her, and almost surely in the other AM Her sources, one is not in the optically thick regime. That gives some satisfaction, in the sense that calculations don't have to deal with the complexities of degradation and of absorption, which can be rather difficult.

ROBINSON: Is the soft X-ray component seen also in the longer period rotators like 2252 and 2251?

LAMB: No. There is a lot of flux missing in the AM Her stars. In the long period DQ Her stars, even more flux is missing.

CHANMUGAM: You had a figure in which you divided the cyclotron cooling sources from the Bremsstrahlung cooling sources and you had a line which divided this. Is that line obtained by equating optically thin cyclotron cooling to optically thin Bremsstrahlung?

LAMB: No. It is obtained by taking into account that the cyclotron emission is optically thick below a certain frequency.

ASTRONOMY IN ISRAEL: FROM OG'S CIRCLE TO THE WISE OBSERVATORY

Yuval Ne'eman[#+]
Tel Aviv University, Tel Aviv, Israel

Supported in part by the U.S.-Israel Binational Science Foundation
+ Wolfson Chair Extraordinary, Tel Aviv University

INTRODUCTION

Colloquium IAU 72 is the first IAU-sponsored activity in this country, and it therefore seemed appropriate to include a paper on Astronomy in Israel. With the highly singular history of this nation, the scene will move from Eretz-Israel* to Babylon (present day Iraq) and to Europe before returning to this area.

MEGALITHS

Prehistoric astronomical activity is represented by a Stonehenge-like megalithic circle and "Observatory" at Rujm-el-Hiri, near Yonathan in the Golan, the Westernmost sector of the historical Bashan plateau[1)] dating from the IIIrd Millenium BC. Star worship is mentioned in the Old Testament[2)] as being common among the Canaanites*, but the Bashan inhabitants who built that Golan megalithic circle antedate the Canaanites. Very little is known about them and the presumably religious role of their edifice. To the XIIIth Century BC Israelites, they appeared as the work of giants (Refa'im, also Anakim, Emim, Zuzim), and this is probably the source of the legends about races of giants that had lived in Eretz-Israel prior to the Israelite conquest - including the characterization "a remnant of the giants" for Og, King of Bashan, in Deuteronomy and Joshua[3)]. Indeed, the Rujm-el-Hiri circle is just one among many megalithic remains in the Bashan, probably at the origin ofthese tales. Notice that megaliths have given birth to just such stories in Greece and England (the "Giants' Dance" = Stonehenge).

ASTRONOMY IN THE MISHNA

The Israelites' abstract monotheism and their centering of intellectual creativity on ethical issues were detrimental to a natural development of observational science, as did happen in Sumeria or Greece. However, the requirements of agriculture induced a cycle of holidays that were incorporated in time into Judaism and were given Ethnical or National religious significance. There thus developed a need for an understanding of the recurrence of seasons and for a synchronized calendar fitted to Solar, Lunar and Sidereal time[4-6)]. Several of the Mishnaic scholars** were versed in Astronomy, such as the "Tannaim"

M. Livio and G. Shaviv (eds.), Cataclysmic Variables and Related Objects, 323–329.

Yehohanan ben Zakkai[7], the Patriarch Gamliel II and in particular Yehoshua (=Joshua) ben Hananiah[8]. In the tractate Horayoth, dealing with errors of Justice, the following anecdote is related[9]:
"Rabbi Gamliel and Rabbi Yehoshua went together on a voyage at sea. Rabbi Gamliel carried a supply of bread, Rabbi Yehoshua carried a similar amount of bread and in addition a reserve of flour. At sea, they used up the entire supply of bread and had to utilize Rabbi Yehoshua's flour reserve. Rabbi Gamliel then asked Rabbi Yehoshua - "Did you know that this trip would last longer than usual, when you decided to carry this flour reserve?" Rabbi Yehoshua answered - "There is a star that appears every 70 years and induces navigational errors. I thought it might appear and cause us to go astray." Rabbi Gamliel then exclaimed "You are so knowledgeable and you nevertheless have to travel to make a living?" Rabbi Yehoshua then answered bitterly - "How come you are so surprised? Don't you know that two of your own students Rabbi Eliezer Hisma and Rabbi Yehohanan ben Gudgada who are so smart that they can tell you how many drops there are in the ocean, have neither bread to eat nor clothes to wear?"

This observation is generally interpreted as relating to Halley's Comet[10], with a period approximating 76 years. Observing a comet's periodicity, with such a long period, requires records covering many centuries; it is possible that the Mishnaic scholars did inherit such records from the Great Knesset scholars (before 300 BC) who received them during the Babylonian exile (586-537 BC) from the "Chaldeans" (i.e. from Sumer, Akad etc. going back to the IIIrd Millenium BC). Indeed, the Hebrew agricultural names of the months*** were replaced by Sumerian and Akadian names after the Babylonian captivity. However, there is perhaps a difficulty with the dates: Rabbi Yehoshua was born in 35 AD and died in 117 AD. Rabbi Gamliel died in 115 AD. The only appearance of Halley's Comet in the interval 55-115 (when Yehoshua was older than 20), is in 66 AD. Rabbi Gamliel was younger than Rabbi Yehoshua and it has been argued that if he was 20-25 years old, it is doubtful whether he could have had students at the time[10]. Still, this is not a strong argument, as Rabbi Gamliel II was also the hereditary Patriarch, and may have had students attached to him formally from the moment he was appointed head of the Sanhedrin.

Philippe Veron has, however, come up[11] recently with a different identification of Rabbi Yehoshua's star, and argues that this was the variable Mira Ceti.

Mar Samuel, who became around 220 AD the Dean of the Talmudic Academy of Nehardea in Babylonia, was an astronomer who could calculate and adjust the calendar with great precision, intercalating an extra month or reassigning the length of a month. The prescriptions for the calendar adjustments were written down in a special Baraita**. They include the 19 years synchronization cycle used to this very day in the Jewish Calendar.

SPAIN AND PROVENCE[12)]

Science spreads by convection. When Khushru Anushirvan, Sassanid Emperor of Persia, signed a ten year truce with Justinian of Byzance, he ensured the continuity of science by requesting that the teachers of the recently abolished Academy of Athens be transferred to Persia. In his fanaticism, Justinian was then trying hard to put an end to Judaism at the sametime eradicating Neo-Platonicism and what was left of Greek science. The Neo-Platonicists settled in Mesopotamia under Persian rule, and their school was already flourishing when Persia was conquered by Islam. Mathematics and Astronomy thrived (e.g. Omar Khayyam, whose Rubayat were just a hobby, his professional creativity having yielded methods for solving factorizable cubic equations etc.) and spread all over the new Mohammedon Empire. The Jews were active participants, and the first Arabic-language treatise on the Astrolabe was written by the Jew Joel, known as Masha-Allah of Basra (Iraq) around the year 800. This is the treatise that was translated into English by Geoffrey Chaucer ("The Treatise on the Astrolabe") around 1380, from a prior Latin translation. Masha-Allah also wrote a book on Lunar and Solar Eclipses, that was later translated into Hebrew by Abraham Ibn Ezra ("sefer be kadrut ha levana ve ha shemesh").

Sind ben Ali, a heretic Jew, was the main contributor (∿830) to the astronomical tables of the Caliph Maimun. The scene now shifts to Spain, where Abraham bar Hiyya Hanasi ("The Prince") of Barcelona (d.1136) improved on these tables,using calculations performed by the Arab astronomer Al-Battani (d. 929). Abraham bar Hiyya was a prominent mathematician and astronomer, and wrote famous textbooks in both fields. He introduced Europe to (Arab) trigonometry in his "Treatise on Mensuration and Calculations". The Hebrew was translated by Plato of Tivoli into Latin in 1145 and the book served as main source material for the later work of Leonardo Fibonacci of Pisa. In Astronomy, his book "The Shape of the Earth" is based upon the Ptolemaic system, contains a roughly correct estimate of the distance to the Moon (but the wrong distance to the Sun). The principles for Calendar intercalation make up yet another book.

His student, Abraham Ibn Ezra (1089-1164), poet, philosopher, Biblical commentator and Astronomer, spent the last part of his life travelling in Italy and France, ending up in Eretz-Israel. He continued the publication of tables, mostly on the movement of the planets. The "Toledo Tables" were compiled by 12 Jewish astronomers led by the Cordovan Arab astronomer Ibn Arzarkali ("Azarchel"). The Latin version (translated by John of Brescia and Jacob Ibn Tibbon) was further improved in 1272 by a group of astronomers led by Isaac Ibn Said, and is known as the "Alphonsine Tables".

Rabbi Moshe ben Maimon's(=Maimonides) main contribution to Astronomy is his complete rejection of Astrology (1194). He is unique, throughout the Centuries, in making this clear-cut. Remember that Kepler was still drawing horoscopes! Perhaps this should justify a visit to Maimonides' tomb in Tiberias.......

Rabbi Levi ben Gershom (= Gersonides, also Maestre Leo de Bagnols, Maestre Leo Hebraeus; 1288-1344), was one of the greatest of Medieval astronomers. He lived in Provence, mostly at Orange. As a mathematician, he rediscovered the law of sines and published a sine table, correct to the 5th decimal. As an astronomer (he wrote 136 "chapters"!), he is the first to have relied on his own observations (in his studies of eclipses) rather than on Ptolemy's. He invented "Jacob's staff", a navigational instrument which was widely used for 3 centuries, and was the first person known to have used a Camera Obscura for his observations.

Rabbi Levi is also the first scientist to derive more realistic estimates of the distance to the fixed stars. Ptolemy's estimate was of the order of 10^{-5} light years (a million times smaller than the distance to the nearest star), whereas Rabbi Levi reached a figure of about 10^{5} light years, 10 times our present estimate for the distance to an average star in the Galaxy. Gersonides was also one of the greatest Medieval philosophers and published Commentaries to the Bible.

The Zohar, a compilation of Jewish mystic writings drawn in Spain in the XIIIth Century anticipates Copernicus by stating that "the whole earth spins in a circle like a ball; the one part is up when the other part is down; the one part is light when the other is dark; it is day in the one part and night in the other".

Jewish astronomers played a key role in the theoretical preparation of the great voyages of discovery in the XVth Century. Judah Cresques, forced to adopt Christianity in the massacres of 1391, later became the Director of the Prince Henry of Portugal's ("The Navigator") Nautical Academy at Sagres. Abraham Zacuto ("Zacut", 1452-1515) worked first at Salamanca but moved to Portugal after the expulsion from Spain. As Court Astronomer to Kings John II and Manuel I, he prepared the voyage of Vasco da Gama (1496) and supplied instrumentation (including his newly perfected copper astrolabe), improved tables, charts, instruction and briefs. He developed the first copper astrolabe. His very precise predictions of eclipses were used by Columbus to threaten the natives at a dangerous moment. Like all Jews, Zacut had to flee Portugal in 1497 and went to Tunis. He died in Eretz-Israel.

MODERN TIMES

The XVI-XVIIth Centuries were centuries of Jewish sufferings, and contributions to Astronomy are less prominent, except perhaps for the Herschel family, of Jewish origin. It was only when Alexander von Humboldt became President of the Prussian Academy of Sciences, that he abolished the requirement of a Christian Oath by a Professor at his ordination. Karl-Gustav Jacobi was the first Jew who did not have to abjure his faith to become a Professor. The oath was re-established by von Humboldt's successor, but it was no more in existence when Einstein, Minkowski and Schwarzschild were ordinated.

MODERN ISRAEL

Until 1965, observational astronomy was entirely a matter of amateurs. In that year we started astronomical research at Tel Aviv University, and by 1969 we had a Department of Physics and Astronomy, still the only one in the country in 1982. In 1971 we inaugurated, with the help of the Smithsonian Astrophysical Observatory, the Florence and George Wise Observatory[13] at Mitspe-Ramon (altitude 950m.) in the Negev. The telescope is a 40" Wide Angle Ritchie-Chretien reflector. European astronomers have often used the instrument. The more important observations at the Wise Observatory have been (1) the first optical identification[14] of an X-ray pulsar, Hercules X-1 (J. and N. Bahcall, 1972);(2) the first direct verification[15] of Whipple's conjecture about comet-tails being composed of water (P. Wehinger and S. Wycoft, together with G. Herbig at Lick Observatory and G. Hertberg and H.Liu in Canada, 1974); (3) the discovery of Sodium[16] and Sulphur[17] clouds around Jupiter, along the orbit of its satellite 10. (Y. Mekler, A. Eviathar and I. Kupo, 1974); (4) the precise observation[18] of a recurrent Nova leading to a better understanding of the Nova mechanism (N. Brosh, I. Leibowitz, 1978); (5) the discovery[19] of the 13-day period in SS433 (I. Leibowitz, T. Mazeh, in coll. with the University of Oregon, 1979).

We now have theoretical groups in almost all institutions in Israel. The late G. Racah was honoured by the IAU for his work in Spectroscopy by giving his name to a crater on the Moon, where he has thus joined the company of Ibn Ezra, Rabbi Levi, Zacut, Jacobi and Einstein.

BIBLIOGRAPHY

1. Z. Vilnay, Ariel - an Eretz-Israel Encyclopoedia, p. 7344.
2. See for example, Deuteronomy 4, 19.
3. See for example, Genesis 14, 5; ibid 15,20; Numbers 13, 33; Deuteronomy 2,10-11; ibid 3,11-13; ibid 9,2; Joshua 12,4; ibid 13,12; ibid 17,15; etc...
4. See for example, Jeremiah 33,25; Job 38,33; Ecclesiastes 1,5-6.
5. See for example, in the Apocrypha, the Book of Enoch.
6. In the Mishna, see for example, the tractate Shabath 75a.
7. See the tractate Sukkoth 28a.
8. J. Podro "The Last Pharisee: the life and times of Rabbi Joshua ben Hananyah", Valentine Mitchell, London (1969).
9. Horayoth, 10a.
10. S. Brodetsky, "Astronomy in the Babylonian Talmud", Jewish Review May 1911, p. 60.
11. P. Veron, L'Astronomie (1982), 351.
12. For the astronomers mentioned in this section, see the relevant entries in Encyclopaedia Judaica 13.
13. "The Florence and George Wise Observatory" - Tel Aviv University, Department of Information, P.R., 1978.
14. J.N. and N.A. Bahcall, "The period and light curve of HZ Herculis", Ap. J. 178 (1972), L1.
15. P.A. Wehinger et al., "Identification of H_2O^+ in the tail of comet Kohoutek", Ap. J. 190 (1974), L43.
16. Y. Mekler & A. Eviatar, "Spectroscopic observations of Io", Ap. J. 193 (1974), L151.

17. I. Kupo, Y. Mekler & A. Eviatar, "Detection of ionized sulphur in the Jovian magnetosphere", Ap. J. 205 (1976), L51.
18. N. Brosch et al., "A spectroscopic study of WZ Sagittae during the 1978 outburst", Ap. J. Lett. 236 (1980), L29.
19. J.C. Kemp et al., :SS 433 "Intensive 4 month Light Curve", Ap. J. Lett. 238 (1980), L133.

FOOTNOTE

*This is the name of this country since the XIIIth Century BC. Outside of the Bible, it is first mentioned in a 1225 BC monument put up by Pharaoh Merneptah at Beit-She'an in the Jordan Valley (now at the Jerusalem Rockefeller Museum). Before, it was known as Cana'an. The Cana'anites, called Phenicians by the Greeks, spoke the same language as the Hebrews, but worshipped Baal and other deities instead of Yahveh. They continued to live along the coast of present Lebanon to Roman times. Cana'anites who settled in present-day Tunisia at the end of the VIIIth Century BC built Keret-Hadashat (the new city), pronounced Carthage by the Romans. Hannibaal (or Hananbaal) thus spoke a dialect of Hebrew. His name means "Baal has forgiven" just as John (Johanan= Yehohanan also Hanan-Yahu=Hananiah means "Yahveh has forgiven". The magistrates of Carthage were known as Shofetim ("Suffetes"), "Judges". ("Patriarchs", then Shofetim=Judges and then Kings).

The first Jewish Commonwealth lasted in various forms between 1700-586 BC. King David installed his capital in Jerusalem around 1000 BC. The second Commonwealth started as an autonomous province of the Persian Empire, Judea with Cyrus the Great's declaration in 537 BC. The Hashmonean (or Maccabean, then Herodian) independent state of Judea lasted from 167 BC to 44 AD (91 AD in the Golan), when it was replaced by the Roman province of Judea, of New Testament renown. After the Bar-Kochba rebellion of 135 AD, the Emperor Hadrian changed the name of the province to Palestine, to erase its connections with the Judean (= Jewish) nation. Palestine and Judea appear as synonims during the Crusades. Jewish autonomy in Galilee (the "Patriarchate") lasted till 425, when the Byzantine Emperor refused to recognize a successor to Gamliel VI. It was renewed for a short spell in 607-614. Galilee with its schools and religious leadership was again central to the Jewish world in the VII-Xth Centuries AD, and again in the XVI Century. Modern Jewish resettlement started in 1777 and accelerated after 1870. Political Zionism was created by Herzl in 1897 and received its Charter over the land in the Balfour Declaration (1917) and the League of Nations Mandate (1921). The Third Commonwealth became in 1948 the State of Israel.

**The Bible (in particular the Old Testament) represents a selection of Hebrew literature, poetry, history, philosophy, etc. between ∿ 1300-400 BC. The Apocrypha are works that existed around 100 AD, when the Old Testament was edited, and were not included. The Mishna is an extensive legal code, (divided into six "Orders"), edited $\sim$ 200 AD by the Patriarch Yehuda the Prince in Tiberias, and representing a selection from the Proceedings of the Rabbinical academies in Eretz-Israel between 300 BC and 200 AD. Its authors are known as Tannaim. The Gmara is a similar but much more extensive code, organized as a Commentary on the Mishna. It was written in the academies of Galilee ("The Jerusalem Talmud" stopping, though unedited, around 350 AD) and Babylonia ("The Babylonian Talmud" edited by Rav Ashi around 400 AD). Its authors are known as Amoraim. The Term Talmud covers both Mishna and Gmara but often denotes the latter. The Talmud is divided into "Tractatea". Discussions that were left out when the Mishna was edited are known as Baraitoth, "externals".

***The months in the Gezer (Halfway between Tel Aviv and Jerusalem) calendar (1000 BC, now in the Istanbul museum) carry names such as "the month of the oat harvest" etc. The months of the present Jewish calendar represent the Babylonian Zodiac appelations, sometimes named after Sumerian or Akadian deities (Tammuz, etc.).

PRE-CONCLUDING REMARKS

G.T. Bath
Department of Astrophysics, Oxford.

In the same spirit as the local organising committee's "post-final announcement" I should like, as chairman of this session, to extend an invitation to myself to present a few pre-concluding remarks before asking Brian to summarise his view of the highlights of the past few days. My comments will be eclectic, coloured by a theorist's wish to impose pattern on chaos; order on the anarchy of observations.

It seems to me that one of the central themes which has run through this meeting, and which marks a new departure in research on cataclysmic variables, has been the spectroscopic, polarimetric and photometric evidence for anisotropic radiation patterns about the binary rotation axis, and their interpretation in terms of a whole range of antisymmetric physical processes. The variety of these suggestions rightly depends only on the limits of the imagination, but if we move out through the gravitational potential well of the accreting star, I noted the following processes under discussion at the meeting:

(1) Magnetic accretion columns on the white dwarf or neutron star.
(2) Localised spots of nuclear burning.
(3) Disc structure and "blobs" orbiting within the disc.
(4) Disc anisotropy over extensive disc regions due to stream penetration or non-circular orbits.
(5) The classic picture of hot-spot (or, rather, bright-spot) radiation from stream impact at the disc edge.
(6) Active regions (excited by bursting mass transfer) on a non-synchronously rotating companion star.
(7) Absorption in material orbiting the whole system.

This is clearly not an exhaustive list of possibilities, and each of us, theorist and observer alike, has his own partiality towards favourites that have served him well in the past. In several cases it is still unclear on simple observational grounds just what is the

M. Livio and G. Shaviv (eds.), Cataclysmic Variables and Related Objects, 331–332.

dominant process at any time, and the theorist's role must be to develop means of distinguishing more rigorously between these possibilities. In this regard let me "blow my own recorder" and remind you of the possible utility of "peapod" diagrams in providing a link between theory and observation, and mapping the structure of these sources. I sense that one of my favourites, the suggestion that Brian Warner was right in 1975 with his original model for the origin of the superhump phenomena, has been coolly received, even by Brian himself. Why I have not fathomed, but there are observational tests, so you observers should get out and prove me wrong, if possible by substituting what is right. Until then I would remind you that some ten years ago when Doug Hall was advocating the star spot model of RS CVn stars, it was spurned by the establishment and even such rebels of the time as John Whelan described it as "Doug's square star model".

Before I hand over to Brian for the observer's view of the reality, rather than my theorist's vision of the ideal, I'd like to read you an excerpt from one of my favourite pieces of poetry, by Rilke - perhaps the greatest poet of the past hundred years. It comes from the first elegy of the Duino Elegies.

"Many a star
was waiting for you to perceive it. Many a wave
would rise in the past towards you; or else, perhaps,
as you went by an open window, a violin
would be utterly giving itself. All this was commission
But were you equal to it? Were you not still
distraught by expectancy, as though all were announcing
some beloved's approach? (As if you could hope
to house her with all those great strange thoughts
going in and out and often staying overnight!)"

Before we go on our various routes to be distraught by expectancy I'm sure I speak for us all in expressing our debt to Mario Livio and Giora Shaviv for the commission we've enjoyed throughout this meeting.

CONCLUDING REMARKS

Brian Warner
Department of Astronomy, University of Cape Town

The great success of this Colloquium has been in a large part due to the high quality of the results presented. Not all of the results have been understandable; neither have all topics been covered. In picking out a few on which to comment I run the risk of offending those whom I omit and possibly offending those whom I include. Nevertheless, these are my thoughts.

Firstly, I want to elaborate on something that Ed Sion did not say. In showing us the IUE spectra of accretion discs he pointed out the recently discovered P Cygni profiles and estimated the mass outflow. What has not yet been satisfactorily modelled is the concomitant angular momentum outflow. It is possible that we see here for the first time direct evidence of significant angular momentum loss from cataclysmic variables. This may be useful in solving what I consider to be the greatest unsolved problem in the CV's: the driving mechanism for the mass transfer. The observationally inferred rates of mass transfer are greatly in excess of what is available from the canonical theory of gravitational radiation. No other quantitative theory has yet been propounded; a subject studiously avoided at last year's Santa Cruz Workshop and again at this Colloquium.

Perhaps the most perplexing result to be launched at this Colloquium was Bob Williams's analysis of the C, N and O emission lines from CV discs. Last year it was He, now it is C, N and O as well that are hugely overabundant. Furthermore, there is rough agreement between the composition of novae ejecta and the CV discs; a conclusion derived from rather different physical conditions in the two locations. If these abundance anomalies are verified by alternative and independent procedures, the implications for evolution of the secondary components of the CVs are profound. As I pointed out in discussion, K-type secondary spectra are observable for several of the longer period CVs. At this temperature, the simultaneous occurrence of low opacity (from H-depletion) and overabundance of C and N should make the CN band at 3883Å of much greater strength than in K stars of normal abundance. The CH band, on the other hand, may not be so anomalous. A direct measure of the C and N abundances would therefore appear feasible and urgent.

M. Livio and G. Shaviv (eds.), Cataclysmic Variables and Related Objects, 333–335.

One of the objects for which a secondary spectrum is accessible is GK Per. Bianchini and Crampton presented new observations, the latter showing that the orbital period is 2 days: considerably longer than previously thought. As the longest period system among the classical novae, and one for which the secondary must be in an advanced stage of evolution, a more concerted effort - employing Williams's techniques, abundance analysis of the secondary spectrum and evolutionary interpretation of the secondary - is overdue.

Rob Robinson started our Colloquium on a rather downbeat note by pointing out the hazards of interpreting emission line profiles for the purpose of deriving radial velocity amplitudes. His apprehensions were borne out by Gilliland who showed that in WZ Sge, despite a careful analysis of the highest quality data, there remains an anomalous phasing of the radial velocity curve and the eclipse - indicative that näive models cannot give reliable results.

This same effect showed up at the end of Jim McDonald's contribution: having presented a simplified but comprehensive theory of nova eruptions, in which he indirectly derived masses for the white dwarf components, his attempt to compare such masses with those "directly" observed from analysis of eclipsing systems, met with the conclusion that, for DQ Her, any mass in the range $0.4 < M/M_{\odot} < 1.4$ is available in the literature. This situation, where an honest theoretician is confronted by observers who speak with many tongues, is a disgraceful one and it is high time that the large range of interpretations of observational material are analysed for their discordancies.

Friedjung and Evans unveiled an unexpected problem with Nova Aquila 1982, which has the appearances of an intermediate supernova. Messrs. Colvin and Schatzman considered supernovae explosions from a theoretical viewpoint, employing various brands of catastrophysics, and Sugimoto presented a study of mass loss from X-ray bursters employing physical theory that would have delighted Sir Arthur Super Eddington. Geoff Bath, in contriving to teach us the α to ω of discs, has got as far as $\alpha\beta$. Peter Eggleton, observing the literature with a critical eye, discovered a large number of exotic beasts that seem never to have ridden on the Common Envelope Binary bandwaggon. All of these individual contributions raised new problems or made progress in the solution of old ones.

Probably the most coherent set of presentations concerned the observation and implications of large magnetic fields in the degenerate components of the cataclysmic variables. The Polars (AM Her stars), reviewed by Santiago Tapia, appear more and more complex and difficult of interpretation. Both Tapia and Piirola find large night-to-night variations in polarisation, not necessarily accompanied by total flux changes. No matter where Tapia observes - from Arizona, Chile or South Africa - the rotation axes seem to point towards him! Whether this is a selection effect or too näive an interpretation of the observations is not yet evident, but the result contains a warning that in most cases where radial velocity measurements have been made in the polars, it is not certain that it is orbital motion (rather than effects of rotation of the white dwarf) that has been measured. In other words, the claim of phase-locked rotation, although probably correct, needs to be carefully re-examined: the evidence for such synchronous rotation in the polars is at present still tenuous.

The intermediate polars, reviewed by Mlle Mouchet and myself, have emerged as a distinct class of asynchronously rotating magnetic white dwarfs that shows great potential for future study. No sooner had the observers presented their results than Don Lamb took the floor and presented a more-than-complete theory that gave not only a quantitative explanation of the observations but provided estimates of magnetic field strengths with such narrow limits of uncertainty that further observations almost seemed unnecessary!

In looking at the detailed structure of accreting magnetic white dwarfs, however, Messrs. King and Chanmugam showed how tremendously complicated even the so-called simplest models really are. Perhaps the intermediate polars, in which we are able to observe the angular distribution of the beamed radiation as the beam sweeps across the secondary, will introduce sufficient new information to assist the theoretical discussion of emissions from accretion columns.

In thanking the Local Organizing Committee, on behalf of the participants, for their sterling work, I must point out that Mario Livio, as well as having had a major share in organizing local affairs, the SOC and apparently most of Haifa, found time to prepare a contribution which turned out to be probably the major announcement at the Colloquium: the discovery of non-radial oscillations in neutron stars.

Although I formally opened the meeting as "The 72nd Colloquium of the International Astronomical Union", it is clear from the events of the week that it will henceforth surely be looked back on with affection as "The Haifa Meeting".

AUTHOR INDEX

Only authors explicitly named in the text and figures were indexed. Names mentioned in acknowledgements were often omitted. Co-authors are not included when not explicitly mentioned.

SUBJECT INDEX

Includes texts, figures and tables, discussions are excluded. The page references are not necessarily complete.

ASTROPHYSICS AND SPACE SCIENCE LIBRARY

1. C. de Jager (ed.), *The Solar Spectrum, Proceedings of the Symposium held at the University of Utrecht, 26–31 August, 1963.* 1965, XIV + 417 pp.
2. J. Orthner and H. Maseland (eds.), *Introduction to Solar Terrestrial Relations, Proceedings of the Summer School in Space Physics held in Alpbach, Austria, July 15–August 10, 1963 and Organized by the European Preparatory Commission for Space Research.* 1965, IX + 506 pp.
3. C. C. Chang and S. S. Huang (eds.), *Proceedings of the Plasma Space Science Symposium, held at the Catholic University of America, Washington, D.C., June 11–14, 1963.* 1965, IX + 377 pp.
4. Zdeněk Kopal, *An Introduction to the Study of the Moon.* 1966, XII + 464 pp.
5. B. M. McCormac (ed.), *Radiation Trapped in the Earth's Magnetic Field. Proceedings of the Advanced Study Institude, held at the Chr. Michelsen Institute, Bergen, Norway, August 16–September 3, 1965.* 1966, XII + 901 pp.
6. A. B. Underhill, *The Early Type Stars.* 1966, XII + 282 pp.
7. Jean Kovalevsky, *Introduction to Celestial Mechanics.* 1967, VIII + 427 pp.
8. Zdeněk Kopal and Constantine L. Goudas (eds.), *Measure of the Moon. Proceedings of the 2nd International Conference on Selenodesy and Lunar Topography, held in the University of Manchester, England, May 30–June 4, 1966.* 1967, XVIII + 479 pp.
9. J. G. Emming (ed.), *Electromagnetic Radiation in Space. Proceedings of the 3rd ESRO Summer School in Space Physics, held in Alpbach, Austria, from 19 July to 13 August, 1965.* 1968, VIII + 307 pp.
10. R. L. Carovillano, John F. McClay, and Henry R. Radoski (eds.), *Physics of the Magnetosphere, Based upon the Proceedings of the Conference held at Boston College, June 19–28, 1967.* 1968, X + 686 pp.
11. Syun-Ichi Akasofu, *Polar and Magnetospheric Substorms.* 1968, XVIII + 280 pp.
12. Peter M. Millman (ed.), *Meteorite Research. Proceedings of a Symposium on Meteorite Research, held in Vienna, Austria, 7–13 August, 1968.* 1969, XV + 941 pp.
13. Margherita Hack (ed.), *Mass Loss from Stars. Proceedings of the 2nd Trieste Colloquium on Astrophysics, 12–17 September, 1968.* 1969, XII + 345 pp.
14. N. D'Angelo (ed.), *Low-Frequency Waves and Irregularities in the Ionosphere. Proceedings of the 2nd ESRIN-ESLAB Symposium, held in Frascati, Italy, 23–27 September, 1968.* 1969, VII + 218 pp.
15. G. A. Partel (ed.), *Space Engineering. Proceedings of the 2nd International Conference on Space Engineering, held at the Fondazione Giorgio Cini, Isola di San Giorgio, Venice, Italy, May 7–10, 1969.* 1970, XI + 728 pp.
16. S. Fred Singer (ed.), *Manned Laboratories in Space. Second International Orbital Laboratory Symposium.* 1969, XIII + 133 pp.
17. B. M. McCormac (ed.), *Particles and Fields in the Magnetosphere. Symposium Organized by the Summer Advanced Study Institute, held at the University of California, Santa Barbara, Calif., August 4–15, 1969.* 1970, XI + 450 pp.
18. Jean-Claude Pecker, *Experimental Astronomy.* 1970, X + 105 pp.
19. V. Manno and D. E. Page (eds.), *Intercorrelated Satellite Observations related to Solar Events. Proceedings of the 3rd ESLAB/ESRIN Symposium held in Noordwijk, The Netherlands, September 16–19, 1969.* 1970, XVI + 627 pp.
20. L. Mansinha, D. E. Smylie, and A. E. Beck, *Earthquake Displacement Fields and the Rotation of the Earth, A NATO Advanced Study Institute Conference Organized by the Department of Geophysics, University of Western Ontario, London, Canada, June 22–28, 1969.* 1970, XI + 308 pp.
21. Jean-Claude Pecker, *Space Observatories.* 1970, XI + 120 pp.
22. L. N. Mavridis (ed.), *Structure and Evolution of the Galaxy. Proceedings of the NATO Advanced Study Institute, held in Athens, September 8–19, 1969.* 1971, VII + 312 pp.

23. A. Muller (ed.), *The Magellanic Clouds. A European Southern Observatory Presentation: Principal Prospects, Current Observational and Theoretical Approaches, and Prospects for Future Research, Based on the Symposium on the Magellanic Clouds, held in Santiago de Chile, March 1969, on the Occasion of the Dedication of the European Southern Observatory.* 1971, XII + 189 pp.
24. B. M. McCormac (ed.), *The Radiating Atmosphere. Proceedings of a Symposium Organized by the Summer Advanced Study Institute, held at Queen's University, Kingston, Ontario, August 3–14, 1970.* 1971, XI + 455 pp.
25. G. Fiocco (ed.), *Mesospheric Models and Related Experiments. Proceedings of the 4th ESRIN-ESLAB Symposium, held at Frascati, Italy, July 6–10, 1970.* 1971, VIII + 298 pp.
26. I. Atanasijević, *Selected Exercises in Galactic Astronomy.* 1971, XII + 144 pp.
27. C. J. Macris (ed.), *Physics of the Solar Corona. Proceedings of the NATO Advanced Study Institute on Physics of the Solar Corona, held at Cavouri-Vouliagmeni, Athens, Greece, 6–17 September 1970.* 1971, XII + 345 pp.
28. F. Delobeau, *The Environment of the Earth.* 1971, IX + 113 pp.
29. E. R. Dyer (general ed.), *Solar-Terrestrial Physics/1970. Proceedings of the International Symposium on Solar-Terrestrial Physics, held in Leningrad, U.S.S.R., 12–19 May 1970.* 1972, VIII + 938 pp.
30. V. Manno and J. Ring (eds.), *Infrared Detection Techniques for Space Research. Proceedings of the 5th ESLAB-ESRIN Symposium, held in Noordwijk, The Netherlands, June 8–11, 1971.* 1972, XII + 344 pp.
31. M. Lecar (ed.), *Gravitational N-Body Problem. Proceedings of IAU Colloquium No. 10, held in Cambridge, England, August 12–15, 1970.* 1972, XI + 441 pp.
32. B. M. McCormac (ed.), *Earth's Magnetospheric Processes. Proceedings of a Symposium Organized by the Summer Advanced Study Institute and Ninth ESRO Summer School, held in Cortina, Italy, August 30–September 10, 1971.* 1972, VIII + 417 pp.
33. Antonin Rükl, *Maps of Lunar Hemispheres.* 1972, V + 24 pp.
34. V. Kourganoff, *Introduction to the Physics of Stellar Interiors.* 1973, XI + 115 pp.
35. B. M. McCormac (ed.), *Physics and Chemistry of Upper Atmospheres. Proceedings of a Symposium Organized by the Summer Advanced Study Institute, held at the University of Orléans, France, July 31–August 11, 1972.* 1973, VIII + 389 pp.
36. J. D. Fernie (ed.), *Variable Stars in Globular Clusters and in Related Systems. Proceedings of the IAU Colloquium No. 21, held at the University of Toronto, Toronto, Canada, August 29–31, 1972.* 1973, IX + 234 pp.
37. R. J. L. Grard (ed.), *Photon and Particle Interaction with Surfaces in Space. Proceedings of the 6th ESLAB Symposium, held at Noordwijk, The Netherlands, 26–29 September, 1972.* 1973, XV + 577 pp.
38. Werner Israel (ed.), *Relativity, Astrophysics and Cosmology. Proceedings of the Summer School, held 14–26 August 1972, at the BANFF Centre, BANFF, Alberta, Canada.* 1973, IX + 323 pp.
39. B. D. Tapley and V. Szebehely (eds.), *Recent Advances in Dynamical Astronomy. Proceedings of the NATO Advanced Study Institute in Dynamical Astronomy, held in Cortina d'Ampezzo, Italy, August 9–12, 1972.* 1973, XIII + 468 pp.
40. A. G. W. Cameron (ed.), *Cosmochemistry. Proceedings of the Symposium on Cosmochemistry, held at the Smithsonian Astrophysical Observatory, Cambridge, Mass., August 14–16, 1972.* 1973, X + 173 pp.
41. M. Golay, *Introduction to Astronomical Photometry.* 1974, IX + 364 pp.
42. D. E. Page (ed.), *Correlated Interplanetary and Magnetospheric Observations. Proceedings of the 7th ESLAB Symposium, held at Saulgau, W. Germany, 22–25 May, 1973.* 1974, XIV + 662 pp.
43. Riccardo Giacconi and Herbert Gursky (eds.), *X-Ray Astronomy.* 1974, X + 450 pp.
44. B. M. McCormac (ed.), *Magnetospheric Physics. Proceedings of the Advanced Summer Institute, held in Sheffield, U.K., August 1973.* 1974, VII + 399 pp.
45. C. B. Cosmovici (ed.), *Supernovae and Supernova Remnants. Proceedings of the International Conference on Supernovae, held in Lecce, Italy, May 7–11, 1973.* 1974, XVII + 387 pp.
46. A. P. Mitra, *Ionospheric Effects of Solar Flares.* 1974, XI + 294 pp.
47. S.-I. Akasofu, *Physics of Magnetospheric Substorms.* 1977, XVIII + 599 pp.

48. H. Gursky and R. Ruffini (eds.), *Neutron Stars, Black Holes and Binary X-Ray Sources.* 1975, XII + 441 pp.
49. Z. Švestka and P. Simon (eds.), *Catalog of Solar Particle Events 1955–1969. Prepared under the Auspices of Working Group 2 of the Inter-Union Commission on Solar-Terrestrial Physics.* 1975, IX + 428 pp.
50. Zdeněk Kopal and Robert W. Carder, *Mapping of the Moon.* 1974, VIII + 237 pp.
51. B. M. McCormac (ed.), *Atmospheres of Earth and the Planets. Proceedings of the Summer Advanced Study Institute, held at the University of Liège, Belgium, July 29–August 8, 1974.* 1975, VII + 454 pp.
52. V. Formisano (ed.), *The Magnetospheres of the Earth and Jupiter. Proceedings of the Neil Brice Memorial Symposium, held in Frascati, May 28–June 1, 1974.* 1975, XI + 485 pp.
53. R. Grant Athay, *The Solar Chromosphere and Corona: Quiet Sun.* 1976, XI + 504 pp.
54. C. de Jager and H. Nieuwenhuijzen (eds.), *Image Processing Techniques in Astronomy. Proceedings of a Conference, held in Utrecht on March 25–27, 1975.* 1976, XI + 418 pp.
55. N. C. Wickramasinghe and D. J. Morgan (eds.), *Solid State Astrophysics. Proceedings of a Symposium, held at the University College, Cardiff, Wales, 9–12 July, 1974.* 1976, XII + 314 pp.
56. John Meaburn, *Detection and Spectrometry of Faint Light.* 1976, IX + 270 pp.
57. K. Knott and B. Battrick (eds.), *The Scientific Satellite Programme during the International Magnetospheric Study. Proceedings of the 10th ESLAB Symposium, held at Vienna, Austria, 10–13 June 1975.* 1976, XV + 464 pp.
58. B. M. McCormac (ed.), *Magnetospheric Particles and Fields. Proceedings of the Summer Advanced Study School, held in Graz, Austria, August 4–15, 1975.* 1976, VII + 331 pp.
59. B. S. P. Shen and M. Merker (eds.), *Spallation Nuclear Reactions and Their Applications.* 1976, VIII + 235 pp.
60. Walter S. Fitch (ed.), *Multiple Periodic Variable Stars. Proceedings of the International Astronomical Union Colloquium No. 29, held at Budapest, Hungary, 1–5 September 1976.* 1976, XIV + 348 pp.
61. J. J. Burger, A. Pedersen, and B. Battrick (eds.), *Atmospheric Physics from Spacelab. Proceedings of the 11th ESLAB Symposium, Organized by the Space Science Department of the European Space Agency, held at Frascati, Italy, 11–14 May 1976.* 1976, XX + 409 pp.
62. J. Derral Mulholland (ed.), *Scientific Applications of Lunar Laser Ranging. Proceedings of a Symposium held in Austin, Tex., U.S.A., 8–10 June, 1976.* 1977, XVII + 302 pp.
63. Giovanni G. Fazio (ed.), *Infrared and Submillimeter Astronomy. Proceedings of a Symposium held in Philadelphia, Penn., U.S.A., 8–10 June, 1976.* 1977, X + 226 pp.
64. C. Jaschek and G. A. Wilkins (eds.), *Compilation, Critical Evaluation and Distribution of Stellar Data. Proceedings of the International Astronomical Union Colloquium No. 35, held at Strasbourg, France, 19–21 August, 1976.* 1977, XIV + 316 pp.
65. M. Friedjung (ed.), *Novae and Related Stars. Proceedings of an International Conference held by the Institut d'Astrophysique, Paris, France, 7–9 September, 1976.* 1977, XIV + 228 pp.
66. David N. Schramm (ed.), *Supernovae. Proceedings of a Special IAU-Session on Supernovae held in Grenoble, France, 1 September, 1976.* 1977, X + 192 pp.
67. Jean Audouze (ed.), *CNO Isotopes in Astrophysics. Proceedings of a Special IAU Session held in Grenoble, France, 30 August, 1976.* 1977, XIII + 195 pp.
68. Z. Kopal, *Dynamics of Close Binary Systems.* XIII + 510 pp.
69. A. Bruzek and C. J. Durrant (eds.), *Illustrated Glossary for Solar and Solar-Terrestrial Physics.* 1977, XVIII + 204 pp.
70. H. van Woerden (ed.), *Topics in Interstellar Matter.* 1977, VIII + 295 pp.
71. M. A. Shea, D. F. Smart, and T. S. Wu (eds.), *Study of Travelling Interplanetary Phenomena.* 1977, XII + 439 pp.
72. V. Szebehely (ed.), *Dynamics of Planets and Satellites and Theories of Their Motion. Proceedings of IAU Colloquium No. 41, held in Cambridge, England, 17–19 August 1976.* 1978, XII + 375 pp.
73. James R. Wertz (ed.), *Spacecraft Attitude Determination and Control.* 1978, XVI + 858 pp.

74. Peter J. Palmadesso and K. Papadopoulos (eds.), *Wave Instabilities in Space Plasmas. Proceedings of a Symposium Organized Within the XIX URSI General Assembly held in Helsinki, Finland, July 31–August 8, 1978.* 1979, VII + 309 pp.
75. Bengt E. Westerlund (ed.), *Stars and Star Systems. Proceedings of the Fourth European Regional Meeting in Astronomy held in Uppsala, Sweden, 7–12 August, 1978.* 1979, XVIII + 264 pp.
76. Cornelis van Schooneveld (ed.), *Image Formation from Coherence Functions in Astronomy. Proceedings of IAU Colloquium No. 49 on the Formation of Images from Spatial Coherence Functions in Astronomy, held at Groningen, The Netherlands, 10–12 August 1978.* 1979, XII + 338 pp.
77. Zdeněk Kopal, *Language of the Stars. A Discourse on the Theory of the Light Changes of Eclipsing Variables.* 1979, VIII + 280 pp.
78. S.-I. Akasofu (ed.), *Dynamics of the Magnetosphere. Proceedings of the A.G.U. Chapman Conference 'Magnetospheric Substorms and Related Plasma Processes' held at Los Alamos Scientific Laboratory, N.M., U.S.A., October 9–13, 1978.* 1980, XII + 658 pp.
79. Paul S. Wesson, *Gravity, Particles, and Astrophysics. A Review of Modern Theories of Gravity and G-variability, and their Relation to Elementary Particle Physics and Astrophysics.* 1980, VIII + 188 pp.
80. Peter A. Shaver (ed.), *Radio Recombination Lines. Proceedings of a Workshop held in Ottawa, Ontario, Canada, August 24–25, 1979.* 1980, X + 284 pp.
81. Pier Luigi Bernacca and Remo Ruffini (eds.), *Astrophysics from Spacelab.* 1980, XI + 664 pp.
82. Hannes Alfvén, *Cosmic Plasma,* 1981, X + 160 pp.
83. Michael D. Papagiannis (ed.), *Strategies for the Search for Life in the Universe,* 1980, XVI + 254 pp.
84. H. Kikuchi (ed.), *Relation between Laboratory and Space Plasmas,* 1981, XII + 386 pp.
85. Peter van der Kamp, *Stellar Paths,* 1981, XXII + 155 pp.
86. E. M. Gaposchkin and B. Kołaczek (eds.), *Reference Coordinate Systems for Earth Dynamics,* 1981, XIV + 396 pp.
87. R. Giacconi (ed.), *X-Ray Astronomy with the Einstein Satellite. Proceedings of the High Energy Astrophysics Division of the American Astronomical Society Meeting on X-Ray Astronomy held at the Harvard-Smithsonian Center for Astrophysics, Cambridge, Mass., U.S.A., January 28–30, 1980.* 1981, VII + 330 pp.
88. Icko Iben Jr. and Alvio Renzini (eds.), *Physical Processes in Red Giants. Proceedings of the Second Workshop, helt at the Ettore Majorana Centre for Scientific Culture, Advanced School of Agronomy, in Erice, Sicily, Italy, September 3–13, 1980.* 1981, XV + 488 pp.
89. C. Chiosi and R. Stalio (eds.), *Effect of Mass Loss on Stellar Evolution. IAU Colloquium No. 59 held in Miramare, Trieste, Italy, September 15–19, 1980.* 1981, XXII + 532 pp.
90. C. Goudis, *The Orion Complex: A Case Study of Interstellar Matter.* 1982, XIV + 306 pp.
91. F. D. Kahn (ed.), *Investigating the Universe. Papers Presented to Zdenek Kopal on the Occasion of his retirement, September 1981.* 1981, X + 458 pp.
92. C. M. Humphries (ed.), *Instrumentation for Astronomy with Large Optical Telescopes, Proceedings of IAU Colloquium No. 67.* 1981, XVII + 321 pp.
93. R. S. Roger and P. E. Dewdney (eds.), *Regions of Recent Star Formation, Proceedings of the Symposium on "Neutral Clouds Near HII Regions - Dynamics and Photochemistry", held in Penticton, B.C., June 24–26, 1981.* 1982, XVI + 496 pp.
94. O. Calame (ed.), *High-Precision Earth Rotation and Earth-Moon Dynamics. Lunar Distances and Related Observations.* 1982, XX + 354 pp.
95. M. Friedjung and R. Viotti (eds.), *The Nature of Symbiotic Stars,* 1982, XX + 310 pp.
96. W. Fricke and G. Teleki (eds.), *Sun and Planetary System,* 1982, XIV + 538 pp.
97. C. Jaschek and W. Heintz (eds.), *Automated Data Retrieval in Astronomy,* 1982, XX + 324 pp.
98. Z. Kopal and J. Rahe (eds.), *Binary and Multiple Stars as Tracers of Stellar Evolution,* 1982, XXX + 503 pp.
99. A. W. Wolfendale (ed.), *Progress in Cosmology,* 1982, VIII + 360 pp.
100. W. L. H. Shuter (ed.), *Kinematics, Dynamics and Structure of the Milky Way,* 1983, XII + 392 pp.